The Chemistry of Cosmic Dust

The Chemistry of Cosmic Dust

David A. Williams
University College London, UK
Email: daw@star.ucl.ac.uk

and

Cesare Cecchi-Pestellini
Observatory of Palermo, Sicily, Italy
Email: cecchi-pestellini@astropa.unipa.it

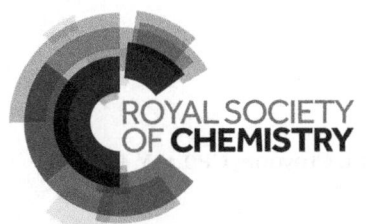

THE QUEEN'S AWARDS
FOR ENTERPRISE:
INTERNATIONAL TRADE
2013

Print ISBN: 978-1-78262-047-1
PDF eISBN: 978-1-78262-369-4

A catalogue record for this book is available from the British Library

Published by The Royal Society of Chemistry,
Thomas Graham House, Science Park, Milton Road,
Cambridge CB4 0WF, UK

Registered Charity Number 207890

Visit our website at www.rsc.org/books

Printed in the United Kingdom by CPI Group (UK) Ltd, Croydon, CR0 4YY, UK

Preface

Although the existence of molecules in interstellar space was recognised as long ago as three quarters of a century, the true astronomical and chemical significance of molecular clouds was only established several decades later with the opening up of the millimetre and submillimetre wavebands for astronomy. It was soon discovered through mapping in lines of carbon monoxide that the Milky Way galaxy included a previously unknown component: interstellar molecular clouds, rich in molecules and in dust. Some of these clouds are giants of their species and may contain up to a million solar masses of gas, although many of them are much less massive. Molecular clouds were found to be present in many galaxies. The total mass in molecular clouds can be an appreciable mass of the galaxy where they are found. This mass is the reservoir of material from which new stars are being formed, and so molecular clouds came to be recognised as an essential component controlling—through the star formation that these clouds promote—the evolution of galaxies.

Molecular clouds, and particularly the star-forming regions they contained, were found to include a wide variety of molecular species, from diatomics to molecules containing a dozen or so atoms, while species possibly containing up to a hundred atoms or so are strongly suspected to be present. The known number of molecular species is around a couple of hundred. However, it is clear that the total number of molecular species must be much larger, as some species that are believed on theoretical grounds to be present are spectroscopically very difficult to detect, while other probable species have abundances below the level of detectability with current instruments. The challenge of understanding the origin of these species attracted intense interest from laboratory and theoretical chemists. Many astronomers also recognised that emissions from these molecular species are valuable new probes of regions of space that are otherwise hidden from view. Over the last

The Chemistry of Cosmic Dust
By David A. Williams and Cesare Cecchi-Pestellini
© David A. Williams and Cesare Cecchi-Pestellini 2016
Published by the Royal Society of Chemistry, www.rsc.org

four decades, the subject of astrochemistry has become very well-established, and the understanding of interstellar physical processes that emerged from such studies showed emphatically how some new astronomical discoveries had their foundations in this new subject of astrochemistry.

Many of the molecules discovered to be present in molecular clouds were found to be formed in complex gas-phase chemical networks. These networks at their most extensive now typically contain thousands of gas-phase reactions. Enormous efforts have been (and continue to be) made in the laboratory and in computational studies to supply accurate rate coefficients and other data required to evaluate the effectiveness of these networks in providing molecular species in the observed abundances. The story of gas-phase interstellar chemistry has been an astounding success, bringing deep insights both to chemistry in exotic environments, and to astronomy in dark regions of space.

Unfortunately, it became unavoidably clear that gas-phase routes alone could not supply the whole range of molecular species in the abundances observed to exist in the interstellar medium. Gas-phase routes for the formation of the most abundant molecule of all, molecular hydrogen, were found to be grossly inadequate to account for molecular abundances obtained from observations in the Milky Way galaxy. Indeed, some other species seemed to be pathological cases for which no effective gas-phase formation routes could be devised. Rather reluctantly, astrochemists began to accept in principle some ideas from the mid-20th century that chemical reactions may occur on the surfaces of dust grains. The reluctance to leave the well-ordered world of gas-phase binary collisions for the relative lack of clarity in surface and solid-state processes was fully understandable, but it was necessary to take this step. Indeed, the discovery that in some regions of space dust grains are coated with icy mantles introduced the idea that reactions could occur within and on the ices. It seemed plausible that surface reactions, or reactions on and in ices, might be able to account of the failures of the well-established gas-phase schemes.

However, it was not until the late 20th century that laboratory techniques and computational facilities improved sufficiently for serious studies of dust-related chemistries to be performed with the required accuracy. Over the last couple of decades there has been a huge effort in the laboratory and in computational modelling, the results of which have made clear that surface and solid-state processes are indeed operating abundantly in interstellar space. Surface and solid-state astrochemistry is now well-established, is developing rapidly, and making substantial progress.

It seems, therefore, a suitable moment to write this book. Our intention has been to review the ideas that currently drive this subject, to assess the current achievements, to demonstrate that this is a research area of intense activity and great promise, if still one of "work in progress", and to try to indicate topics of future study. However, a comprehensive account of the hundred thousand or so of published articles in this field is clearly an impossible task, and we have not attempted to do that. Instead, we have concentrated on modern themes driving surface and solid-state astrochemistry. In this book, we give some astronomical background to provide motivation, and indicate where further information may be obtained. However, the book is aimed primarily at those who might wish to enter research in the murky world of surface and solid-state astrochemistry. We hope that established workers in this exciting and challenging field will also find this book a useful repository of current knowledge and ideas for future work.

David A Williams
University College London

Cesare Cecchi-Pestellini
Observatory of Palermo

We dedicate this book to the memory of Alex Dalgarno (1928–2015). His leadership of the field for many decades has been an inspiration and example to all of us who try to contribute to progress in astrochemistry.

Contents

Section I: Defining the Chemical and Physical Nature of Interstellar Dust

The Chemistry of Cosmic Dust
By David A. Williams and Cesare Cecchi-Pestellini
© David A. Williams and Cesare Cecchi-Pestellini 2016
Published by the Royal Society of Chemistry, www.rsc.org

Section II: The Formation of Dust and Its Evolution in the Interstellar Media of Galaxies

Section III: Chemically Active Interstellar Dust

Section IV: Roles of Dust in the Universe

CHAPTER 1

Dust-Related Chemistry in Space

1.1 Introduction

This is a book about dust and chemistry in space. The space we mean is *inter-stellar space*, that is, the space between the stars in our galaxy, the Milky Way, and the interstellar space between the stars in other galaxies. The chemis-try we shall describe is the chemistry associated with dust particles that are observed to be mixed with the interstellar gas.

We are not concerned here with *gas-phase* chemistry in interstellar space. That subject has been investigated intensively for more than half a century through laboratory experiments and *ab initio* computation; it has been driven by the results from astronomical molecular line observations, and extensive chemical networks have been explored and tested through many astrophys-ical models. Interstellar gas-phase chemistry is well described in many texts (see Further Reading).

In this book, we shall describe *the chemistry of interstellar dust* and *the chemistry initiated by interstellar dust.* These dust-related astrochemical topics are currently being very actively pursued, but in some aspects are less well-established than interstellar gas-phase chemistry. In some areas, there is less of a consensus about the nature of dust and dust-related chemistry than in gas-phase chemistry. This makes dust-related chemistry a very lively area of current research. We do not have all the answers! This book is intended for active researchers, and will highlight areas of uncertainty as well as areas of solid achievement.

Dust enables two kinds of interstellar chemistry to occur through its inter-actions with the gas. Firstly, the presence of dust promotes reactions between

The Chemistry of Cosmic Dust
By David A. Williams and Cesare Cecchi-Pestellini
© David A. Williams and Cesare Cecchi-Pestellini 2016
Published by the Royal Society of Chemistry, www.rsc.org

atoms and molecules adsorbed on the surfaces of bare dust grains. Surface reactions of this type turn out to be very important because they provide, in particular, almost all the hydrogen molecules that are essential partners in the gas-phase network. In that sense, even gas-phase interstellar chemistry depends fundamentally on the presence of dust, through the need for molecular hydrogen. Undoubtedly, interstellar species other than molecular hydrogen are also formed in reactions at the surfaces of dust particles.

Secondly, observations of some particular astronomical locations lead to the conclusion that the dust grains in them accumulate mantles of ices containing a limited variety of fairly simple molecules, mainly water, carbon monoxide and carbon dioxide, with smaller amounts of some other simple molecules. While these unprocessed ices are certainly interesting in themselves, their chemical complexity can be dramatically enhanced, apparently in some form of solid-state chemistry, with the more complex products enriching the gas phase. These more complex product species, such as ethanol, acetic acid, and glycolaldehyde, are detected at relatively high abundances in various interstellar locations, especially in regions of star formation, and are considered by some astrochemists to be related to the emerging subject of astrobiology.

While these two topics—surface chemistry on bare grains and solid state chemistry in ices—are the main targets for discussion for which the material in this book is organised, the chemical nature of the dust itself is also important. Where can solid particles, the dust grains, form from the gas? By what chemical pathways can this occur? How does the chemical and physical nature of dust affect surface chemistry and the deposition of ices? Obviously, atoms locked up in dust are not available in the gas and do not take part in gas-phase reactions, so the existence of dust grains can affect chemistry involving those atoms in a passive and negative way. If almost all the atoms of one particular kind of atom, say, silicon, are incorporated in dust (silicon atoms in silicates, for example), then any silicon-related gas-phase chemistry will be very significantly suppressed – unless, of course, those atoms can be released from the dust. It usually takes rather extreme circumstances for that release to be possible, for example, to return silicon atoms from silicate dust to the gas phase in the interstellar medium.

1.2   Interstellar Dust – A Brief History

1.2.1   Observations

It has been clear for a long time that the interstellar space of our galaxy, the Milky Way, contains material that tends to obscure starlight. Perhaps the first suggestion that such material exists in interstellar space arose from observations made by William Herschel in his 1784 telescope surveys of rich star fields in the Milky Way. He noticed, in particular, that there was a small well-defined region in Ophiuchus that was apparently devoid of stars even though it was embedded in a very rich field of stars, and he remarked "Hier

ist wahrhaftig ein Loch in Himmel" ("Here is indeed a hole in the Heavens").
While Herschel may have believed that there was indeed a true absence of
stars in such regions, an alternative explanation is simpler: that something
in the foreground is obscuring the optical light of the background stars. Late
19th century astrophotographic studies by Edward Emerson Barnard were
much more sensitive than Herschel's naked-eye observations. Barnard's
work showed that a variety of structures existed in interstellar space, and
supported the view that material in those structures was capable of obscuring starlight.

Was the obscuring material confined to relatively small regions of interstellar space such as those observed by Herschel and Barnard, or was it more
generally distributed? In the mid-19th century Wilhelm Struve analysed star
counts, *i.e.* the distribution of stars with distance in particular directions. He
showed that these data seemed to imply that the number of stars per unit volume of space declines with distance from the Sun. This apparently privileged
position for the Solar System was untenable. Therefore, Struve proposed
the existence of some obscuring material that was uniformly distributed in
space, and that this material reduced the measured intensity of starlight.
The effect of such obscuration is that these stars would be perceived to be
less bright and therefore apparently further away, so that the number density
of stars at a particular distance would be lower than was actually the case.
However, Struve's hypothesis was largely ignored until Jacobus Kapteyn and
others returned to the issue of star counts in the early 20th century.

In a related study, Robert Trumpler in the 1930s made studies of large,
spherical associations of stars known as globular clusters and found that
the size of these globular clusters, like stellar number densities, appeared
to increase with distance from the Sun. Following Struve, and assuming that
this anomalous behaviour, too, could be understood as a result of a general
interstellar extinction, Trumpler used this information to compute the average amount of extinction per unit distance that would remove this illogical-
ity. The amount of extinction that Trumpler deduced is, in fact, close to the
value obtained from modern measurements. But Trumpler also went further:
from observations made at different visual wavelengths he showed that there
is a wavelength dependence of interstellar extinction in such a way that blue
light is more heavily extinguished than red. Therefore, the spectrum of starlight is "reddened" as it passes through the interstellar medium. The amount
of reddening is easy to measure, and can be used as an alternative measure of
the amount of obscuring material on a particular line of sight.

Many astronomers in the first half of the 20th century wished simply to
obtain clear optical images and spectra of distant stars, to observe their properties so that they could then speculate about stellar origin and evolution
(the main focus of astronomy at that time). For astronomers such as these,
the obscuring material was regarded only as a major impediment to the
detailed observational study of stars. Therefore, most of the early observational work related to this obscuration was aimed at measuring its strength
and its variation with optical wavelength. Then, if extinction properties were

assumed to be the same on all lines of sight, the effects of extinction on the light received from any particular star could effectively be allowed for by a simple calculation, and the observed stellar spectrum could be corrected and the true spectrum obtained. Of course, it was soon recognised from observational studies that the basic assumption of uniformity of extinction on all sight-lines was seriously flawed.

A curious and unexpected observational result related to extinction in the visual region was obtained independently by Hall and by Hiltner in 1949. These authors found that visual starlight is often slightly linearly polarized, at a level of a few percent. The polarization was interpreted as a differential extinction, in which material in the interstellar medium apparently slightly favours the extinction of starlight in one plane of polarization over another.

The lack of uniformity in extinction became even more obvious as measurements were extended into the ultraviolet, when these became possible from (first) rocket-borne and (later) orbiting satellite-borne telescopes. Results obtained by Ted Stecher in 1965 from rocket-borne observations first established the existence of a broad "bump" in interstellar extinction in the near ultraviolet, peaking at a wavelength of about 220 nanometres (nm). Later, satellite-borne observatories showed that the amount of extinction in the far-ultraviolet, beyond the "bump" and less than a wavelength of about 160 nm, rose relative to the visual extinction. On some lines of sight the far-ultraviolet extinction rose very strongly indeed. The variation of extinction with frequency along any line of sight to a star in the Milky Way is called the interstellar extinction curve (or ISEC). Although astronomers referred to an *average* ISEC for the Milky Way, taken over many sight-lines, it was recognised that there is a wide scatter about this average.

Finally, astronomical observations in the infrared became possible from the 1960s, as detectors developed initially for military use became more widely available. Infrared observations have had a profound effect on our understanding of interstellar extinction. Infrared astronomy is carried out either at telescopes that are at high altitude and above much of the atmospheric water vapour, or from telescopes carried outside the Earth's atmosphere on orbiting satellites. To reduce background noise, infrared detectors are normally required to be cooled to very low temperatures. Extinction measurements extended into the infrared showed that the dependence of extinction on wavelength first discovered in the visual region by Trumpler tended to continue into the infrared, but that extinction in the infrared was much less than in the visual region. This was an important result: it meant that infrared observations could penetrate regions that were opaque in the visual region. For example, observations in the infrared confirmed that Herschel's "holes in the sky" were simply regions where foreground obscuration extinguished the visual light of background stars, while infrared radiation was much less impeded, so that those optically "absent" stars could readily be detected as infrared objects.

Infrared continuum emission, and line emission and absorption are also detected from the interstellar medium. Since by the mid-20th century it was already well established that a population of dust grains was the source of interstellar obscuration, much of the observed continuum emission was attributed to thermal radiation from dust grains heated by stellar radiation. The line emissions (from hot material) and absorptions (from cold material) were regarded as signatures of various components of the interstellar dust.

1.2.2   Inferences

While many astronomers wished simply to employ a formula to remove the effects of obscuration from their observed stellar spectra, nevertheless, there were others who sought to understand the source of that obscuration. Trumpler's determination that interstellar extinction (quantitatively defined by astronomers as a logarithmic measure of stellar intensity) was inversely proportional to wavelength in the visual region of the spectrum was perhaps the single most important factor in determining the origin of interstellar obscuration. If the interstellar extinction was caused by particles much smaller than the wavelength of visual light, such as atoms or molecules, then the scattering would be so-called Rayleigh scattering, in which the intensity of scattered light varies inversely as the fourth power of the wavelength. That very strong behaviour was excluded by the observational data. Therefore, extinction must be mainly caused by some other carrier, and attention was turned to solid particles – although it seemed rather unlikely in the early part of the 20th century that such entities could exist in interstellar space.

Calculations of the scattering of electromagnetic radiation by solid spheres, using a theory first developed by Gustav Mie in 1908, showed that it was possible to account for the inverse linear dependence of extinction on wavelength if the size of the spheres was comparable to the wavelength of the radiation. This was a strong indication that interstellar extinction was caused (at least, in large part) by macroscopic particles of a size around a few hundred nm. A range of particle sizes was probably required to match precisely the wavelength dependence of the visual interstellar extinction along any particular line of sight. The extinction was caused because the macroscopic particles were capable of *absorbing* starlight as well as *scattering* it. The starlight absorbed by a particle is a source of energy that tends to raise the temperature of the particle. But the particle also tends to cool by radiating classically, so that an average grain temperature can be established.

Early ideas as to the nature of these postulated particles were guided by the fact that the two most abundant chemically-active elements in the interstellar medium were hydrogen and oxygen. This led to the idea that the particles were ices that somehow nucleated and grew in the interstellar gas. By the mid-20th century, the idea that the interstellar medium was populated by dust grains composed of ice became widely accepted – although that particular model was soon to be rejected.

The discovery of interstellar polarization in 1949 also strongly supported the idea that solid macroscopic particles really do exist in the interstellar medium. But polarization required that some of the particles (at least) must be asymmetric (rather than spherical) and capable of being partially aligned by some external process. The alignment mechanisms discussed included magnetic relaxation effects in which some components of a dust grain's angular momentum are suppressed by the dissipative interaction with an external magnetic field. Alignment of dust grains in gas streaming by supersonic flows was also considered and may be viable in suitable circumstances.

The computation of the properties of extinction and polarization by interstellar dust particles was, nevertheless, poorly defined. It was relatively easy to adjust the many and various parameters involved in a model (*e.g.* dust size distribution, maximum and minimum grain sizes, the dust refractive indices, the total dust: gas ratio, the dust anisotropies, *etc.*) to obtain a good fit to visual observational data of extinction and polarization along any particular line of sight. Mathematically, the problem was poorly constrained, and modelling fits to the observational data were merely illustrative, and could not be regarded as giving definitive information on dust grain populations and properties.

Tighter constraints on the nature of interstellar dust were therefore required. When observations provided data other than information on visual extinction and polarization, the fitting procedures were much tighter, and the fits to the data became much more convincing. For example, light scattered from dust grains near bright stars in optical nebulae is a useful constraint. It provides direct information on the component of starlight that is scattered rather than absorbed by dust, and scattering and absorption are affected by the refractive indices of the dust materials and the grain size distributions. In another example, measurements of extinction in the ultraviolet provide data that seem to require many grains of a much smaller size than those responsible for visual extinction and polarization; therefore ultraviolet observations help to constrain the dust grain size distribution. In a further example, the icy nature of the dust was challenged by the detection of the so-called extinction "bump" at a wavelength near 220 nm, and it was argued that this feature could be attributed to graphite. It was also suggested that graphite grains, by the anisotropic nature of that material and assuming some partial alignment, might be responsible for the observed visual polarization. These ideas opened the door to consideration of alternative materials to ice as the nature of dust grain materials. In any case, it was becoming accepted that serious problems existed with the nucleation of ice particles in the interstellar medium. In fact, the ultraviolet extinction "bump" continues to be associated in current models with sp^2 bonding between carbon atoms that is present in graphite and to varying extents in other carbonaceous materials, including hydrogenated amorphous carbons (HACs) and free-flying polycyclic aromatic hydrocarbons (PAHs).

When infrared observations became available, they began to place even tighter constraints on models of interstellar dust. Continuum emission from dust is an indirect measure of energy absorbed by dust from starlight, so—taken together with measures of the scattered starlight—can, in principle, provide a complete description of the extinction (*i.e.* absorption and scattering) mechanism. Line emission and absorption in the infrared is also very helpful—if not absolutely definitive—in determining the chemical nature of the dust material. Absorptions from cool dust at wavelengths near 10 and 20 micrometres (µm) were assigned to Si–O stretch and bend vibrations in silicates, indicating the presence of a silicate component in the dust. Emissions at a series of wavelengths in the near-infrared (3.3, 6.2, 7.7, 8.7, 11.3, 12.0, and 12.7 µm) were detected from material in the environment of hot stars; these features are considered to be characteristic of the spectra of organic material. While no accurate fit has been obtained of all the observational data with specific organic species or with a mixture of species, these spectral features are taken as indicating that molecular structures similar to conventional PAHs exist in the interstellar medium, and that they radiate when excited by the absorption of starlight. It is likely that the harsh interstellar environment modifies the structure and spectrum of conventional PAHs. It is less certain that observations require PAHs to populate the entire interstellar medium; however, they are commonly assumed to be widespread and to occupy the small size limit of the grain size distribution. In addition to these emission features a weak absorption near 3.4 µm on long low density paths is regarded as characteristic of sp^3 hydrocarbon material.

Deep infrared searches were made for absorption at wavelengths near 3 µm, corresponding to the O–H stretch in water ice, along low density paths in the interstellar medium, initially with negative results. This result supported the arguments that ice would not be able to nucleate in the low density interstellar medium. However, infrared searches in denser regions of the interstellar medium showed that—in those denser regions—water ice could be very abundant, taking up a significant fraction of the available oxygen. It is curious that although the original widely-held suggestion of the nature of dust as ice had been rejected, the suggestion that ice is indeed an abundant component of interstellar dust has returned: ice is now fully accepted as a chemically-important interstellar material. The ices observed to exist in denser regions of interstellar space are now believed to be formed directly when incident oxygen atoms are hydrogenated on the surfaces of dust grains of silicates or carbon. The difficulty of nucleating pure ice particles directly from the gas phase is therefore overcome.

There is a further constraint that can be placed on the dust composition and size distribution. Atoms contained in the dust are obviously not in the gas phase, so if one knows the *total* elemental composition of the interstellar medium and also can determine abundances of the relevant elements present in *gas phase* atoms and molecules, one can deduce very easily the abundances of elements unaccounted for and assumed to be locked up in the dust. While such a calculation does not determine the actual chemical composition of the dust, it provides a useful limiting constraint. For example, it was once suggested that interstellar dust grains were composed of bacteria and viruses. However, these entities contain phosphorus atoms, and the number of phosphorus atoms that would be required for all grains to be bacteria and viruses exceeds the number available in the total composition of the interstellar medium by a factor of the order of one hundred. Evidently, if interstellar bacteria and viruses exist, they can only be a very minor component of interstellar matter.

In practice, an important uncertainty in this calculation is the total composition. While the Sun was once considered a suitable standard, it is now more usual to adopt as standard the relative elemental composition measured in the atmospheres of hot stars. Then by determining the relative elemental abundances in the interstellar medium and subtracting from the hot star relative elemental abundances, one find that roughly half of the carbon atoms, one tenth of the oxygen atoms and almost all of the silicon atoms are locked in the dust. These elements, with some hydrogen and minor amounts of other species, seem to comprise the bulk of material in the dust in the Milky Way.

Many grain models have been proposed. Most current models that are consistent with the observational constraints necessarily adopt a fairly similar composition: they propose a population of dust grains consisting of magnesium/iron silicates, solid carbons and/or hydrogenated carbons, together with a collection of free-flying PAHs, each PAH molecule containing up to around a hundred carbon atoms. The adopted dust grain sizes usually range from values on the order of 1 nm to values on the order of 1 μm, with a size distribution that gives very many more small grains than large. The size distribution conventionally adopted assumes that the number of dust grains per unit volume with radius a is proportional to $a^{-3.5}da$, although there seems to be no *a priori* reason for this choice.

1.3   Why is Dust Important?

In a little over half a century, the way that we perceive dust as a component of the Universe has changed significantly. Up to the mid-20th century, dust was regarded by most astronomers merely as an annoying fog that prevented a clear view of the stars and galaxies that—self-evidently, it seemed—were the most important items in the Universe. The important task for astronomers was to know how to "disperse" that fog (in a theoretical sense) so as to reveal the Universe in its unshielded glory.

The modern view of dust as a component of the Universe is almost entirely the reverse of that earlier view. We now know that almost every aspect of the formation of planets, stars and galaxies is influenced in some way by interstellar dust. A description of modern astronomy that omits the roles of dust is not merely incomplete, it is simply incorrect.

From a chemical point of view, the interstellar medium is a giant laboratory in which the interaction of gases with tiny solid particles at ultra-low pressures and temperatures for unattainably long timescales, enables surface- and solid-state-chemistry to be explored in extreme regimes inaccessible in terrestrial experiments. These cosmic experiments are being performed while the materials are being irradiated at all frequencies from radio to γ-rays. It is not surprising that the study of dust-related interstellar chemistry attracts the interest of chemists as well as astronomers.

The most obvious important role of interstellar dust is a direct result of the extinction caused by dust. Dense interstellar regions (such as that causing Herschel's "hole in the skies", for example) are almost totally shielded from ambient ultraviolet and visual radiation. This extinction permits the rich chemistry of dark clouds to develop, unhindered by the photoionization and photodissociation caused by starlight. Unshielded regions of the interstellar medium, by contrast, have a rather limited chemistry.

To be effective in converting much of the available material into molecular form, interstellar gas-phase chemistry requires the presence of abundant molecular hydrogen, a product of surface reactions on dust grains. This is a hugely important role for interstellar dust.

The astronomical importance of interstellar chemistry is that simple product molecules such as carbon monoxide and water radiate efficiently at low temperatures. An interstellar gas cloud that is collapsing under its own weight will heat up and therefore re-expand unless there is a means of cooling available. Radiation from interstellar molecules and from dust provides that cooling mechanism. The thermal control exerted by interstellar molecules is what permits the formation of galaxies and stars from very tenuous intergalactic and interstellar material in the Universe in its present state of evolution. The whole edifice of galactic evolution in the nearby Universe ultimately relies on the presence of interstellar dust, and as we will see in Chapter 11 it seems inescapable that dust has a crucial role to play even in the very distant Universe.

Through the polarization of starlight, one can map the interstellar magnetic fields that give rise to partial alignment of anisotropic dust grains. The strength and orientation of this field also affects star formation, because a partially ionized gas is closely linked to the magnetic field. Therefore, a magnetic field may provide additional support to a collapsing cloud, in addition to thermal pressure.

The ices that accumulate on the surfaces of dust grains in dense gas in star-forming regions are known to undergo a transformation from fairly simple molecular ices to relatively complex gas phase species, under the influence of heat and radiation from the central star. These molecules are of considerable interest to the emerging subject of astrobiology.

Star formation is the end-point of the collapse of an interstellar gas cloud. Therefore, newly-formed stars are embedded in dense clouds of gas and dust, and largely invisible at visual wavelengths. The intense ultraviolet from the young stars is absorbed by dust and re-radiated in the infrared. Intense infrared emission is therefore a signature of star formation.

During star formation, dust grains accumulate in disks in orbit around the central protostar. Planets form in these disks from the gradual accumulation of dust grains into larger and larger entities. Grain charge may affect the way that the accumulation occurs.

1.4   What is in This Book?

This book is divided into several parts. In the first, Section I, we lay a foundation for the remainder of the book, and discuss in more detail the foundations of the chemical and physical nature of interstellar dust that has been merely sketched out in this introductory chapter. In Chapter 2 we discuss in some detail remote observations of interstellar dust, and then in Chapter 3 we describe the modelling that attempts to interpret the observational data. Extensive laboratory studies that have been made of candidate interstellar dust materials are described in Chapter 4.

In Section II, we discuss the life cycle of dust in the interstellar medium. How is dust formed (Chapter 5) and how does it evolve (Chapter 6)? What is the chemistry of these events? In Chapter 7 we describe dust in the Solar System; this contains dust that is older than the Solar System, *i.e.* is *presolar*. This presolar dust differs markedly from interstellar dust. Why?

The main active chemical roles of dust are discussed in Section III. Chapter 8 describes the chemically-active role of dust in promoting reactions on the surfaces of bare dust grains, here assumed to be either silicates or carbons. This chapter covers both laboratory work and theoretical studies. In Chapter 9, we describe the ices that may be deposited on dust in denser regions, and the processes involved in that deposition. Reactions similar to those described in Chapter 8 may also take place on ice surfaces; they are discussed here. Then we consider the chemical complexity that observations tell us exists in star-forming regions. How can this complexity be developed in simple ices, under interstellar conditions? Laboratory experiments are helpful in showing that such transformations can occur; but the precise mechanisms by which this "cooking" occurs are as yet unclear.

Finally, in Section IV we discuss some of the roles of dust in the Universe at large. In Chapter 10 we consider the roles of dust in star- and planet-forming regions and how dust-promoted chemistry and dust-charging are influential in these roles. Dust is known to exist in galaxies at high redshifts. Is it possible to infer the properties of dust in high redshift galaxies, where physical conditions may be very different from those in the local Universe? Is surface chemistry on dust likely to occur in galaxies near the "observational edge" of the Universe? We discuss some of these issues in Chapter 11. In Chapter 12 we speculate about the possible connection between dust-related chemical complexity found in star-forming regions and the hugely more complex chemistries involved in astrobiology.

Chapter 13 attempts to sum up current achievements and identify areas where further work is needed.

1.5   The Interstellar Medium of the Milky Way

Planet Earth is in orbit around the Sun, a modest star among about a hundred billion stars in the Milky Way. The Milky Way galaxy is but one of many billions of galaxies in the Universe. As we shall describe further in Chapter 11, there is a great variety of galaxies in the Universe: some are dominated by a high formation rate of massive stars that generate intense fields of electromagnetic and particle radiation, while others may be relatively dim and quiescent. However, all galaxies possess, to greater or lesser degrees, gas and dust in the space between the stars. This material, the gas and the dust taken together, is the interstellar medium of that galaxy. It may amount to a significant fraction by mass of the total mass of a particular galaxy. The material of the interstellar medium is important; it is a reservoir of matter from which new stars may be formed: the larger the reservoir, the greater the potential for future star formation. The details of this star formation process (see Chapter 10) can be followed to a great extent through the detection of millimetre-wave emissions from molecules in the star-forming gas. This is a major research area of modern astronomy and one to which the young interdisciplinary subject of astrochemistry contributes enormously.

The nature of the interstellar medium varies from galaxy to galaxy. We shall now describe very crudely the nature of the interstellar medium in the Milky Way, as revealed by observations at wavelengths from radio to X-rays. To some extent, the interstellar medium of the Milky Way is representative of the variety of interstellar structures that one finds in all galaxies. Interstellar gas is almost entirely hydrogen with a significant fraction of helium (about 9% by number of atoms). The chemically-important elements oxygen, carbon, nitrogen are present at a total abundance relative to hydrogen of about 0.1%, while other elements have even lower relative abundances. Table 1.1 shows the relative abundances of some elements found in the interstellar medium of the Milky Way. These abundances can vary from place to place, because stellar nucleosynthesis that powers the stars creates more massive nuclei at the expense of hydrogen and helium and injects those products into interstellar space through stellar winds and stellar explosions such as novae and supernovae.

The interstellar medium is flooded with electromagnetic and particle radiations. The electromagnetic radiation is mainly starlight, and is dominated by the radiation from massive stars. These are relatively few in number, and relatively short-lived, but on average maintain intense sources of ultraviolet and visible light in interstellar space. Visible and ultraviolet radiation absorbed by dust is re-radiated in the infrared, so interstellar space is pervaded by a strong infrared field. Under the influence of a rather tangled

Table 1.1 Approximate relative elemental abundances of some elements relative to hydrogen, by number.

Element	Abundance	Element	Abundance
H	1	Si	3.16×10^{-5}
He	9.77×10^{-2}	Mg	3.63×10^{-5}
O	5.75×10^{-4}	Fe	3.31×10^{-5}
C	2.14×10^{-4}	S	1×10^{-5}
N	6.2×10^{-5}	Na, Ca	2×10^{-6}

galactic magnetic field, ions and electrons generate a radio background. At the other wavelength extreme, powerful galactic and extragalactic sources generate X-rays and γ-rays. Transient events such as supernovae and novae explosions drive interstellar gas dynamics and are sources of radiation flashes.

The whole Milky Way—and probably the entire Universe—is permeated by cosmic rays. These are fast-moving atomic nuclei (mainly protons) and electrons with energies ranging up to $\sim 10^{20}$ eV. However, cosmic rays with energies of a few MeV are very much more numerous and much more effective in ionizing atoms and molecules. Cosmic rays penetrate almost everywhere in the interstellar medium, whereas ultraviolet and visible radiation is excluded from denser dusty regions. Cosmic rays ionize hydrogen molecules inside dark clouds where stellar ultraviolet does not penetrate. In the subsequent recombinations of ions with electrons, emitted photons create a comparatively weak ultraviolet radiation field. This field turns out to have important consequences for both gas-phase and surface and ice chemistries in dark clouds.

The denser, cooler parts of the interstellar medium have an interesting and astronomically important complex chemistry that is generated in part through gas-phase reactions. The chemistry is generally initiated by reactions of gas-phase atoms and ions with molecular hydrogen. Cosmic rays and stellar ultraviolet both have an important role by generating ions which may enter a chemical network of ion-molecule, atom-exchange, radiative association and other types of gas-phase reaction. These gas-phase networks are now fairly well understood, and have been successful in accounting for much of interstellar chemistry. However, as noted earlier, they fail in many important ways and it is now accepted that they must be accompanied by chemical reactions on and in dust grains. The ultraviolet and cosmic rays also play a role in the heterogeneous chemistries, both in initiating the network by creating suitable ions and also by tending to destroy product molecules.

Much of the volume of interstellar space is occupied by a very hot ($\sim 10^6$ K) and low density ($\sim 10^{-3}$ H^+ cm^{-3}) fully ionized gas (mainly protons). The presence of this gas is confirmed, for example, in X-ray emission lines of highly ionized Fe atoms. Ultraviolet absorption lines from highly ionized atoms of O, C, and N also arise in this very hot gas. The ultraviolet absorption lines are also found in the solar corona, so this interstellar component is often (and rather misleadingly) called the *coronal gas*. The gas is heated by the shock waves that supernova explosions send into the gas; it takes a very long time to cool. From a chemical perspective, this gas is not important, as the temperatures are simply too high for molecules to exist. A roughly similar volume of space in the Milky Way is occupied by *warm ionized regions*, detected in atomic hydrogen Hα emission (at 656 nm), in which the number density of protons is much larger than in the coronal gas, ~ 0.3 H^+ cm^{-3}, so that cooling is faster and the temperature is much lower (~ 8000 K). Again, these regions are chemically unimportant. Ionized gas at a similar temperature is found around bright massive stars whose intense ultraviolet radiation maintains a zone of ionized hydrogen (the *HII regions*) within a relatively small volume of interstellar space around the star. These are also unimportant from an astro-chemical point of view. At a roughly similar temperature and density to the warm ionized regions are the *warm neutral regions* but in which conditions are sufficiently different to allow much of the ionized hydrogen to recom-bine; the neutral atoms emit the famous 21 cm wavelength radio radiation. These regions have the potential for chemistry to be established, but forma-tion routes are very weak and destruction through ultraviolet photodissocia-tion is very fast, so chemistry is heavily suppressed. The regions mentioned so far (see Table 1.2) occupy more than 90% of the volume of interstellar space – but only a tiny fraction of the interstellar mass.

The even denser *cold neutral regions* are also much cooler than the warm neutral regions. They are also almost entirely atomic, for the same reasons. However, still denser and cooler gas exists, and it is here that interstellar chemistry begins to be active and to play a role in modifying the thermal properties of the gas. *Diffuse clouds* are denser and cooler than cold neutral regions. The larger density is important, because the dust grains carried with the gas now provide some modest defence against the interstellar radiation field and some chemical processes—involving both surface chemistry on the grains and purely gas-phase chemistries—can generate observable conse-quences. In particular, the gas in diffuse clouds contains a significant amount of molecular hydrogen, and this is overwhelmingly important in promoting an efficient chemistry. A large number of molecular species (more than 30) have been identified in diffuse clouds of the Milky Way. Thus, the transition for significant chemistry to occur requires a favourable density/temperature regime (*i.e.* from warm neutral region to diffuse cloud) and is quite abrupt. *Translucent clouds* have larger densities and consequently the embedded dust grains provide even greater shielding ability against starlight from massive stars. In them, the hydrogen is almost entirely molecular.

A range of *molecular clouds* have densities ranging upwards from a few thousand H_2 molecules cm^{-3}. The external radiation field from bright mas-sive stars is almost totally extinguished by the concentration of dust grains in the gas. Molecular clouds are almost entirely molecular. In many of the dense regions, absorptions from molecules in ices can be detected and the possibilities of chemical processing of ices through solid-state chemistry become available. Extensive relatively dense regions known as *infrared dark clouds* are believed to be the sites where massive stars are formed; these are

Table 1.2 Hot regions of the interstellar medium: very approximate characteristics.

Name	Filling factor	Dimension (pc)	Number density (cm^{-3})	Temperature
Coronal gas	~50%	1000	10^{-3}	10^6
Warm ionized	~40%	1000	0.3	8000
HII	<1%	1	10^2–10^4	8000
Warm neutral	~10%	100	0.3	8000

Table 1.3   Cool regions of the interstellar medium: very approximate characteristics.

Name	Size (pc)	Number density (cm^{-3})	Temperature (K)	Nature of hydrogen	Chemistry present?	Ice present?	Dynamical age (My)
Cool neutral	10	30	100	H	No	No	10
Diffuse	10	100	100	H/H$_2$	Yes	No	3
Translu-cent	1	1000	30	H$_2$	Yes	No	3
Molecu-lar	0.1	10^4	10	H$_2$	Yes	Yes	1
Hot core	0.01	10^7	300	H$_2$	Yes	No	0.01
Warm core	0.01	10^7	100	H$_2$	Yes	No	0.1

currently regions of intense study by astronomers. The densest parcels of gas observed in the interstellar medium are the so-called *molecular cores*, seen near very recently-formed stars of high or low mass (*hot* and *warm cores*, respectively). The densities in these cores are thousands of times greater than in the molecular clouds, and the temperatures are much higher because of the heating caused by the radiation from the very nearby star. The cores are, famously, the sites of the richest molecular distributions and locations of the largest detected interstellar species. Table 1.3 lists some characteristics of the cooler regions of the interstellar medium. Figure 1.1 shows images and false-colour images of a few of these regions in which the effects of dust are prominent.

We use these names—diffuse cloud, hot core, *etc.*—as a convenience. In fact, we should properly think of the interstellar medium as a dynamical system in which atoms may move from one phase, *i.e.* one type of region, to another, as time passes. For example, a hydrogen atom might at one time be in a warm neutral region and at another in a hot core. Gas is moving and evolving under the action of shocks and flows generated by the rotation of the galaxy, by supernova and other explosions, and by the presence of gravity that strives constantly to compress diffuse gas to denser forms. A crude timescale for the existence of each cool phase is indicated in Table 1.3. If chemical effects are more rapid than this timescale, then one may safely ignore evolutionary considerations. Otherwise, one must couple the descriptions of chemistry and of dynamics together.

Throughout this discussion, we have assumed that gas and dust are closely coupled, so that, for example, when the gas density changes, the number density of dust grains changes in proportion. Thus, in denser regions, there are more dust grains per unit volume of space, so that the grains cause a greater extinction of external starlight, and provide a greater amount of surface area per unit volume of space. On a large scale, this constancy between gas and dust is observed to be approximately conserved. However, the ratio of grains to gas may change from place to place, under the actions of, for example,

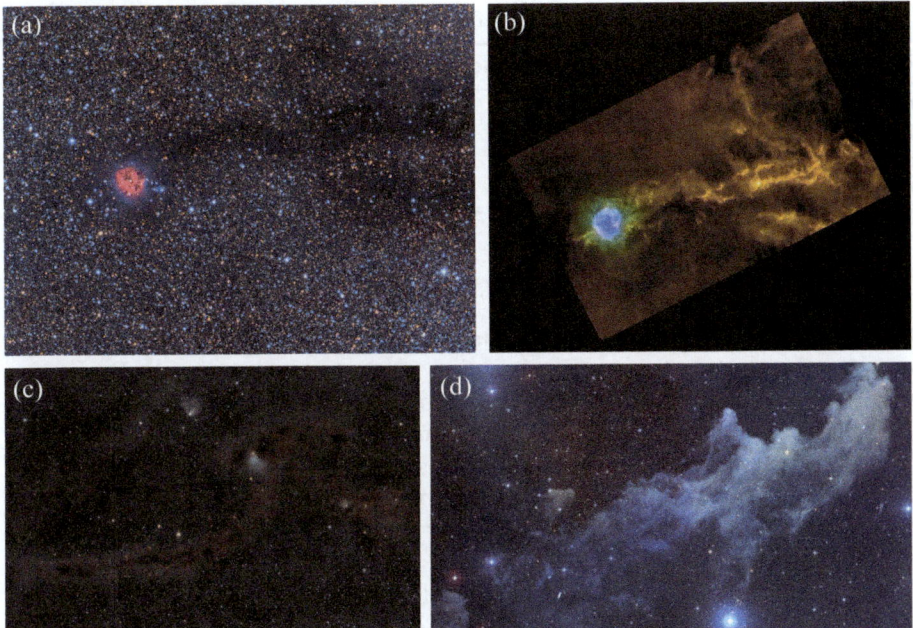

Figure 1.1 (a) Cocoon Nebula and trail of dark interstellar dust clouds, in the visible (credit and copyright: Tony Hallas); (b) as (a) but in the infrared, showing emission from the dust (credit: ESA, SPIRE & PACS Consortia, Doris Arzoumian [CEA, Saclay] *et al.*); (c) Dark Markings catalogued by E. E. Barnard (credit and copyright: Steve Cannistra [Starry Wonders]); (d) Witch Head Nebula, a faint Reflection Nebula, in which grains reflect starlight from Rigel (just outside this image) in the constellation of Orion (credit NASA/STScI Digitized Sky Survey/Noel Carboni).

radiation or magnetic pressures. Also, since grains are in a real sense the "smoke" formed in the envelopes of cool stars of modest mass and from supernovae and novae, a population of these sources of dust may create an enhancement of dust relative to gas in certain regions. Those galaxies with physical conditions that favour the formation of dust-forming stars may have higher dust-to-gas ratios. Very distant galaxies, representing an epoch much earlier in the evolution of the Universe, may have lower dust-to-gas ratios because there has been insufficient evolutionary time to allow the dust-forming stars to evolve and to pollute their environment with dust (see Chapter 11).

There are also various regions in near-stellar environments in which chemistry may be important. These include the envelopes of cool stars (in which dust grains may form), planetary nebulae, the ejecta of novae and supernovae, and circumstellar disks. In some of these regions, grains may have an important chemical and/or physical role, and some of them are sources of dust (see Chapter 6). Some information about these objects is given in Table 1.4.

Table 1.4   Near-stellar environments of chemical importance.

Name	Number density (cm^{-3})	Temperature (K)
Cool stellar envelopes	10^{12}	2000
Planetary nebulae	10^6	200
Supernova ejecta	10^{12}–10^5	4000
Circumstellar disks	10^{10}–10^{12}	10–100

1.6   Physical Units in This Book

Chemists are notoriously punctilious in their use of SI units. Astronomers, on the other hand, are cavalier in using whatever units seem to be convenient. To some extent, this attitude can be excused, because of the remarkable range of quantities with which astronomers must deal. It is of little value to define interstellar distances or dimensions in terms of metres, and it is acceptable to introduce some more appropriate measure. In the case of distance, this is the parsec, the light year, or the Astronomical Unit. These and other units are defined in Table 1.5.

Astronomers conventionally use cm^3 rather than m^3 as a unit volume when describing number densities of gases. There is little justification for this, except perhaps that the average number density of hydrogen atoms in the interstellar medium of the Milky Way is about 1 cm^{-3}. It seems more cumbersome to write 10^6 m^{-3}. In any case, it is useful to refer to a molecular cloud as having, say 10^4 H$_2$ molecules cm^{-3}, and recognising at once that this means that it is about 10^4 times as dense as the mean interstellar density in the Milky Way.

In measuring wavelengths of radiation, astronomers have moved partly into the modern era, using SI units such as nanometre, micrometre (or micron), millimetre, metre and kilometre where appropriate. At the same time, astronomers retain a deep-rooted affection for the Ångström (= 0.1 nm) and the centimetre (= 0.01 m).

Astronomers are also shockingly indiscriminate in their use of energy units adopting, as seems appropriate, the electron volt, the inverse centimetre or inverse micron, the Kelvin, and Joules (and some astronomers still even prefer ergs). We shall follow conventional astronomical practice. Approximate conversion factors for energy units are given in Table 1.6.

Finally, we need to discuss the units used by astronomers to measure the amount of extinction caused by dust in interstellar clouds. Extinction is the loss of light from a background star caused by the absorption and scattering of light by dust grains along the line of sight through the interstellar cloud. Extinction is a useful measure of the amount of dust in the cloud, and—by extension—of the amount of gas.

Astronomers measure extinction in units known as *magnitudes*. The term goes back to ancient astronomy when the optical brightness could only be estimated by eye; the brightest stars were said to be of first magnitude, while less bright stars were said to be of second or third, *etc.*, magnitude.

Table 1.5 Some astronomical data.

Units of length	
Parsec	1 pc = 3.086×10^{16} m
Light year	1 l.y. = 9.461×10^{15} m
Astronomical unit	1 AU = 1.496×10^{11} m

Mass of object	In solar masses	In kilogrammes
Milky Way galaxy	$1.0–1.5 \times 10^{12}$	$2.0–3.0 \times 10^{42}$
Sun	1	1.989×10^{30}
Jupiter	9.54×10^{-4}	1.897×10^{27}
Earth	3.00×10^{-6}	5.97×10^{24}
Moon	3.695×10^{-8}	7.35×10^{22}

Table 1.6 Some energy units.

	J	eV	cm^{-1}	K
1 Joule	1	6.24151×10^{18}	4.8675×10^{22}	7.24297×10^{22}
1 electronvolt	1.6022×10^{-19}	1	8065.5	11604
1 inverse cm	2.0544×10^{-23}	1.2398×10^{-4}	1	1.4388
1 Kelvin	1.38065×10^{-23}	8.6173×10^{-5}	0.69502	1

The faintest stars that can be detected by eye are of sixth magnitude. In the 19th century, photometric measurements became possible and showed that a first magnitude star was one hundred times brighter than a sixth magnitude star. Thus, a five magnitude change corresponds to a brightness change of a factor of one hundred, and this relationship became the basis of the modern magnitude scale. With this definition (eqn (1.1)), the magnitude is approximately equivalent to the optical depth, a term in more general use in other branches of science.

$$A_\lambda = 2.5 \log_{10} (I_0/I) \qquad (1.1)$$

Here, A_λ is the extinction in magnitudes at wavelength λ, caused by dust along the line of sight through an interstellar cloud, of the radiation from a background source whose measured brightness emerging from the cloud is I but whose brightness in the absence of extinction would have been I_0.

Extinction measurements caused by dust along the line of sight through an interstellar cloud to a background star require the background star to have a known brightness at a particular wavelength. Background sources are therefore selected, where possible, to have a well-characterized spectrum and spectral energy distribution. Of course, allowance must be made for the distance between the star, the cloud, and Earth. Extinction measurements may be made at any wavelength. Extinction curves (see Chapter 3) are usually curves showing extinction as a function of inverse wavelength, with the values of A_λ normalized to the value at a specified wavelength – often taken to be the visual wavelength, $\lambda_V = 0.55$ μm. With this normalization, *i.e.* A_λ/A_V, one may compare extinction on one line of sight with another, regardless of how much dust is on the various lines of sight.

Interstellar dust is mainly in clouds, but on sufficiently long paths in the Milky Way galaxy the amount of dust increases linearly with path length at a rate of about 1.8 magnitudes per kiloparsec. On the large scale, dust and gas are well mixed and in the Milky Way the average column density of hydrogen (in both atoms and molecules), N_H, and the average visual extinction, A_V, taken over many lines of sight, are related as in eqn (1.2).

$$(N_H/A_V) = 1.9 \times 10^{21} \text{ cm}^{-2} \text{ per magnitude} \qquad (1.2)$$

Further Reading

The following texts give a general background for the study of gas and dust in the interstellar medium.

1. B. T. Draine, *Physics of the Interstellar and Intergalactic Medium*, Princeton Series in Astrophysics, 2011.
2. J. E. Dyson and D. A. Williams, *Physics of the Interstellar Medium*, Institute of Physics Publishing, 1997.
3. S. Kwok, *Organic Matter in the Universe*, Wiley, 2011.
4. A. G. G. M. Tielens, *The Physics and Chemistry of the Interstellar Medium*, Cambridge University Press, 2005.
5. D. C. B. Whittet, *Dust in the Galactic Environment*, Institute of Physics Publishing, 2003.
6. D. A. Williams and T. W. Hartquist, *The Cosmic-Chemical Bond: Chemistry from the Big Bang to Planet Formation*, RSC Publishing, 2013.
7. D. A. Williams and S. Viti, *Observational Molecular Astronomy*, Cambridge University Press, 2013.

Section I

Defining the Chemical and Physical Nature of Interstellar Dust

CHAPTER 2

Remote Observations of Interstellar Dust

2.1 Introduction

In this book we are concerned with the chemical and physical nature of inter-stellar dust grains and their ability to promote surface and solid-state chem-istries that contribute to the wider network of interstellar chemistry. The nature of the dust is obviously crucial to these active chemical roles. How do astronomers obtain information about interstellar dust?

Most of the information that we have about interstellar dust is obtained remotely, by the influence of dust on various kinds of astronomical obser-vations. These observations may be carried out at a very wide range of wave-lengths, from X-rays to radio, but traditionally the most important and defining data have come from the infrared, visible, and ultraviolet parts of the spectrum. These remote observations were at one time the only sources of information about interstellar dust. However, it is now also possible to find samples of interstellar dust grains within interplanetary dust particles collected by high-flying aircraft and by satellite experiments, and in other col-lected samples of Solar System material. Thus, some interstellar dust grains can now be examined in the laboratory. We shall describe such "hands-on" studies in Chapter 7.

As we mentioned in Chapter 1, the existence of interstellar dust was inferred from the extinction of starlight in dark regions revealed against the background star fields, and also as a general background of extinc-tion throughout interstellar space. The dependence of extinction on wave-length indicated the presence of small solid particles of a size comparable with the wavelength of visible light. Extinction studies still remain crucial

The Chemistry of Cosmic Dust
By David A. Williams and Cesare Cecchi-Pestellini
© David A. Williams and Cesare Cecchi-Pestellini 2016
Published by the Royal Society of Chemistry, www.rsc.org

in constraining the physical and chemical nature of interstellar dust, as we will see in Chapter 3. Any new ideas about the nature of dust must be immediately tested by comparing the extinction predictions of an appropriate model with observational data. There are detailed extinction measurements along hundreds of lines of sight in the Milky Way, and less accurate extinction data for the interstellar media of external galaxies. Another phenomenon, related to extinction and mentioned in Chapter 1, is interstellar polarization: observations show that partially extinguished visible starlight reaching Earth is often slightly linearly polarized, implying that radiation in one plane of polarization is more severely extinguished than radiation in the orthogonal plane. This is often attributed to a partial alignment of dust grains that are (also) of a size comparable to the wavelength of visible light. Polarization measurements can constrain the shape and—through the various alignment mechanisms—also the grain composition.

As we will see in Chapter 3, the procedure to fit an interstellar extinction curve is not a severe constraint on a model, because there are a significant number of free parameters defining the dust grain size distribution, the grain structures, and the dielectric constants of the materials involved. Fortunately, there are many other observational constraints on possible models of dust grains, perhaps the most important of which is infrared spectroscopy. This technique identifies individual bonds in grain material (such as Si–O, or C–H, as we will see), rather than complete structures. These features can be found in emission or absorption, and appear in the near- and mid-infrared. Other phenomena throwing light on the physical and chemical nature of dust are the extended red emission (ERE)—a photoluminescence in the visible and near-infrared—and light scattered by dust grains either in the general interstellar medium to form the diffuse galactic light (DGL) or—locally—in a particular reflection nebula (an association of a bright star with a nearby dust cloud, from which scattered starlight can be detected, see Figure 1.1d). Observations of scattered light are important constraints on the optics of grains; while extinction is a result of both absorption and scattering, the scattering observations in principle allow contributions from both effects to be identified. The scattering by dust of X-rays through small angles leads to the appearance of a diffuse "halo" around the source, and this also provides a constraint on models.

As mentioned above, extinction by dust grains arises from both absorption and scattering. What happens to the absorbed radiation? It is a heat source for the dust grains, and the energy it represents is radiated away in cooling radiation in the infrared, at wavelengths of a few microns to about one millimetre. This radiation is readily detected, and observations provide a further constraint on dust models.

Finally, there is emission at centimetre wavelengths emitted from grains rotating with rates ~5–100 GHz. The high rotation rate is maintained by kinetic collisions with gas molecules, and the emission arises when the grains have electric dipole moments.

In addition to these direct observations involving dust grains, there is (as we described in Section 1.2.2) another important body of work that provides an indirect constraint on the composition of dust. Atoms in the interstellar medium can either be in the gas phase or in the dust. There is no other alternative. So if the total abundance of a particular element in the interstellar medium is known (as given in Table 1.1), and the gas phase abundance of that element can be measured by conventional spectroscopy, then the difference between the two must be the abundance of that element that is locked up in dust grains. For example, if gas-phase silicon in the interstellar medium is only 1% of the total silicon abundance in the interstellar medium then—assuming that there are no silicon-bearing molecules present (which may be the case in diffuse clouds)—we must infer that 99% of the available silicon is in the dust grains. Of course, this method does not determine the chemical nature of the dust, but it does constrain the chemical composition.

This is a formidable list of observational constraints (but they are not all equally stringent)! Nevertheless, there are viable models of interstellar dust grains that meet these constraints (we will see them in Chapter 3), and it is therefore possible—as we shall see in Sections III and IV of the book—to explore in great detail the chemical consequences of dust in the Universe. In the remainder of this chapter, we will look in slightly more depth at some of the observational phenomena that we summarized above. Additional information may be found in the indispensable and comprehensive textbook on Milky Way interstellar dust by Whittet (2003),[1] in the superb review by Draine (2003)[2] and in Draine's (2011)[3] textbook (see Further Reading).

2.2 Observational Phenomena Related to Interstellar Dust

2.2.1 Interstellar Extinction

The extinction of starlight by interstellar dust at wavelength λ is written A_λ and is measured in units of magnitudes (see Chapter 1). The variation of interstellar extinction with wavelength is often expressed as A_λ/A_V where V corresponds to the wavelength 550 nm. The variation of A_λ/A_V with wavelength λ as deduced from observations (the interstellar extinction curve, or ISEC) is used to test the validity of any proposed model of interstellar dust. By convention, the curves are usually plotted with respect to $1/\lambda$, (measured in inverse micrometres, μm^{-1}), effectively an energy measurement, in which measurements at long wavelengths (the infrared) are on the left and at short wavelengths (the ultraviolet) are on the right.

ISECs have been measured along many lines of sight through the diffuse gas of the interstellar medium in the Milky Way, and at wavelengths from the infrared to the ultraviolet. These lines of sight pass through diffuse clouds and are such that A_V is usually less than about one magnitude per cloud.

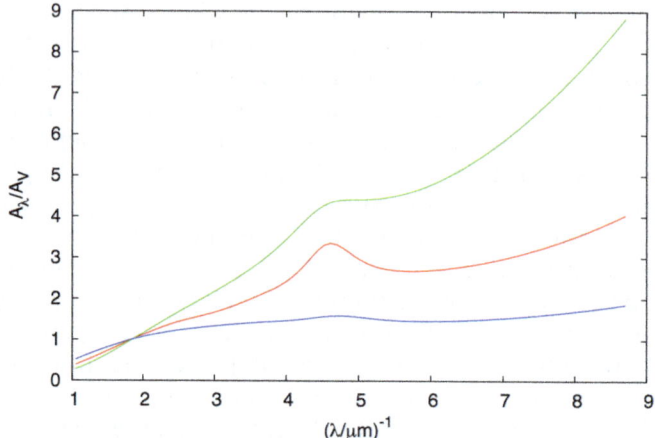

Figure 2.1   Normalized interstellar extinction curves from observations of the Milky Way galaxy: lines of sight toward stars HD37130 (green line) and HD210121 (blue line) provide ISECs that have extreme departures from the average for the Milky Way ISECs, also shown (red line). Data taken from Fitzpatrick & Massa (2007).[4]

Figure 2.1 (ref. 4) shows three ISECs (A_λ/A_V plotted as a function of λ^{-1}) for the Milky Way; one is the ISEC for an average of results for many lines of sight in the Milky Way, and two for ISECs that are extreme outliers from that average.

The three curves are different but are clearly members of the same family of ISECs. The family characteristics are (i) low extinction in the infrared and declining with increasing wavelength, (ii) a near-linear rise in the region 1–2 μm^{-1} (*i.e.* wavelengths ~1000–500 nm), (iii) a reduction of slope near 2–3 μm^{-1} (sometimes called the "knee"), (iv) a pronounced and fairly symmetric "bump" near 4.6 μm^{-1}, (v) all ISEC bumps have the same central wavelength of 217.5 nm, (vi) a minimum around 6 μm^{-1} (1670 nm), and finally (vii) a steep rise into the far ultraviolet (10 μm^{-1} corresponds to a wavelength of 100 nm). The curves differ mainly in having weaker or stronger bumps, and in having steeper or shallower rises towards the far ultraviolet. The visual and infrared regions give less information because the normalization at wavelength *V* means that the curves are somewhat similar in that vicinity. However, there appears to be no common behaviour in the infrared.

The implication to be drawn seems to be that grain models must be complex. There may be different "carriers" for each of the different components of the ISECs, so that differences in the abundances of the different carriers produce the variations on a standard curve. We shall see in Chapter 3 that this approach is indeed adopted by current models of interstellar dust. No single-component grain model can currently account for the details of ISECs and the variations from one line of sight to another. In some models, the various components are essentially independent, but in Chapter 3 we shall meet one grain model in which the various components are related in a "unified model".

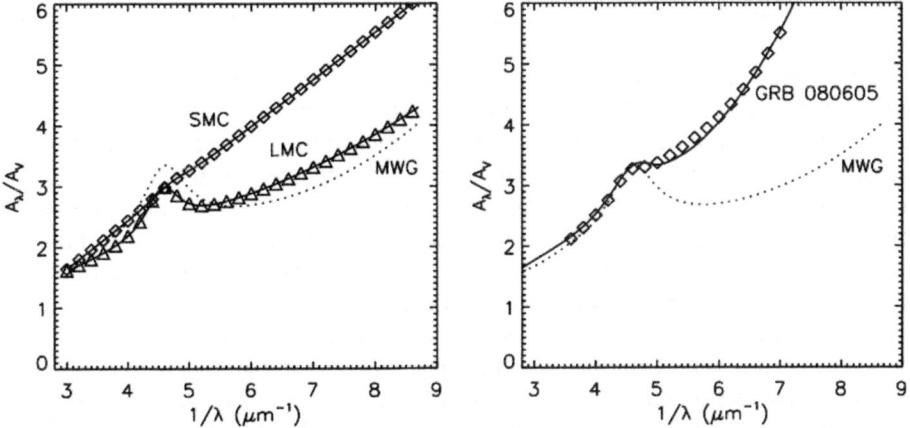

Figure 2.2 Normalized ISECs observed for some nearby and distant galaxies, with the Milky Way ISEC for comparison. Data for AzV 18 (SMC) and Sk-68 129 (LMC) from Cartledge *et al.* (2005),[5] for GRB 080605 from Zafar *et al.* (2012).[6]

Is the Milky Way rather special, or is dust in external galaxies generally similar to that in the Milky Way? We show in Figure 2.2 (ref. 5 and 6) some interstellar extinction curves for some nearby and distant galaxies compared to the average curve for the Milky Way. It should be remembered, however, that determinations of interstellar extinction in external galaxies are extremely difficult and are attended by significant errors.

The curves shown in Figure 2.2 indicate that ISECs for external galaxies generally belong to the same family as the Milky Way curves. Although the curve for the Small Magellanic Cloud (SMC, a near neighbour of the Milky Way) looks rather different, it is merely because the bump is especially weak in that galaxy, compared to that in the Milky Way. The Large Magellanic Cloud (LMC) has an ISEC that is closer to the average ISEC for the Milky Way. An ISEC for a quite distant galaxy (GRB 080605) and that for the line of sight to the Milky Way star HD210121 seem rather similar, both to each other and to the ISEC for the Small Magellanic Cloud.

2.2.2 Interstellar Polarization

A partial alignment of asymmetric dust grains can arise through the action of a variety of physical processes. Where this occurs, some linear polarization of starlight (typically, up to a few percent) passing through a diffuse interstellar cloud containing partially aligned grains can be created and is detected on many lines of sight in the Milky Way. In fact, linear and circular polarization of radiation is detected from a wide variety of astronomical sources including active galactic nuclei, circumstellar disks, circumstellar envelopes, and even comets, and quite large degrees of polarization can be detected in some objects.

There are various mechanisms by which alignment can be achieved, and these are discussed in Draine's (2011)[3] textbook and at some length in the review by Lazarian (2007).[7] The various mechanisms may operate in regions with different physical conditions. In principle, each mechanism also depends on the physical and chemical nature of dust grains and therefore the models they generate should be capable of providing information about the grains, when confronted with the relevant observations. In fact, because of uncertainties in the astronomical data and in the complexities of the models, it is difficult to gain much new information about dust.

As mentioned in Chapter 1, alignment was traditionally thought to be achieved by the dissipation of a component of a grain's rotational energy through paramagnetic or superparamagnetic coupling to the magnetic field. Superparamagnetic grains are those containing ferromagnetic inclusions; as we will describe in Chapter 7, collected interplanetary particles called GEMS (*g*lasses with *e*mbedded *m*etals and *s*ulfides) do contain inclusions of iron sulfide (FeS), a ferromagnetic material. In general, the alignment created by these and other processes affects the larger grains, but not the smaller grains. If superparamagnetic alignment is occurring, then polarization gives some information about ferromagnetic inclusions in the grain material.

However, several other alignment mechanisms are now believed to be important. Mechanical alignments may occur when grains are affected by gas flows; both supersonic and subsonic flows may generate partial alignments. Torques introduced by the impact of an anisotropic radiation field can both spin grains and align them. Systematic torques introduced by H_2 formation at specific surface sites, by photoelectric emission, and by variation in adsorption accommodation coefficients over the surface of a grain may all affect the degree of alignment achieved by a grain. It appears the alignment by these mechanisms takes place with respect to the local magnetic field, even if the field is not involved in the alignment mechanism; thus, the field direction is indicated by the polarization generated by the dust grain alignment. Radiative torque alignment is now considered to be an important mechanism and appears to be significant in a variety of interstellar situations. It depends on the grains being irregular and rotating.

Linear polarization in the diffuse interstellar medium can be measured at various wavelengths through the visual from near-infrared to near-ultraviolet and shows a broad asymmetric curve peaking at a wavelength in the visual of λ_{max}. It is a remarkable result that the dependence of the polarization, p_λ, on wavelength, λ, is generally well-fitted by one of a number of empirical curves, such as

$$p_\lambda = p_{max} \exp\left\{-K \ln^2 (\lambda_{max}/\lambda)\right\} \qquad (2.1)$$

where p_{max} is the peak polarization and K is a constant close to unity. Optical theory suggests that for dielectric grains the value of λ_{max} is determined by the grain size and the refractive index of the grain material. Thus, the wavelength of maximum polarization gives some information about the grain material and the size distribution; unfortunately, this information is not very precise.

2.2.3   Spectroscopy in the Interstellar Medium

2.2.3.1   *Absorption Features in the Interstellar Extinction Curves in Diffuse Interstellar Gas*

The dominant feature in the ISEC is the bump at a wavelength of 217.5 nm in the near-ultraviolet (or $\lambda^{-1} = 4.6\ \mu m^{-1}$). It is present in many ISECs. Its central position is fixed, although the width of the feature can vary significantly from one line of sight to another. As we shall discuss further in Chapter 3, this is attributed to $\pi* \leftarrow \pi$ transitions in graphite or in PAHs. In the near-infrared, there is a weak absorption feature at a wavelength of 3.4 μm present in the ISECs, detectable on long paths through diffuse gas. It is attributed to absorption in the sp^3 C–H stretching bond. Taken together, these two features strongly support the presence of some form of carbon/hydrocarbon as a component of interstellar dust. Further into the infrared are two absorption features in ISECs, at 9.7 μm and 18 μm. Features near 10 and 20 μm are known to arise in silicate materials, from Si–O stretching and O–Si–O bending modes, respectively. These features are attributed to some form of silicate in the dust.

There is a large set of unidentified absorption bands arising in the diffuse interstellar medium of the Milky Way and also in nearby galaxies: these features are the diffuse interstellar bands (DIBs). The lowest wavelength feature is a very broad feature and centred at 443 nm, and the set has several hundred members in the visible into the near-infrared. It is embarrassing to report that even after about a century of study, these features remain unidentified. There are two schools of thought about the carriers of these features; firstly, that they are solid-state features arising in dust grains, and secondly, that they are associated with large molecules flying freely in the interstellar medium. Current opinion favours the latter view, and that the features belong to poorly-characterized complex structures, probably of an organic nature.

2.2.3.2   *Absorption Features in the Interstellar Extinction Curves in Dense Interstellar Gas*

Along lines of sight through denser gas, on which the extinction to background stars is larger than a few magnitudes, additional absorption features are found in the interstellar extinction curves. The most pronounced feature is centred at a wavelength of 3.1 μm and is very broad (typical full width at half the maximum of the feature, ~0.3 μm). This feature is attributed to the

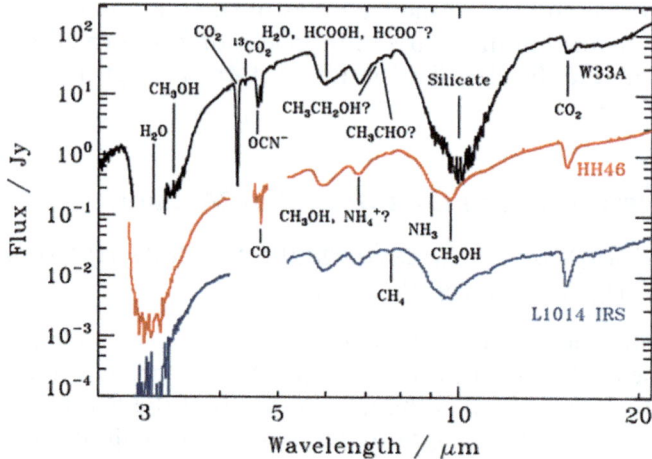

Figure 2.3 Infrared absorption spectra from cold dust along the lines of sight toward three protostars, W33A, HH46, and L1014 IRS. Assignments of the specific spectral features are indicated. Reproduced with permission from Öberg *et al.* (2011).[8]

O–H stretching mode of water ice, assumed to be deposited on the surface of dust grains. Absorption at 3.1 μm is accompanied by other H_2O–ice absorptions and by features associated with ices of some other simple molecular species, including CO, CO_2, CH_4, H_2CO, CH_3OH, and NH_3 (see Figure 2.3).[8] These ices are discussed further in Chapter 9.

2.2.3.3   Emission Bands in the Interstellar Medium

The temperatures of dust grains are typically very low, ~10 K. However, if interstellar material is relatively close to a bright star, then dust (and other materials) may become hot enough to emit in spectral features that are diagnostic of the dust material. The silicate absorption features at wavelengths of 9.7 and 18 μm found in cold dust (discussed above) are also found in emissions from dust that has a temperature of more than a few hundred K.

There is also a set of detected infrared features variously known as the aromatic infrared bands (AIBs), alternatively known as the unidentified infrared bands (UIBs), or the PAH features. The main features occur at 3.3, 6.2, 7.7, 8.6, and 11.3 μm (see Figure 2.4).[9] Although no perfect match has been obtained with features from any organic molecule, the AIBs are commonly accepted as arising in free-flying interstellar PAHs; however, these interstellar PAHs may not be exactly similar to laboratory samples since they exist in relatively harsh conditions. A single PAH is assumed to be heated by a single ultraviolet photon, and in relaxing from an excited state emits in various C–H and C–C stretching and bending modes. A more detailed discussion of these features is given in Chapter 4.

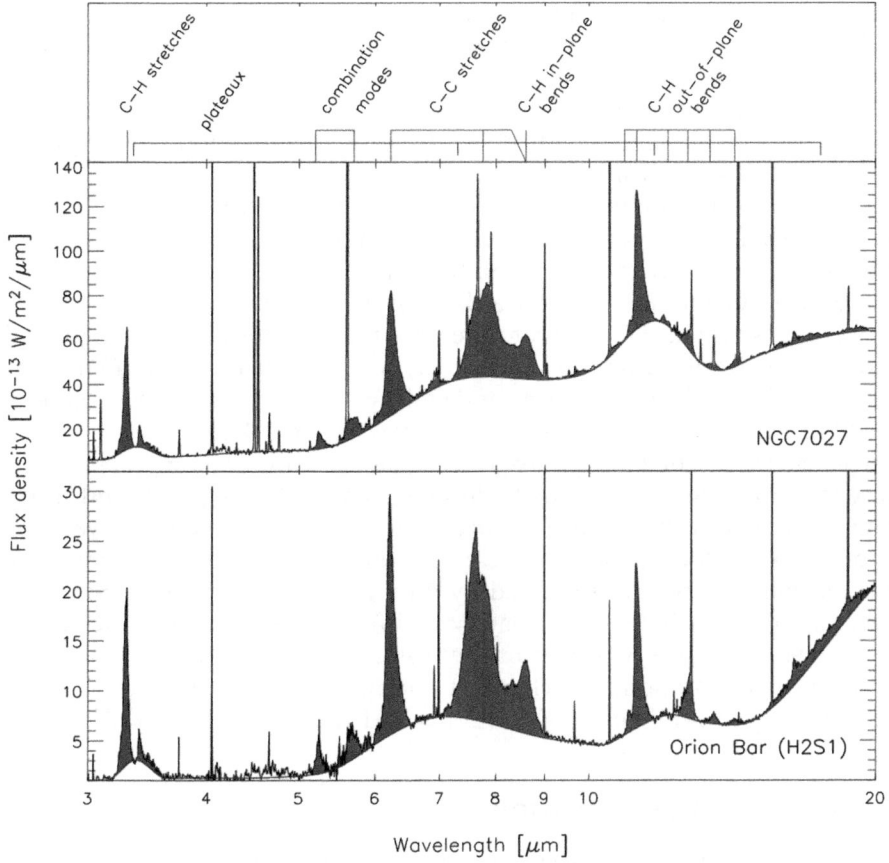

Figure 2.4 Infrared emission spectra from warm dust associated with the planetary nebula NGC 7027 and with the Orion Bar. Assignments of the features are indicated. Reproduced with permission from Peeters *et al.* (2004).[9]

2.2.4 Scattering by Dust Grains

Measurements of scattered light are usually expressed in terms of the effective albedo (*i.e.* the reflectivity, with values in the range 0–1) of the grains, and the scattering asymmetry factor, g, computed at various wavelengths. The factor g describes the tendency of the grains to scatter in forward ($g = 1$) or backward ($g = -1$) directions, isotropically ($g = 0$), or in some intermediate way. Measurements of light scattered by dust can best be made in reflection nebulae where the location of the star (the source of the radiation) relative to the nearby scattering dust is reasonably clear, and the spectral energy distribution of the stellar radiation is also known. It is also possible, but more difficult, to measure the effective albedo from the DGL, *i.e.* radiation from the mean interstellar radiation field that has been scattered by the dust grains. The DGL is weak and is contaminated by local sources of radiation such as faint stars in the field and by other sources such as fluorescence from inter-stellar gas phase H_2. Nevertheless, measurements from both reflection neb-ulae and the DGL give rather similar results for the effective albedo of Milky Way dust, within the rather substantial error bars. Obviously, these mea-surements give an effective albedo for the whole population of dust grains, including grains of all sizes.

The inferred albedos in the near-infrared and visible are typically fairly high at about 0.6, falling through the blue to a minimum of about 0.4 in the near-ultraviolet (~200 nm), then rising to high values (~0.8) in the mid-ultraviolet (~160 nm) but falling again towards shorter wavelengths (~100 nm). The scattering asymmetry factors for reflection nebulae are generally high at all wavelengths from infrared to ultraviolet, typically ~0.8. For the DGL, how-ever, there is considerable scatter, and while g is large in the visible it appears to fall in the ultraviolet at wavelengths below about 160 nm.

The chemistry of dust grains is accessible to direct study using X-ray absorption features. Soft X-ray features—such as the oxygen K and iron L edges—are currently accessible. Future instrumentation should give direct information on Mg and Si edges near 2 keV and on Fe edges at 7.1 keV. X-rays from a point source are scattered by dust grains so that the point source appears to have a halo which may extend for nearly 1° around the source. Spatially-resolved spectroscopy of the scattering halo will be possible using future instrumentation. The chemistry of dust grains can be directly deter-mined using a combination of scattering and absorption of X-rays.[10]

2.2.5   Continuum Emission

2.2.5.1   *Continuum Emission in the Far-Infrared*

The absorption of starlight by dust grains is a significant heating source for the grains. A realistic steady-state temperature can be defined for all but the smallest grains. For a particular grain, this temperature takes a value that achieves a balance between the energy input rate and the rate of cooling by radiation. For typical grain temperatures of, say 10–30 K, then this emitted radiation peaks in the range 300–100 μm. However, smaller grains, with radii less than or about 5 nm, do not have a steady-state temperature; their temperature fluctuates, rising rapidly to a peak when a ultraviolet photon is absorbed and falling more slowly as this energy is distributed among various bending and stretching modes of the material. The energy is ultimately radi-ated away in discrete near-infrared lines, as discussed above, together with an underlying continuum. Thus, the infrared emission spectrum from dust grains may be crudely considered as arising in two parts: the small grains contribute a spectrum dominated by lines, extending in wavelength from about 3 μm up to about 20 μm, while a smooth continuum extends to around a thousand μm, peaking at a few hundred μm. Most of the energy is in this smooth continuum.

From an astronomical perspective, the far-infrared emission is a useful tracer of recent star formation, since newly-formed stars, powerful sources of ultraviolet radiation, are embedded within the dense and dusty clouds in which they were formed. In these clouds, the dust transforms the energy that was in ultraviolet and visual radiation into the far infrared, so that observations of the long wavelength emission give a measure of the amount of star formation. However, the observations do not discriminate sensitively between different models of interstellar dust. The far infrared emission is determined by the size distribution and the refractive indices of the grain material.

2.2.5.2 *Continuum Emission in the Visual and Near-Infrared*

Photoluminescence is a commonly observed phenomenon in the laboratory; in this process a higher energy photon is absorbed by a material and re-emitted at lower energy. It is not surprising, therefore, that a phenomenon of this kind is observed in the interstellar medium. Emission in a very broad, featureless band is observed in some sources to extend from about 500 nm to about 900 nm, peaking at about 700 nm; the actual values vary from one source to another (ERE). The emission is often found in regions in which the AIBs are prominent. Since the AIBs are almost certainly due to carbonaceous material, it is reasonable to suppose that the ERE is photoluminescence from some form of carbon. Small particles of HAC with mixed sp^2/sp^3 bonding have been shown in the laboratory to reproduce the ERE profile and the necessary photon efficiency, but also to generate an additional peak at 1 μm that is not observed.[11] No generally accepted identification of ERE with a carbonaceous solid has yet been made. Models based on silicon-based solids have been unsuccessful.

In principle, the ERE should be a powerful test of dust grain models, but further work is required before the ERE can be used to constrain models of interstellar dust.

2.2.5.3 *Continuum Emission in the Centimetre Wavelength Range*

Continuum emission in the frequency range 5–100 GHz has been observed in many sources in the Milky Way and has been attributed to emission from rotating dust grains. Small grains rotate at frequencies in this range, and if they possess dipole moments will certainly radiate. Larger grains with magnetic inclusions (as found in interplanetary particles, see above and Chapter 3) will also radiate magnetic dipole radiation. Smaller grains will generally rotate at higher frequencies than larger grains. The emission from both large and small grains is controlled largely by the grain size distribution but gives little information about the grain composition. However, it appears to be necessary that large grains contain inclusions of magnetic material. This is an important constraint.

2.2.6 Depletions

The elemental depletion is a measure of the underabundance of that element in the gas phase of interstellar medium with respect to an adopted standard. This standard is often taken to be the list of relative elemental abundances measured in the atmospheres of hot stars whose temperature is such that any dust grains that were in the gas are now returned to the gaseous state. Other choices for the adopted standard might be the list of measured relative elemental abundances in stars hotter than the Sun. The depletions are usually given for lines of sight through diffuse clouds. The assumption is made that atoms missing from the expected standard are incorporated in dust grains. Dust grains in these regions do not have coatings of ice, so the depletions give a measure of the relative abundances of elements in refractory grains.

We show in Table 2.1 (ref. 12) depletions in the diffuse interstellar medium for a range of elements. Some, such as nitrogen, (shown in the table), sulfur, and the noble gases (not shown) have depletions that are near zero, meaning that these elements are not heavily depleted and are not significantly present in the dust material in diffuse clouds (but see also Chapter 7). Others, such as iron or magnesium, have depletions that indicate much of that element is missing from the gas and is—presumably—a significant component of the dust grain material in diffuse clouds. Of course, in these considerations we have to take account of the particular element in all forms. For example, in diffuse clouds carbon may exist as C^+ ions, C atoms, CO molecules, and various hydrocarbons, as well as being in a solid form as part of the dust.

While depletions are helpful in constraining the chemical composition of dust, some anomalies remain. Whittet (2010)[13] has noted that oxygen atoms in known gaseous species and locked in solid silicates and metal oxides do not account for all the oxygen assumed to be present. Evidently, another interstellar component contains a significant amount of oxygen. Since cometary particles are found to contain oxygen-bearing carbonaceous matter,

Table 2.1 Depletions of the elements in diffuse interstellar gas. The elemental abundances are presented as the numbers of atoms of a particular element relative to 10^6 H atoms. The standard abundance is measured in the atmospheres of hot stars, in which dust is not present. The abundance in the interstellar gas (measured in the same way) is, apart from nitrogen, much less. The difference shown in the final column is assumed to be present in dust. It gives an approximate relative composition of interstellar dust. Data from Nieva & Przybilla (2012).[12]

Element	Standard abundance	Interstellar gas	Interstellar dust
C	214	91	123
N	62	62	0
O	575	389	186
Mg	36.3	1.5	34.8
Si	31.6	2.2	29.4
Fe	33.1	0.3	32.8

Table 2.2 Sources of information about dust grain properties from astronomical observations.

	Extinction	Polarization	Absorption lines (diffuse)	Absorption lines (dense)	AIBs	Scattered light	Continuum emission	Depletions
Size distribution	x	x				x	x	
Asymmetry		x						
PAHs			x		x			
Elemental composition								x
Chemical composition	x		x	x	x		x	(x)
Inclusions		x					x	(x)
Ices	x			x				

he suggests that oxygen-containing "organic refractory" material[14] may be present in interstellar clouds. Such material could be an important chemically-active component in the interstellar medium.

2.3   Conclusions

Information constraining models of interstellar dust grains comes from a variety of observational sources. Table 2.2 summarizes the situation.

All of the information is useful. A single type of observation—such as traditional extinction measurements from the infrared to ultraviolet—can in general be fitted quite well because there are a significant number of free parameters available. It is in the intersection of fitting of different kinds of observations that tighter constraints put models under closer scrutiny.

In Chapter 3 we shall describe current models of dust grains. From the chemical point of view taken by this book, it is particularly important to know the physical and chemical nature of the surfaces and of the bulk. There is a significant measure of agreement on these issues, so that it is reasonable to build theories of dust-related chemistry on the basis of what is known about interstellar dust.

The emphasis over the last century has been to fit with reasonable models the variety of ISECs and other data from observations of different lines of sight in the Milky Way and other galaxies. The fitting procedures have been largely successful, so that one might say with confidence that along one line of sight we find more of one type of grain than another. Much less attention has been given to understanding why these variations occur: it is important that we move on from merely fitting observational data to a position in which we understand why those variations occur and what they may tell us. We begin to address these questions by considering the life cycle of dust in Section II, and in particular recognising (in Chapter 6) that dust is not immutable but responds physically and chemically to the local physical conditions. Thus, the implication for the topic of this book is that the chemical properties of dust grains may also change from place to place and time to time within any galaxy.

Further Reading

1. D. C. B. Whittet, *Dust in the Galactic Environment*, Institute of Physics Publishing, Bristol and Philadelphia, 2003.
2. B. T. Draine, *Annu. Rev. Astron. Astrophys.*, 2003, **41**, 241.
3. B. T. Draine, *Physics of the Interstellar and Intergalactic Medium*, Princeton University Press, Princeton and Oxford, 2011.
4. E. L. Fitzpatrick and D. Massa, *Astrophys. J.*, 2007, **663**, 320.
5. S. I. B. Cartledge, G. C. Clayton, *et al.*, *Astrophys. J.*, 2005, **630**, 355.
6. T. Zafar, D. Watson, *et al.*, *Astrophys. J.*, 2012, **753**, 82.
7. A. Lazarian, *J. Quant. Spectrosc. Radiat. Transfer*, 2007, **106**, 225.

8. K. I. Öberg, A. C. A. Boogert, *et al.*, *Astrophys. J.*, 2011, **740**, 109.

9. E. Peeters, L. J. Allamandola, D. M. Hudgins, S. Hony and A. G. G. M. Tielens, *ASP Conf.*, 2004, **309**, 141.

10. E. Costantini, M. J. Freyberg and P. Predehl, *Astron. Astrophys.*, 2005, **444**, 187.

11. S. S. Seahra and W. W. Duley, *Astrophys. J.*, 1999, **520**, 719.

12. M.-F. Nieva and N. Przybilla, *Astron. Astrophys.*, 2012, **539**, A143.

13. D. C. B. Whittet, *Astrophys. J.*, 2010, **710**, 1009.

14. A. Li and J. M. Greenberg, *Astron. Astrophys.*, 1997, **323**, 566.

CHAPTER 3

Models of Interstellar Dust

3.1 From Data to Models

How do astronomers use the information described in Chapter 2 to deduce a description of interstellar dust grains? For the purposes of this book, we need the best possible description of the physical and chemical nature of interstellar dust. Then we can discuss in Section II of the book the chemistry of dust formation and describe the reasons for believing that the nature of dust is not fixed but evolves in the interstellar medium. We can then go on in Section III to explore the chemistry that occurs on the dust, and consequently in Section IV we can outline the extensive roles played by the chemistry of dust in the Universe.

The traditional approach, since the groundbreaking work of van de Hulst (1946; 1957)[1,2] and Oort and van de Hulst (1946),[3] has been to use the interstellar extinction curve to constrain the possible grain models, and then to subject the resulting chosen model to other observational constraints (as discussed in Chapter 2). In those early days, observations were confined to the visual region of the spectrum, and there was no spectral information about the chemical nature of the dust. The only feature that had to be explained was the near-linear portion of the ISEC in the visual region of the spectrum. Using Mie theory[4] to solve the interaction of electromagnetic radiation with a spherical grain, those early workers showed that this very restricted piece of the interstellar extinction curve could be explained by a wide range of materials in the form of particles with a size comparable with the wavelength of visual light. Evidently, without further information this approach was not very discriminating.

However, this traditional approach remains the most successful technique today. Nowadays, however, the extinction measurements extend from the

The Chemistry of Cosmic Dust
By David A. Williams and Cesare Cecchi-Pestellini
© David A. Williams and Cesare Cecchi-Pestellini 2016
Published by the Royal Society of Chemistry, www.rsc.org

infrared to the ultraviolet, and the range of potential dust materials is closely guided by infrared spectroscopy from the laboratory and from astronomical data, and by information from collected interplanetary particles. The resulting model can then be applied to other astronomical scenarios and tested against observations of scattering, emission, *etc.*, discussed in Chapter 2.

Mathis *et al.* (1977)[5] demonstrated how successful this approach could be, by developing a model (the MRN model from the initials of the authors Mathis, Rumpl & Nordsieck) that has been the basis of much further work. Mathis *et al.* (1977)[5] assumed that there were either two or three separate components in interstellar dust grains, selected from a list including silicates (enstatite and olivine), graphite, silicon carbide, iron, and magnetite (these listed materials being consistent with depletion data). Graphite was assumed to be the carrier of the 217.5 nm feature (as had been originally proposed by Stecher & Donn in 1965 [6]) and therefore an essential component. These component populations were entirely separate and were assumed to consist of spherical grains of radius *a*. Each population was assumed to have a power law form of the grain size distribution noted in Chapter 1: $n(a) \propto a^{-3.5}$ (where $n(a)\, da$ is the number of grains per unit volume with radii in the range $a \rightarrow a + da$, implying that there are very many more small grains than large.

Theoretical MRN ISECs involving either two or three components from the list were obtained using Mie theory by a simple minimization procedure; all of the choices gave good agreement with the average ISEC for the Milky Way for a wide range of wavelengths from the ultraviolet (~110 nm) through the visual to the near-infrared (~1 μm). Graphite, in the form of spheres of radii from 5 nm–1 μm, was essential, while the other component(s) had a rather more restricted range: 25 nm–0.25 μm.

While this curve-fitting approach did not discriminate well between the various potential dust models considered by Mathis *et al.*, the MRN paper has nevertheless been very influential because it defined a useful procedure that many other investigators have since followed. But like all curve-fitting with many free parameters available, it cannot give conclusive results. The MRN results were not the end of the story; there were many issues not considered in the MRN model.

For example, should a population of PAHs be included? The evidence—*i.e.* the AIBs—for their existence in the interstellar medium is strong; so do they contribute to extinction and—particularly—to the 217.5 nm bump and to extinction in the far-ultraviolet? Secondly, it is possible that the larger grains would be formed by aggregation of smaller grains, and therefore be composites of various materials. Perhaps distinct populations of grains of particular materials do not occur. How would composite grains affect the grain optics? Thirdly, in the case of larger grains at least, is porosity in the grain material important? After all, if large grains are simply aggregations of smaller grains that probably fit together rather poorly, then cavities within them would be very likely to occur. Fourthly, should the only form of solid carbon included be graphite? Is it not more likely that an amorphous form of carbon—possibly hydrogenated in the hydrogen-rich interstellar medium—is present and deposited on other surfaces? Fifthly, could the precise nature of the silicate component be established? These and other issues were considered by many authors in the years following the MRN paper; most of them adopted the MRN procedure.

At the same time, numerical procedures improved. The original Mie theory methods for the interaction of electromagnetic radiation with dielectric spheres were extended to spherical core/multiple mantle grains and to cylinders and to spheroids; these latter procedures were used in discussions of linear polarization. Powerful new methods were introduced to allow the computation of optical properties for grains of arbitrary shape and arbitrary composition,[7] with results of any desired accuracy. These new methods can—in principle—replace various approximations that had been (and still continue to be) widely used. Thus, the necessary tools for the calculation of optical properties of dust grains of any shape, material and composition were available. At the same time, laboratory work on refractive indices of potential materials provided the input data for these calculations.

Several models are currently widely used by astronomers and astrochemists. In the remainder of this chapter we shall describe these models and consider the implications of these models for interstellar surface chemistry.

3.2   The model of Bruce Draine and His Collaborators

The model developed by Draine and his many collaborators is very successful and has been widely applied to many of the dust-related observational phenomena discussed in Chapter 2. It is a direct descendant of the Mathis *et al.* ISEC model, but takes into account the population of PAH molecules inferred to be present in the interstellar medium; these PAHs are considered to give rise to the infrared emission spectrum (the AIBs) observed in the interstellar medium. PAHs are believed to contain a significant fraction of the available carbon in the interstellar medium, and they contribute to interstellar extinction as well as generating the infrared emission spectrum.

Draine and his collaborators extended the MRN model in a straightforward and plausible manner (as described in Draine's comprehensive review of 2003 [8]). Interstellar grains are assumed to be, as for the MRN model, totally distinct silicate and carbonaceous populations. The carbonaceous grains have a size distribution that extends into the PAH regime, and the abundance and size distribution of the PAH molecules are adjusted to reproduce infrared emission detected in the diffuse regions of the interstellar medium.[9] Following the MRN method, Li and Draine (2001)[9] adjusted the size distributions of the silicate and carbonaceous grains to compute an ISEC that fits the average Milky Way interstellar extinction, using a minimization procedure. The dielectric functions used in the Draine model are important refinements on those adopted by Mathis *et al.* In particular, the dielectric functions for silicates have been constructed to reproduce the broader 9.7 and 18 μm observed interstellar features than the narrower features found in terrestrial and lunar silicates, and to take account of structure at X-ray absorption edges.[10,11] This modified dielectric function for silicates is referred to as belonging to "astrosilicate" material, and is designed to allow appropriately for the modified properties associated with silicates in the harsh environment of the interstellar medium as compared to the relatively benign terrestrial environment.

In making fits to the average ISEC for the Milky Way, the best-fitting size distributions obtained using the Draine model[12] are different from the simple $dn \propto a^{-3.5}da$ of the MRN ISEC. For silicates, the best-fitting size range is similar to that of the MRN but the index of the power law of the distribution is smaller. For carbonaceous grains and PAHs, for which graphite dielectric functions are used, a complex tri-modal distribution gives the best fit to the average ISEC. A peak in the carbonaceous grain size distribution at grain radius $a \approx 0.3$ μm is required to fit the average ISEC, a peak at radius 5 nm improves the far-infrared fit, and a peak at radius 0.5 nm is required to fit the observed near-infrared 3–12 μm emission features (see Chapter 2).

One observational feature that is not normally accounted for in the Draine model is the absorption at a wavelength of 3.4 μm on long interstellar paths. The feature is normally attributed to the sp^3 C–H stretching mode in hydrogenated amorphous carbon. The dielectric function used in the Draine model for carbonaceous material is that of graphite which does not possess that mode; it could be included with little modification.

Draine's model has been applied with success to measurements of scattered starlight in studies of the DGL and in reflection nebulae. It gives (as it was designed to do) excellent fits to measurements of the discrete and continuous infrared emission from the Milky Way from 3 to 1000 μm. It is consistent with microwave emission in the range 5–100 GHz, depending on the alignment method adopted. Magnetic inclusions appear to be necessary to match the emission at the lower frequencies.

Two recent space missions—the Spitzer Space Telescope (active cold phase 2003–2009) and the Herschel Space Observatory (active 2009–2013)—have opened up infrared observations of galaxies external to the Milky Way. Between them, they cover the wavelength range 3.6–500 μm, so that the infrared part of the spectrum of any astronomical object observed in the two missions is tightly constrained. The Draine model has been of inestimable value in the interpretation of these observations. The use of this model with the observations from space has allowed the determination of important dust-related parameters in external galaxies, including the dust mass surface density, the gas-to-dust ratio, the PAH mass fraction, the intensities of starlight that is heating the dust, and the total infrared luminosity. The scientific return from the Draine model used with these observations has been excellent, and includes information about galaxies M33,[13] M31,[14] and NGC 628 and NGC 6946.[15]

The success of the Draine *et al.* model has inspired the study of similar models by other authors (*e.g.* Nozawa & Fukugita 2013[16]). However, from the point of view of the chemistry of cosmic dust, the Draine *et al.* model and similar models predict that the surfaces available for chemistry in space are those of silicates, graphite, and of PAHs.

3.3   Composite Dust Grains

3.3.1   The "Unified" Model

It is evident that potential dust grain materials cannot be totally inert in the interstellar medium, but will respond appropriately to the local physical conditions (as we will discuss in more detail in Chapter 6). Some materials will be affected by starlight or by cosmic rays, and some may be modified by chemical reactions with gas phase species. All grains will be eroded and possibly reduced to atoms or small molecular fragments in the intermittent passage of interstellar shock waves, with some materials (such as silicates) being more resistant to this kind of erosion than others (such as hydrocarbon polymers). Therefore, it seems necessary to develop a model of dust that takes these ideas into account. Such a model will have two important characteristics: it explicitly contains the idea of the *evolution* of dust (we shall return to this idea in Chapter 6) and of interstellar extinction that is time-dependent (as explored first by Cecchi-Pestellini & Williams 1998[17] and developed more accurately by Cecchi-Pestellini *et al.* 2010[18]), and it links the various components of dust together – hence, we give it the title of the *unified* model. Neither of these ideas is contained in the MRN approach, which is essentially one of curve-fitting and offers no possible explanation of why one line of sight might have a different ISEC from another.

Many of the extensive laboratory results on which this new approach is based were obtained by Walt Duley and collaborators, and others were gleaned by him from the literature; the overall picture is summarized by Jones *et al.* (1990).[19] The main features of the dust model developed in that paper were that carbon would be slowly deposited on bare silicate grains in the hydrogen-rich interstellar medium in the form of a hydrogenated amorphous carbon, initially mostly with sp^3 bonding; this hydrogenated amorphous carbon would absorb ultraviolet radiation from starlight to be gradually converted to a more graphitic sp^2 structure. The optical properties of the sp^3 and sp^2 structures are distinctly different, and the conversion from sp^3 to sp^2 is termed "photodarkening" to represent the fact that the graphitic material (sp^2) is visually opaque. Thus, there are several important timescales in the development of interstellar extinction according to this model: the timescale for deposition of the carbon layer and the timescale for photodarkening; there is also another—usually much longer—timescale associated with the removal of the carbon layers when the dust grains pass through a shock. The circulation of carbon in the interstellar medium is summarized schematically in Figure 3.1.[20] Extinction characteristics and other dust properties computed using the unified model are inherently time-dependent.

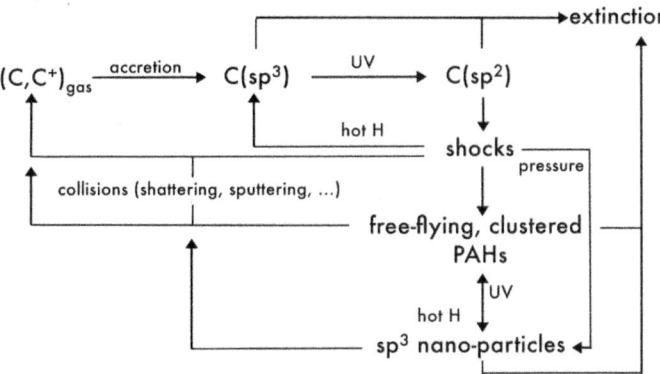

Figure 3.1 A schematic diagram indicating the circulation of interstellar carbon into and out of dust, according to the "unified" model of dust. Reproduced with permission from Cecchi-Pestellini *et al.* (2014).[20]

In this unified model of interstellar dust, the PAHs and other atoms and fragments of erosion are part of the natural circulation of carbon in the interstellar medium between gas and solid phases. Instead of adopting values for the optical constants of PAH component that are identical with the optical constants of graphite (as is done in other models), in the unified model the PAH optical properties are treated in a more fundamental manner. They are computed *ab initio* for a representative sample of 54 different types of PAH in four different charge states.[21,22] The PAH contribution to extinction obtained from these *ab initio* computations is then fitted to two Lorentz profiles. The $\pi* \leftarrow \pi$ resonance transition contributes to the 217.5 nm extinction bump, while the $\sigma* \leftarrow \sigma$ transition contributes to the far-ultraviolet rise in extinction. For the solid grains, the optical constants used are those for Draine's "astrosilicates", and experimental values for sp^2 and sp^3 hydrocarbons.[23,24]

In the unified model, the grains are not separate populations of graphite and silicates, but are mixtures of at least two forms of carbon and one of silicate. In the simplest case, the dust grains are considered as spherical silicate cores, coated with distinct layers of sp^2 (inner) and sp^3 (outer) carbon. The silicate cores may contain vacuum inclusions, and for convenience these voids are placed symmetrically at the centre of the silicate core. Thus, there may be up to four components in each grain, and the optical properties of these structures are computed using the methods described in Borghese *et al.* (2007).[7]

The grain size distribution is chosen to be a power law, $\propto (a + w)^{-q}$, where w is the mantle thickness on a grain of radius a and q is a free parameter. The size distribution is assumed to include separate small grains (somewhere in the range 5–12 nm) and large grains (somewhere in the range 60–100 nm), governed by the same power law. The free parameters are the minimum and maximum sizes in each range, the power law index, q, the vacuum fraction, f_v, in the silicate core, f_{sp2}, the fraction of sp^2 material. Figure 3.2 (ref. 25 and 26)

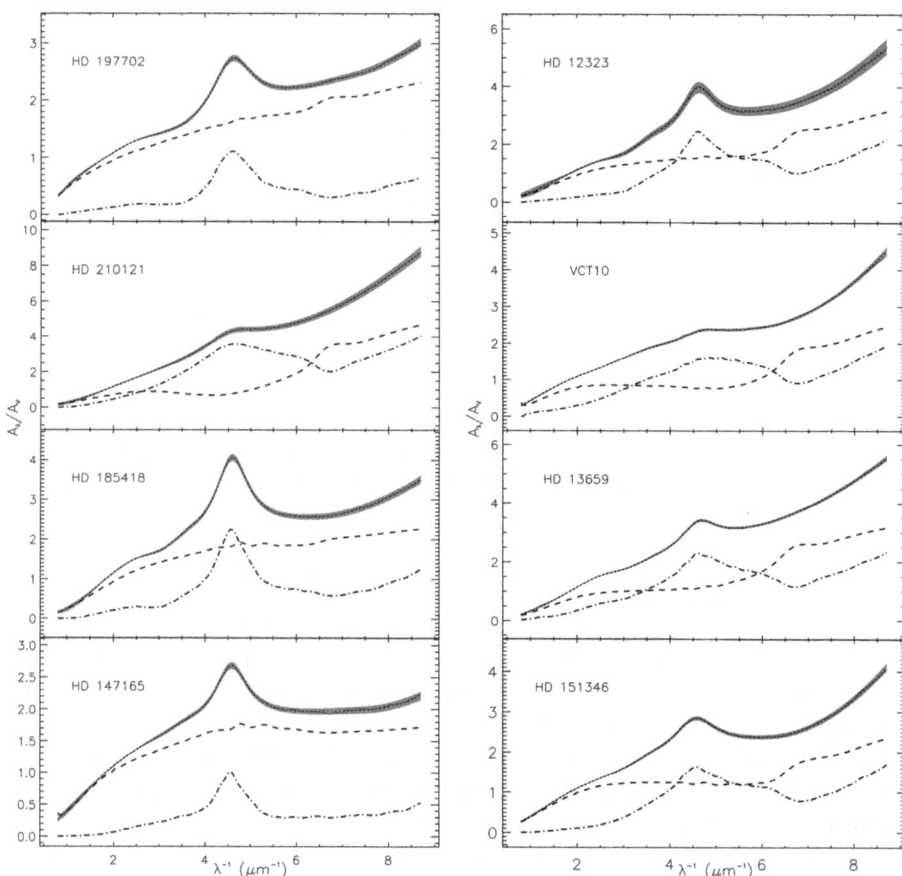

Figure 3.2   Fits to the normalized interstellar extinction curves for eight lines of sight representative of the Milky Way galaxy. Solid lines: fits; dashed lines: classical dust contributions; dot–dashed lines: PAH contributions. The shaded areas are of the observational errors given in Fitzpatrick & Massa (2007).[25] Reproduced with permission from Mulas *et al.* (2013).[26]

shows fits to normalized extinction curves along eight lines of sight in the Milky Way representative of the variety of observed ISECs.[26] The corresponding best fit parameters are shown in Table 3.1.

In general, the curves are well fitted by power law size distributions and the index q is very close to the value of 3.5 used in the MRN curves. Variations in the ranges of small and large grains and in f_{sp2} and f_v account for the variety of ISECs shown in Figure 3.2. The abundances of carbon and silicon required for these dust models are shown in Table 3.2. These values may be compared with the reference values shown in Chapter 1, Table 1.1, and are consistent with those values.

The unified model has not been applied to the wide variety of interstellar observations (discussed in Chapter 2) so successfully achieved for the Draine

Table 3.1 Best fit parameters for the solid dust component of the unified model. Here, f_v is the vacuum fraction of the silicate core, w (in nm) is the total carbonaceous mantle thickness, f_{sp2} is the fraction of w that is sp^2 material, a and b values (in nm) give the minimum and maximum radii of the small and large grain components, respectively, and q is the index of the power law size distribution.

Line of sight	f_v	w	f_{sp2}	a_-	a_+	b_-	b_+	q
Average ISEC	0.45	0.97	0.92	5.0	27.0	70.4	493	3.47
HD 12323	0.48	0.75	0.67	5.0	42.0	133.0	492	3.48
HD 13659	0.36	0.51	0.00	5.0	55.5	129.4	486	3.49
HD 147165	0.37	3.00	0.78	6.4	13.6	65.0	566	3.47
HD 151346	0.27	0.98	0.91	5.0	26.1	107.7	492	3.49
HD 185418	0.52	3.00	0.07	5.0	25.0	85.9	470	3.40
HD 197702	0.09	3.00	1.00	5.1	28.0	45.4	567	3.51
HD 210121	0.46	0.61	0.21	5.0	47.2	252.6	387	3.45
VCT10	0.33	0.45	0.66	5.0	36.3	160.0	561	3.51

Table 3.2 Si and C abundances in the dust per million hydrogen nuclei (ppM) along the line of sight resulting from the unified model.

Line of sight	Si/H (ppM)			C/H (ppM)			
	Small grains	Large grains	Total	Small	Large	PAHs	Total
Average ISEC	6.0	39.0	45.0	14.5	5.7	64.3	84.5
HD 12323	8.6	27.2	35.8	10.3	1.9	281.1	293.2
HD 13659	11.4	26.0	37.5	4.1	0.6	70.5	75.2
HD 147165	1.9	59.3	61.1	21.7	23.3	40.2	85.3
HD 151346	9.4	52.3	61.7	23.3	6.2	71.7	101.3
HD 185418	3.1	37.6	40.8	11.8	7.1	78.6	97.5
HD 197702	8.7	91.8	100.5	72.1	50.4	46.4	168.8
HD 210121	7.7	8.2	15.9	4.6	0.2	257.3	262.1
VCT10	14.0	44.5	58.5	10.9	1.6	67.2	69.6

model. However, the unified model has been applied to the observations of ISECs in external galaxies[20] with some success (as we will see in Chapter 11). It appears that all known types of observed ISECs can be accounted for on the basis of the unified model. We defer to Chapter 11 our discussion of extinction in very distant high red-shift galaxies (for which observational data are extremely sparse).

Finally, from the point of view of the chemistry of cosmic dust, the unified model predicts that the surfaces available for chemistry in space are those of silicates, of amorphous carbon in both sp^2 and sp^3 bonding, and of PAHs.

3.3.2 The "Holistic" Model

Jones *et al.* (2013)[27] have proposed a model of dust, the "holistic" model, which also includes silicates and amorphous carbon, similar in many ways to the "unified" model described in Section 3.3.1, but without an explicit

PAH component. The essential picture follows that outlined by Jones *et al.* (1990).[19] The interstellar radiation field plays a crucial role in converting hydrogenated amorphous carbon to a more graphitic form. Small amorphous carbon particles (radii less than about 20 nm) are assumed to be completely converted to a graphitic form, while larger particles are only partly converted. Grains of silicates with or without carbon are also included; the carbon may be in the form of mantles on the silicate surfaces or arise as a result of the accretion of small carbon grains.

The grains are therefore composed of the two materials, silicates and carbons, but arranged in three dust populations, whose properties are size-dependent. Small amorphous carbon grains (radii about 0.4 to 100 nm) have been processed by the interstellar radiation field to a depth of 20 nm. These are assumed to have a power law distribution with an exponential tail. Large amorphous carbon grains (radii ~200 nm) with a photo-processed outer layer (20 nm thick) are present with a log-normal size distribution. The third component consists of large (radii ~200 nm) amorphous forsterite-type silicates, with Fe nanoparticle inclusions (taking some 70% of cosmic iron); these particles are coated with mantles of amorphous carbon that are 5 nm thick. These grains are assumed to have a log-normal size distribution. There are no free-flying PAHs in this model.

Optical constants for the silicates with Fe inclusions are based on laboratory data, using Effective Medium Theory to obtain average values for the composite silicate/iron mixture. Optical constants for the amorphous carbon are newly-derived based on modelling using random covalent networks for defective graphite. With some "fine-tuning" of laboratory-derived data, this model gives a good fit to astronomical observations, without graphite or PAHs. We show in Figure 3.3 (ref. 27) fits to data for the interstellar extinction curve and to infrared emission from 1 to 1000 μm, obtained using the basic version of this model.

As in the "unified" model, there is a clear implication in the "holistic" model that time-dependence is important. In the "unified" model, this is explicit in the basic mathematical equations of the model, while in the "holistic" model the time-dependence is given implicitly. However, the adopted circulation of carbon in the "holistic" model is shown in Figure 3.4 (ref. 27), where carbon dust emerges from a cool stellar envelope, passes through the diffuse and then the denser regions of interstellar space, and is present in near-stellar environments in star-forming regions. We'll discuss the idea of the time-dependence of dust properties in more detail in Chapter 6.

3.4   Conclusions: Implications for Interstellar Surface Chemistry

Our purpose in the chapter is to determine the physical and chemical properties as predicted by models of dust grains. These properties will determine the chemical activity of dust grains, as we shall discuss in Chapters 8 and 9.

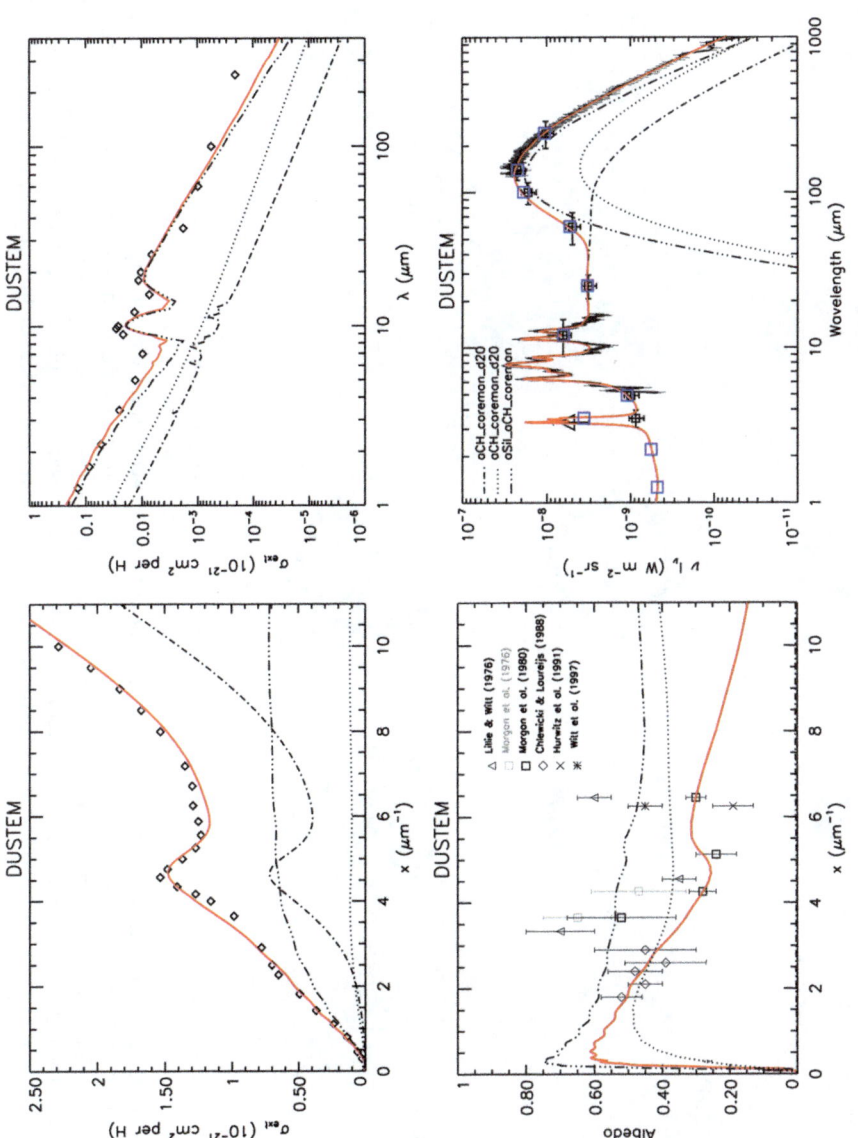

Figure 3.3 (Top) computed ISEC for the "holistic" model compared with the average extinction curve for the Milky Way galaxy (diamonds). The contributions of the amorphous silicate + iron core/amorphous carbon mantle is shown (triple dot–dashed); of the hydrogenated amorphous carbon core/amorphous carbon mantle (dotted); and of the small amorphous carbon grains (dash–dotted). (Bottom) computed emission spectrum, IR–UV, for the "holistic" model, compared with observational data. λ is the wavelength and x the wavenumber ($=\lambda^{-1}$). Reproduced with permission from Jones *et al.* (2013).[27]

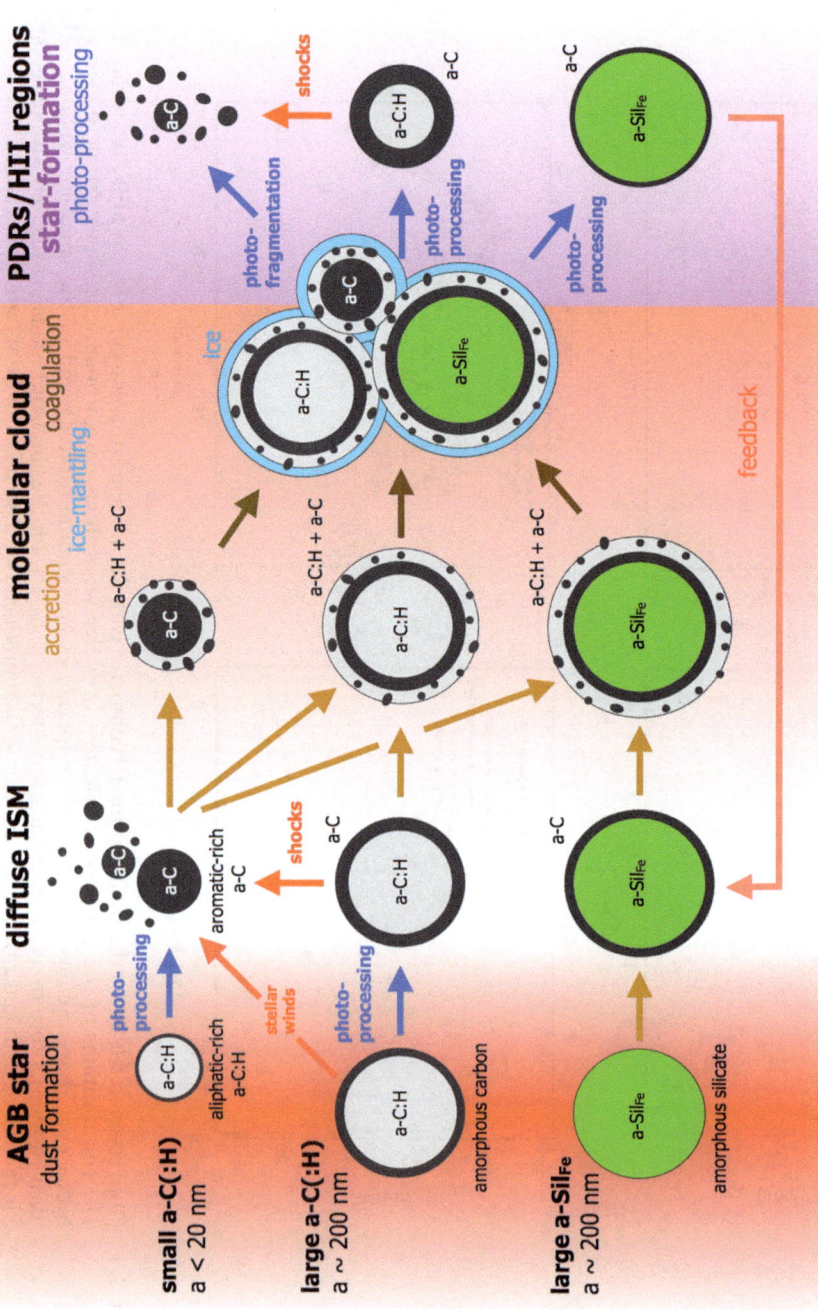

Figure 3.4   A schematic view of the life-cycle of interstellar dust in terms of evolutionary tracks: "parallel-processing" tracks that lead to a smooth evolution are horizontal and "cross-talk" tracks that lead to significant mass-transfer between grain sizes (coagulation and disaggregation) are diagonal and vertical. Qualitatively, the red shading reflects the density, and the violet shading represents high radiation fields. The symbol a-C and a-C:H represent amorphous carbon and hydrogenated amorphous carbon, while a-Sil$_{Fe}$ represents amorphous forsterite-type with iron nano-particle inclusions. Reproduced with permission from Jones *et al.* (2013).[27]

Evidently, there are various ways of approaching the modelling process. Should the materials of the dust grains be discrete or mixed? What are the size distributions of the various components? Are there separate components of large and small grains? To what extent can one use laboratory-derived optical constants for materials that are probably unlike (in size, shape, and structure) any terrestrial materials? How reliable is it to modify these optical constants? Should voids be included? Is shape important, other than for polarization? There are important differences between the various models discussed, although the grain materials adopted in all three models have similarities. All models adopt some form of amorphous silicate and some form of carbon. However, there are differences in how the carbon is perceived to be present. Two of the models described here include free-flying PAHs, but differ in the way that the optics of these molecules is included. The solid carbon is described as graphitic in one model and as hydrogenated amorphous carbon in the other two. The carbon may be distinct from the silicate, or intimately associated with it. The amorphous carbon is recognised as a material that evolves in the interstellar medium, and as it does so its optical properties change. Indeed, gaseous carbon is readily deposited on terrestrial surfaces, and this should also occur in interstellar space. Thus, models involving amorphous carbon as a component of the dust necessarily imply that there is a time-dependence associated with interstellar extinction and other observed signatures of interstellar dust.

There are significant differences in the adopted size distributions of the dust. Nevertheless, in spite of differences in approach and in basic data used, all these models are capable of representing the physics of the situation rather well. Evidently, the physical constraints are sufficiently loose that fitting procedures (with a rather large number of free parameters available) can accommodate a variety of descriptions.

However, from the point of view of astrochemistry occurring on the surfaces of dust grains, the various models of dust in diffuse interstellar clouds are in reasonable harmony. There are surfaces of amorphous silicate (only a small proportion of interstellar silicate is observed to be crystalline) and of amorphous carbon which often has graphitic (sp^2) form but may also take polymeric (sp^3) form. Therefore, when we come to consider surface reactions on bare interstellar dust grains in Chapter 8 and the formation of ice mantles in Chapter 9, these are the kinds of solid surface we need to consider, along with reactions on PAHs.

Further Reading

1. H. C. van de Hulst, *Recherches Astronomique de l'Observatoire d'Utrecht*, 1946, **11**, 1.
2. H. C. van de Hulst, *Light Scattering by Small Particles*, Wiley, New York, 1957.
3. J. H. Oort and H. C. van de Hulst, *Bull. Astron. Inst.*, 1946, **10**, 187; G. Mie, *Ann. Phys.*, 1908, **25**, 377.

4. G. Mie, *Ann. Phys.*, 1908, **25**, 377.
5. J. Mathis, W. Rumpl and K. H. Nordsieck, *Astrophys. J.*, 1977, **217**, 425.
6. T. P. Stecher and B. Donn, *Astrophys. J.*, 1965, **142**, 1681.
7. F. Borghese, P. Denti and R. Saija, *Scattering from Model Nonspherical Particles: Theory and Applications to Environmental Physics*, Springer-Verlag, Berlin, Heidelberg, New York, 2nd edn, 2007.
8. B. T. Draine, *Annu. Rev. Astron. Astrophys.*, 2003, **41**, 241.
9. A. Li and B. T. Draine, *Astrophys. J.*, 2001, **554**, 778.
10. B. T. Draine and H. M. Lee, *Astrophys. J.*, 1984, **285**, 89.
11. B. T. Draine, *Astrophys. J.*, 2003, **598**, 1017.
12. J. C. Weingartner and B. T. Draine, *Astrophys.J.*, 2001, **548**, 296.
13. M. D. Calapa, D. Calzetti, B. T. Draine, *et al.*, *Astrophys. J.*, 2014, **784**, 130.
14. B. T. Draine, *et al.*, *Astrophys. J.*, 2014, **780**, 172.
15. G. Aniano, B. T. Draine, *et al.*, *Astrophys. J.*, 2012, **756**, 138.
16. T. Nozawa and M. Fukugita, *Astrophys. J.*, 2013, **770**, 27.
17. C. Cecchi-Pestellini and D. A. Williams, *Mon. Not. R. Astron. Soc.*, 1998, **296**, 414.
18. C. Cecchi-Pestellini, A. Cacciola, M. A. Iatì, R. Saija, F. Borghese, P. Denti, A. Giusto and D. A. Williams, *Mon. Not. R. Astron. Soc.*, 2010, **408**, 535.
19. A. P. Jones, W. W. Duley and D. A. Williams, *Q. J. R. Astron. Soc.*, 1990, **31**, 467.
20. C. Cecchi-Pestellini, S. Viti and D. A. Williams, *Astrophys. J.*, 2014, **788**, 100.
21. G. Malloci, C. Joblin and G. Mulas, *Astron. Astrophys.*, 2007, **332**, 353.
22. C. Cecchi-Pestellini, G. Malloci, G. Mulas, C. Joblin and D. A. Williams, *Astron. Astrophys.*, 2008, **486**, L25.
23. F. Rouleau and P. G. Martin, *Astrophys. J.*, 1991, **377**, 526.
24. J. Ashok, P. L. H. Varaprasad and J. R. Birch, *Handbook of Optical Constants of Solids II*, ed. E. D. Palick, Academic, New York, 1991, vol. 957.
25. E. L. Fitzpatrick and D. Massa, *Astrophys. J.*, 2007, **663**, 320.
26. G. Mulas, A. Zonca, S. Casu and C. Cecchi-Pestellini, *Astrophys. J., Suppl. Ser.*, 2013, **207**, 7.
27. A. P. Jones, L. Fanciullo, *et al.*, *Astron. Astrophys.*, 2013, **558**, A62.

CHAPTER 4

Laboratory Studies of Candidate Interstellar Dust Materials

4.1 Terrestrial Replicas of Interstellar Dust Grains

The chemical composition of dust particles is, as we have seen in Chapters 2 and 3, mainly determined *via* remote sensing observations over the infrared, optical, and ultraviolet spectral ranges. The composition of interstellar dust grains is, however, not entirely known. Clues about the nature of dust grains are obtained by the numerous astrophysical spectra that invariably identify metal (primarily iron and magnesium) silicate compounds and carbonaceous matter. In the infrared, features related to the abundant silicates are frequently observed, especially the bands around 10 and 18 μm attributed to Si–O stretching and bending vibrations. In the diffuse interstellar medium, silicate materials are predominantly amorphous, while crystalline silicates, whose composition is mostly magnesium-rich (forsterite), are frequently observed in protoplanetary disks; this indicates that *in situ* formation by thermal annealing or shocks may have occurred. Strictly speaking the word "silicate" implies that the material is crystalline and could be a mineral, making the term "amorphous silicate" mineralogically incorrect, and "crystalline silicate" redundant.[1]

Nearly all conceivable forms of carbon, from quenched carbonaceous composite to graphene, have been proposed as carriers of the interstellar features. The AIBs at 3.3, 6.2, 7.7, 8.6, 11.3 and 12.7 μm (see Chapter 2), characteristic of C–H and C–C chemical bonds in materials possessing aromatic structures, suggest (but do not necessarily imply) the existence of a particular family

The Chemistry of Cosmic Dust
By David A. Williams and Cesare Cecchi-Pestellini
© David A. Williams and Cesare Cecchi-Pestellini 2016
Published by the Royal Society of Chemistry, www.rsc.org

of macromolecules, the PAHs. There are also bands observed in absorption (and recently also in emission in the Magellanic Clouds, see Mori *et al.* 2012[2]) with features at 3.4, 6.8 and 7.2 µm. The band at 3.4 µm, with its associated substructures, has been identified with methyl (CH_3-) and methylene ($-H_2C-$) group aliphatic stretching modes in solid carbons. In some evolved circumstellar environments, aliphatic emissions at 8 and 12 µm, together with the 3.4, 6.9, and 7.3 µm features and the AIBs, have been detected. This is a strong indication of the coexistence of both aliphatic and aromatic structures in the circumstellar shells of these evolved stars, suggesting the presence of a solid material with a mixture of aliphatic and aromatic structures such as hydrogenated amorphous carbon, a-C:H, or HAC (*e.g.* Jones *et al.* 1990[3]), or coal (*e.g.* Guillois *et al.* 1996[4]).

As discussed in Chapter 2, depletion arguments also provide indirect support for the presence of such materials in space. Since hydrogen and helium cannot solidify under interstellar conditions, dust must therefore consist mainly of compounds of the more common heavy elements. In oxygen-rich environments we may expect the presence of oxide materials, most of them silicates, and oxygen-containing ices, H_2O, CO, and CO_2. Carbon-rich environments contain carbonaceous solids, carbides, and sulfides. We shall defer our discussion of molecular ices to Chapter 9.

To study dust in space astronomers combine observations and theoretical models with laboratory analyses of materials believed to be similar, or analogous, to interstellar solids. Analogue materials are, therefore, tools that may be used to explore in the laboratory the response of dust grains to the conditions of the regions in which they are embedded, through their interaction with gas ions, atoms, and molecules, and electromagnetic radiation. Optical and spectroscopic properties of these analogue materials are considered a primary goal to study for providing basic data for the interpretation of astronomical observables, such as extinction, thermal emission, and scattering. However, despite the claimed "remarkable" closeness of synthetic spectra to astronomical features, experiments on terrestrial analogues of cosmic solids often do not take into account the specifics of cosmic dust materials—composition, structure, processing, *etc.*—nor have they covered the relevant frequency ranges. As we will see in Chapters 6 and 7, dust grains condense in circumstellar environments, are modified by interstellar processes, and are probably reformed and reassembled in the interstellar medium. We must therefore expect that astronomical dust is likely to be an amalgam of a number of different materials, probably very chaotic in composition and structure, with different individual substances dominating at different wavelengths. Consequently, the spectra of terrestrial materials may be fundamentally different from the spectra of astronomical dust. Experiments on analogue materials are valuable, but cannot be an exact replica of natural environments, and thus they may produce unnatural artefacts. In other words, cosmic dust analogues are excellent working examples whose usefulness decreases with increasing expertise!

4.1.1 Structures of Silicates and Carbons

Glassy solids such as obsidian and basaltic glass have been considered as reasonable approximations of interstellar amorphous silicates (*e.g.* Pollack *et al.* 1973[5]), as opposed to the original suggestion put forward by Woolf & Ney (1969)[6] based on mixtures of spectra of crystalline silicate species predicted to form by theoretical models. However, this is an oversimplification. In dust forming regions, for example in the cooling outflows of late-type stars (see Chapter 5), the fundamental constituents of dust, including water and metals, begin to condense, forming chaotic and disordered structures. Nuth & Hecht (1990)[7] introduced the concept of "chaotic silicates" in which the level of disorder is even greater than for glass. Glass is characterized by the absence of long range order in the atomic arrangement beyond nearest neighbours, while the definition of crystalline is somewhat semantic, including poly-nanocrystalline agglomerates, with a continuum which extends from a true glass to a single crystal grain. Considering also the possibility that many materials may be assembled in the agglomerate, an astronomical silicate cannot be considered a solid with a definite stoichiometric composition. Such materials may also occur in groups that recall solid solutions, in which one or more types of atoms or molecules of the solid may be partly substituted for the original atoms and molecules without changing the structure. A complete solid solution occurs when two elements can substitute for each other completely in a mineral. Olivines are excellent examples of this type of solid solutions. Forsterite, Mg_2SiO_4, and fayalite, Fe_2SiO_4, have identical structures because the ions Mg^{2+} and Fe^{2+} are very nearly the same size and are chemically similar. If both iron and magnesium are present, a single material with a "continuous" composition intermediate between forsterite and fayalite will form. This compositional mixture is considered to be a "solution" of fayalite and forsterite (rather than a compound). Because minerals with all possible intermediate compositions can be formed, the chemical formula is $Mg_{(2-x)}Fe_xSiO_4$ ($x \leq 2$), and the general name, regardless of the Mg:Fe ratio, is olivine. When the substituting cation has a different size, the solid solution is more limited, for example between enstatite ($MgSiO_3$) and diopside ($CaMgSi_2O_6$). Astronomical solids may be porous, and therefore of much lower density than a glass. Thus, the nature of an "astronomical silicate" is rather loosely constrained, and possibly limited to a material whose infrared spectrum is dominated by Si–O stretching and bending vibrations. Thermal annealing (*e.g.* in shocks) of precursor materials that were probably amorphous, may explain the presence of "crystalline" silicates in circumstellar regions and protoplanetary disks, as the exposure to high temperatures causes the condensate to gain order and become more crystalline. The process is by no means simple or well understood (see Section 4.2.2).

The basic building block of silicates is the SiO_4^{4-} tetrahedron. In order to neutralize the +4 charge on the Si cation, one negative charge from each of the oxygen ions will reach the Si cation. Thus, each oxygen atom will be left with a net charge of −1, resulting in a SiO_4^{4-} tetrahedral group that can be bonded to other cations. Since Si^{4+} is a highly charged cation, Pauling's rules state that it should be separated as far as possible from other Si^{4+} ions. Thus, when these SiO_4^{4-} tetrahedrons are linked together, only corner oxygen atoms will be shared with other SiO_4^{4-} groups. In all cases, the non-shared oxygen atoms (non-bridging oxygen atoms) are charge-balanced by other cations (*e.g.* Mg^{2+}, Fe^{2+}, Ca^{2+}). Tetrahedrons can be linked in a framework, and this gives rise to different silicate groups. If the corner oxygen atoms are not shared with other tetrahedrons, each tetrahedron will be isolated. The basic structural unit is then SiO_4^{4-}. In this group, the oxygen atoms are shared with octahedral groups that contain other cations like Mg^{2+}, Fe^{2+}, or Ca^{2+}. The olivine series—from forsterite to fayalite—is a good example. When each oxygen atom is shared between two tetrahedra (*e.g.* SiO_2 minerals and feldspars) the structural unit becomes $Si_2O_7^{6-}$. When linked in single (*e.g.* enstatite) or double chains the basic structural units are $Si_2O_6^{4-}$ or SiO_3^{2-} and $Si_4O_{11}^{6-}$, respectively.

Minerals occur as crystals, whose structure is a continuous ordered arrangement of one or more elements. Thus, they possess long range order. This gives rise to symmetry, with very narrow distributions of bond angles and lengths. This leads both to anisotropy and narrow spectral features. Glasses form before the atoms in a disordered liquid state have time to arrange themselves into a crystal structure. They are therefore both isotropic and have broad spectral features. As in crystals, the basic structural unit of silicate glasses and melts (high temperature liquids) is the SiO_4^{4-} tetrahedron. There are three classes of components for oxide glasses: network formers (*i.e.* silicon for silicates), intermediates, and modifiers. Silicon atoms form a highly cross-linked network of chemical bonds. The modifiers alter the network structure, and coordinate non-bridging oxygen atoms. The intermediates can act as both network formers and modifiers, according to the glass composition. Tetrahedral cations include not only Si^{4+}, but also trivalent cations such as Al^{3+} and Fe^{3+}, that must be charge-balanced by other cations, usually alkalis or alkaline earths (*e.g.* Na^+ and Ca^{2+}). The degree of polymerization of a melt or glass can be summarized by the ratio between the number of non-bridging oxygen atoms and tetrahedral cations, which can range from 0 (fully polymerized, *e.g.* SiO_2) to 4 (fully depolymerized, *e.g.* Mg_2SiO_4). An important parameter is the glass transition temperature, *i.e.* the temperature at which an amorphous solid becomes soft upon heating or brittle upon cooling. If the cooling is fast enough, the network present in a melt is preserved in the glass structure. More de-polymerized melts (*e.g.* olivines and pyroxenes) are more difficult to quench, and form glassy grains below their glass temperatures, because the cooling rate required for quenching to a glass is extremely rapid. From the perspective of interstellar dust formation, the glass transition temperature is essentially the temperature above which a given composition should form as or convert to crystalline solids, whereas solids formed below this temperature will be amorphous if they cool sufficiently fast, depending on the competition with the annealing timescales. Crystalline grains carry the imprinting of extended periods at high temperature, such as in a stellar atmosphere, whereas amorphous materials are often the products of rapid mutations in the thermal conditions at which they were formed, as might found in the outflows around evolved stars (see Chapter 5).

Carbonaceous materials in space are difficult to constrain. A striking example is given by the nature of the carrier of the interstellar ultraviolet extinction bump at 217.5 nm, that was originally attributed to small crystalline graphite particles,[8] followed by a plethora of proposals including mixture of spheres composed of graphite, amorphous carbon, and silicate,[9] irregular or fractal arrangement of graphite and amorphous carbon,[10] PAHs,[11] natural coal,[12] and even electronic transitions of OH^- ions in sites of low coordination in silicates.[13] In general, carbonaceous materials contain greater or lesser hydrogen fractions, varying proportions of different chemical bonding and different degrees of long range order. All these forms of carbon can under suitable conditions be readily converted from one to another. Using an evocative but rather incorrect expression (in strict geological terms) carbonaceous materials may be defined as metamorphic.

In principle, a carbon atom can adopt three different bonding configurations, sp^1, sp^2, and sp^3. In the sp^3 configuration, each of the carbon's four valence electrons is assigned to a tetrahedrally directed sp^3 hybrid orbital, which then forms a strong σ bond with an adjacent atom. At an sp^2 site, only three electrons are used in σ bonds. The three sp^2 hybrid orbitals arrange themselves as far apart as possible while the fourth enters a π orbital which lies normal to the σ bonding plane. The π bond is the weaker bond so that it is closer to the chemical potential of electrons, *i.e.* the measure of the energy of the least tightly held electrons within a solid (the Fermi level). π states will therefore form both the valence- and conduction-band states. In sp^1 hybridization two electrons form σ bonds, while the remaining (two) electrons form π bonds. A hydrocarbon containing only single bonds is called "saturated". Unsaturated systems have double or triple bonds between adjacent carbon atoms, for example a system of separate double bonds in "olefinic" systems such as ethylene, $H_2C=CH_2$, or as delocalized or "conjugated" π bonded systems such as the aromatic six-membered rings in benzene, C_6H_6, and in graphite. The manifold of possible bonding arrangements produces several allotropes of carbon of which the best known are graphite, diamond and amorphous carbon. The physical properties of carbon vary widely with the allotropic form. Diamond consists of sp^3 sites. The saturated bonding produces the wide 5.5 eV band gap and low conductivity, and the isotropy of the bonding gives it its strength. Graphite consists of hexagonal layers of sp^2 sites, weakly bonded together by van der Waals forces into a stacking sequence along an axis normal to the basal plane. The conductivity is high in the basal plane, but rather low across the planes. Graphite shows a semi-metallic behaviour with a band overlap of about 41 meV. As graphite is the stable allotrope of carbon, many disordered forms of carbon have structures based on its lattice, such as glassy and amorphous carbons (a-C) that may be hydrogenated (a-C:H or HAC).

While glassy carbon is essentially metallic, a ribbon-like structure with local graphitic order, a-C:H is truly amorphous and semiconducting, a predominantly sp^2 bonding material with little two-dimensional order. The carbon bonding and the hydrogen content define the short-range order in amorphous carbon. This is not, however, sufficient to describe its structure, since there is a substantial degree of medium-range order on the scale of 1 nm: the sp^2 sites tend to occur in bunch of warped graphite layers at low hydrogen content, while sp^2 and sp^3 bonds are somewhat segregated and clustered with increasing hydrogen concentration. There exists another kind of disordered carbon, the tetrahedral, or diamond-like, amorphous carbon. This material is a metastable form of amorphous carbon containing a significant fraction of sp^3 bonds. Like diamond, it can have a high mechanical hardness, chemical inertness, optical transparency; it is a wide band gap semiconductor, and is much denser than a-C:H. Fullerenes and carbon nanotubes constitute other allotropic forms of carbon. Fullerenes are built up of fused pentagons and hexagons and have a bond hybridization which is intermediate between graphite and diamond due to their spherical shape. The response to this non-planarity is the σ-character of curved π orbitals. Graphene is the most recently discovered allotropic carbon form, a one-atom thick flat sheet of carbon simultaneously the world's thinnest, strongest and stiffest material, as well as being an excellent conductor of both heat and electricity. Smaller subsets of fused aromatic rings constitute the class of PAHs. They have a relatively low solubility in water, but are highly lipophilic.

Amorphous carbonaceous materials cover a wide range of compositions, from wide band gap, H-rich, aliphatic-rich a-C:H to narrow band gap, H-poor, aromatic-rich a-C materials. The properties of a-C:H materials are determined by the sp^3 : sp^2 ratio for the carbon atoms and the hydrogen concentration X_H. A C–H bond contributes to the formation of sp^3 bonding and the reduction of the defects in the amorphous carbon network. Therefore, to tune the properties of the a-C:H films such as hardness, band gap energy, and electrical resistivity, the control of the hydrogen concentration is an important issue. In general, it is found that the (optical) energy gap increases with hydrogen concentration (*e.g.* Higa *et al.* 2006[14]). This is essentially due to two factors controlling the density of states of electrons, which in turn regulate the band gap: the number of both sp^2 carbon atom clusters and unpaired electrons of dangling bonds. In the first case, the gap reduces because the band gap between π and π* bands is lower than that between σ and σ* bands; when X_H increases so does the sp^3 : sp^2 ratio, and consequently the band gap. In the second, since the defects in σ bonds are π-like, the stronger the defects the lower the gap; by increasing X_H we force the unpaired electrons to be terminated by hydrogen atoms, increasing the gap. From the above consideration it appears unlikely that well-ordered crystalline graphite grains are an important component of interstellar dust. We must instead consider interstellar carbonaceous dust to be primarily composed of a suite of materials represented by the collective term a-C:H.

Laboratory studies (in the astrophysical literature) are preferentially oriented towards "crystalline" and "amorphous" silicates, to a-C:H, and to PAHs. The comparison among different analogues requires a detailed structural and chemical characterization of the samples that in the case of amorphous materials may be very complex to perform. Other important factors are morphology, *i.e.* shape, agglomeration, and size, and the temperature dependence of the physical and chemical properties. To simulate what is going on in space, from "birth" to "death" (and rebirth as well) of dust grains, laboratory simulations are performed focusing on three main interconnected topics: synthesis; grain processing through thermal annealing and irradiation by ions, exposure to vacuum and extreme ultraviolet photons and X-rays; spectral properties from the far-ultraviolet to the millimetre-range, with different chemical composition, structure, sizes and shapes of the grains, and agglomeration.

Studies of laboratory-produced analogues of cosmic dust are clearly of great potential value in interpreting the nature of interstellar dust. Evidently, however, results from the laboratory should be used with caution. This chapter is mainly devoted to the laboratory production of analogues of cosmic dust in diffuse interstellar regions. Dust material in these regions is believed to be mainly silicate and carbon.

4.2 Silicates

4.2.1 Synthesis of Silicate Dust Analogues

Perhaps the simplest way to produce amorphous silicate materials is by quenching a melt to glass. Mixtures of specific compositions serve as precursors for the melts. As an example, Mohr *et al.* (2013)[15] exploited magnesium carbonate ($MgCO_3$), iron oxalate ($C_2FeO_4 \cdot 2H_2O$), and silicon dioxide powder (SiO_2) to synthesize a series of Mg/Fe-glasses with pyroxene-like nature ($Mg_xFe_{1-x}SiO_3$). The powders are placed in a platinum crucible within a high temperature oven that slowly warmed them up above 1600 °C producing a homogeneous mixture. The melt must be kept for a suitable time (approximately 1 h) over the melting temperature, and after that cooled rapidly. Very fast cooling can be achieved pouring the samples between two copper cylinders, of which one is rotating. The melt is drawn through the rotating rollers and pressed to a strip whose thickness is a fraction of a millimetre. Although some degree of crystallization is unavoidable, the sharp decrease of temperature preserves the disordered state of the melt.

Two important issues are the role of water and the oxidation state of iron. At low water content, water dissolves in silicate glasses almost exclusively as hydroxyl (OH^-) ions, and acts as a network modifier. As such water has a larger effect than other modifier oxides such as Na_2O and MgO, reducing melt viscosity and glass transition temperatures. Iron is intermediate, playing both as modifier (octahedral Fe^{2+} or Fe^{3+}) or network-forming cation (tetrahedral Fe^{3+}). Consequently, the oxidation state of an iron-bearing glass or melt has a significant effect on its structure and properties. For instance, while iron-free $MgSiO_3$ is almost totally amorphous, a higher iron content tends to result in increased crystallinity.

In olivine stoichiometry (Si : Mg + Fe = 0.5) a high Mg content leads to melting temperatures higher than the platinum melting point, making this method impractical. Such materials may be produced at lower temperatures exploiting a sol–gel technique (*e.g.* Burlitch *et al.* 1991[16]). In the sol–gel process, microparticles or molecules in a solution (sol) agglomerate and under controlled conditions eventually link together to form a coherent network (gel). The starting materials (precursors) are suspended or dissolved in water or alcohol. The precursor is then activated by the addition of an acid or a base. Finally, the activated precursors react together to form a network. To synthesize silicates of astrophysical interest Jäger *et al.* (2003)[17] used as precursors metal organic compounds such as magnesium methylate $Mg(OCH_3)_2$ and tetraethoxysiloxane that were hydrolysed to form hydroxides, $Si(OH)_4$ and $Mg(OH)_2$, followed by their condensation. The process is conducive to the formation of a porous three-dimensional magnesium silicate network. Since bonds are broken down by hydrolysis and built up by the opposite reaction of condensation, to avoid inhomogeneities both silicon and magnesium components should present similar hydrolysis and condensation rates. An important parameter is the water to silica (molar) ratio; a low H_2O : Si ratio (<10) tends to inhibit the condensation reactions, thus favouring the formation of silanol groups (Si–OH) and small pore sizes. Higher H_2O : Si ratios favour condensation reactions and the formation of siloxane bridges (Si–O–Si) and larger pore sizes. The synthesis is facilitated by adding hydrogen peroxide (Burlitch *et al.* 1991[16]). Finally, care must be taken in removing solvents from the gel, as methods involving heating to evaporate the liquid may introduce some degree of crystallinity (if actually not required). A suitable "cold" method is freeze-drying that works by freezing the material and then reducing the surrounding pressure to allow for example, the frozen water in the material to sublimate directly from the solid to the gas phase.

To synthesize ordered materials crystal growth techniques can be exploited. The principle of flux growth does not differ significantly from that of growth in solution: the temperature is higher and related to the melting temperature of the flux (solvent). The components of the desired substance are dissolved in the flux. It takes place in a crucible made of highly stable, non-reactive material. The single crystals are deposited from a high-temperature supersaturated solution, while the temperature of the solution decreases.

All these techniques usually proceed in thermodynamical equilibrium, so that the final products are bulk materials that are very homogeneous in structure and chemical composition.

However, interstellar grains are thought to be formed by gas-phase condensation (see Chapter 5). Gas-phase synthesis techniques for the production of amorphous nanometre-sized silicate grains have been applied by various groups. Such techniques, resembling suggested astrophysical condensation processes, are very useful methods, but quite often not applicable to the production of very homogeneous materials. Condensation of dust proceeds through nucleation and growth. We can, in principle, divide condensation experiments into two types, one where the smoke generated at high temperature collides with a substrate at much lower temperature or is embedded in a quenching atmosphere, and the other one in which the condensation occurs in an environment where gas atoms and molecules interact in cooling conditions. Very accurate studies of non-equilibrium gas-to-solid condensation processes have been performed by the groups of J. Nuth at the NASA Goddard Space Flight Center and of F. Rietmeijer at the Department of Earth and Planetary Science in Albuquerque (*e.g.* Rietmeijer & Nuth 2013[1]). These researchers exploited an apparatus able to produce large quantities of highly amorphous condensates from a well-mixed homogenous vapour phase. While these materials are interesting *per se*, they also constitute the natural starting point for annealing experiments to monitor the emergence of crystallinity in time and temperature. Grains are produced in a hydrogen atmosphere, at temperatures in the range 500–1500 K and pressure of about one tenth of atmosphere, through combustion of gas-phase precursors such as silane, SiH_4, iron pentacarbonyl, $Fe(CO)_5$, trimethyl aluminum, $(CH_3)_3Al$, or titanium tetrachloride, $TiCl_4$, that constitute few percent of the total gas input. The oxidant is introduced separately just before the furnace, and is typically molecular oxygen. One metal at the time can also be added to the mixture. The gas is injected in the chamber flowing through the furnace at high speed, in order to minimize the time spent by a grain close to the heating source, following nucleation and growth. The hot gas and the newly-formed grains are rapidly quenched as they enter in the chamber kept at a temperature of about 300–350 K (see Figure 4.1).[18]

Grains formed in these experiments are frequently fluffy, open aggregates. Although the grain chemical composition is determined stochastically by the composition of the inflowing vapour, evidences are found of a complex chemistry ongoing during the very short duration of the vapour growth phase. In a series of experiments Rietmeijer, Nuth and collaborators found that dust condensed from $Mg-SiO-H_2-O_2$ or $Fe-SiO-H_2-O_2$ atmospheres consisted of highly disordered amorphous Mg-SiO or Fe-SiO nanograins (see Figure 4.2),[19] plus crystalline SiO_2 and the corresponding metal oxides.

A eutectic point in a phase diagram indicates the chemical composition and temperature corresponding to the lowest melting point of a mixture. If the condensation proceeds at thermodynamic equilibrium the compositions of the Mg-SiO condensates—occurring at three stable eutectic points—would be 2% in MgO (and 98% in SiO_2) at 1986 K, 36% MgO at 1816 K, and 63% MgO at 2123 K.[18] The magnesium or iron silicates produced in the experiments show instead three favoured compositions, 25%, 45%, and 85% MgO. While the third condensate composition was not constrained by the existing phase diagrams, the first and second compositions fell in between the first and the second, and the second and the third eutectic points, respectively. These out of equilibrium compositions occur at deep (colder) metastable eutectic points, as that shown in the hypothetic equilibrium phase diagram

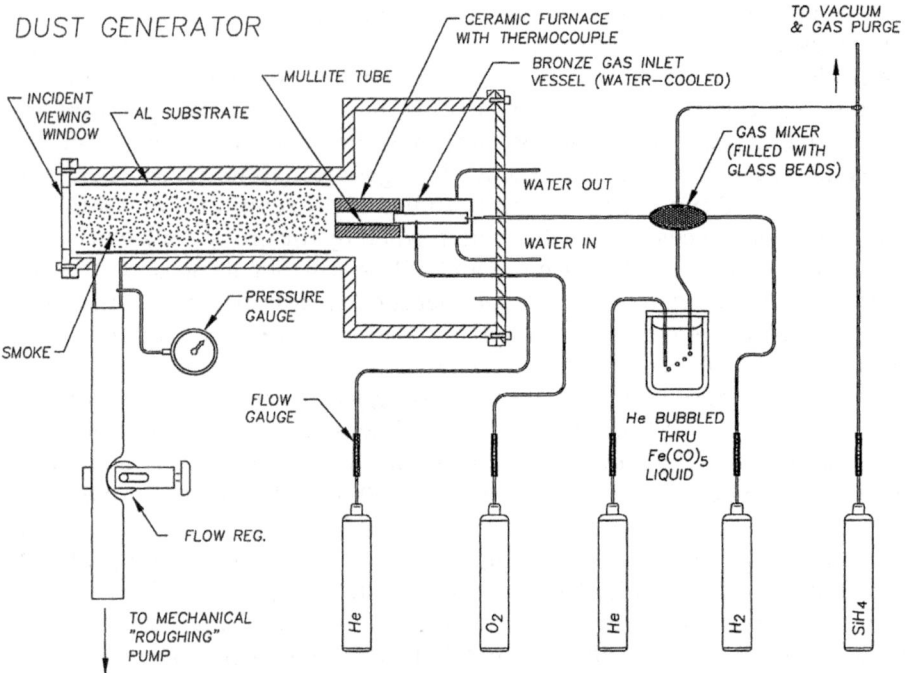

Figure 4.1   Schematic of the dust generator used to manufacture nanometre-sized smoke particulates. The gas is injected into the chamber, flowing through the furnace at high speed in order to minimize the time spent by a grain close to the heating source. Reproduced with permission from Nuth *et al.* (2002).[18]

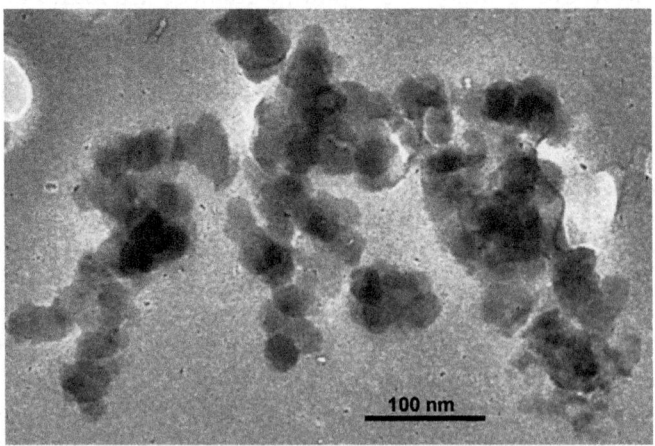

Figure 4.2   Transmission electron microscope image of Si-oxide grain condensates present in grains condensed from the Ca–SiO–H$_2$–O$_2$ vapour. Reprinted with permission from Rietmeijer *et al.* (2008).[19]

depicted in Figure 4.3,[19] in which the phase composition and temperature are reported on the horizontal and vertical axes, respectively.

The stable liquidus (solid) line shows two stable eutectic points where the liquidus has two minima. Extending the stable liquidus line to a lower temperature (dashed lines) from both minima, a deeper, far from thermodynamic equilibrium metastable eutectic point is intersected. Such a point corresponds to a composition "AB" that is reached if the cooling phase is short enough to prevent equilibrium to be achieved. The chemical composition at the deep metastable eutectic point will always be positioned in between two stable eutectic points. Rietmeijer & Nuth (2000)[20] interpret the results of their experiment identifying the deep metastable eutectic points with the emergence of a new state of matter where the extreme disorder becomes a metastable state, *i.e.* a dissipative structure such as that described by I. Prigogine (see for example Prigogine, 1979[21]).

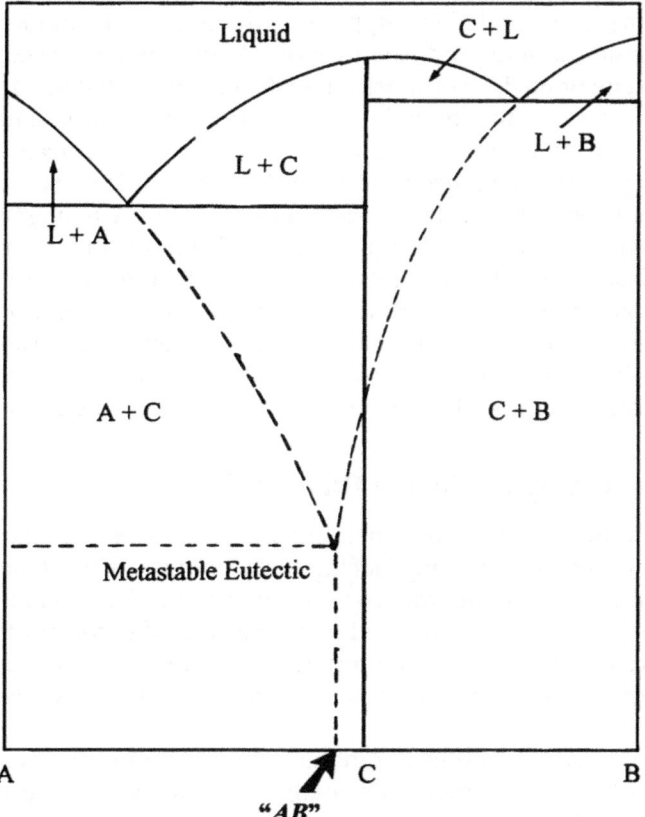

Figure 4.3 The hypothetical binary system A–B with an intermediate stable phase C, showing two stable eutectic points that define a metastable eutectic one (dashed lines) with a composition AB at some quench temperature. Reprinted with permission from Rietmeijer *et al.* (2008).[19]

Laser ablation of target material has also been employed to condense amorphous silicate grains. The condensed particles are extracted from the condensation zone by particle beam extraction and deposited on a KBr substrate, for example, or as made by Sabri *et al.* (2014)[22] by condensation of the evaporated atoms, molecules, and clusters in a quenching gas atmosphere. An advantage of such a technique is the high-power density that can be achieved by using a pulsed laser producing a high concentration of reactive clusters and atoms. Sabri *et al.* (2014) used either Mg-Fe-Si targets with metal ratios expected for olivine or pyroxene stoichiometry (Si : Mg + Fe = 1.0), and amorphous materials of olivine or pyroxene compositions. The newly-formed condensates were allowed to flow adiabatically into a chamber with a pressure as low as 10^{-3} mbar. This produced a decoupling of the particles from the gas, with a corresponding decrease of their density, inhibiting further particle growth and coagulation processes. Finally, the grains are directed into a second chamber (at much lower pressure) where they are deposited as a film. Grain compositions range from non-stoichiometric magnesium iron silicates to pyroxene- and olivine-type stoichiometry. Other amorphous magnesiosilica smokes were produced by laser ablation of glassy silicate targets with a pyroxene or olivine composition by Fabian *et al.* (2000).[28] The smallest grains in these experiments had non-stoichiometric silicate compositions. The largest amorphous magnesiosilica grains showed significant grain coagulation. Large (1–2 μm) amorphous grains had forsteritic and enstatitic compositions and contained scattered (crystalline) forsterite nanograins.

Unfortunately, all these techniques yield different results in terms of optical properties, for example in the line ratio of the 10 and 18 μm features, even if the materials show clearly the same compositions (see Section 4.4). This suggests that the samples do not have the same structure. The sample preparation techniques vary, and evidently this induces a wide range of disordered structures, that frequently lead to complex physical and chemical properties.

4.2.2 Processing of Silicate Dust Analogues

Laboratory studies of thermal annealing processes have been carried out by several groups, most commonly placing samples of condensed materials into a furnace under vacuum in order to monitor specific changes in the solids as functions of the annealing time and temperature. Other forms of processing are exposure of dust analogue materials to radiation and to energetic ions. All these experiments simulate different phases in the life of astronomical silicate grains.

It is well known that a phase transition between amorphous and crystalline silicates leads to a remarkable change of their spectral properties from the ultraviolet through visible to the far-infrared region. During thermal annealing, the variation of transmittance indicates changes in the structure of the dust. Such spectral evolution as a function of temperature and time may be exploited to measure the activation energy intrinsic to each material of the metamorphic transformation from chaotic condensates through to the crystalline phase. The characteristic annealing time τ is marked by the appearance of the first hints of fine structure indicating the increase of structural order. An Arrhenius law relating the annealing time τ to the activation energy E_a *via* the temperature T (K) is usually adopted, $\tau = 1/v_0 \exp(E_a/kT)$, where v_0 is the vibrational frequency of the lattice. In other words, the crystalline activation energy indicates how quick the crystallization process is once the temperature of dust is fixed. The experimental results show that high energy barriers have to be overcome by the atoms of Mg-silicate glasses and smokes to reach a more favourable and ordered configuration, *i.e.* E_a/k ~40 000–50 000 K. The activation energies of magnesium silicates produced by sol–gel synthesis are about two thirds of these values.[17] This behaviour derives from an excess of Si–OH bonds stemming from the sol–gel synthesis. OH groups act as network modifiers introducing non-bridging oxygen atoms, thus decreasing the degree of polymerization. A less polymerized material will have a higher viscosity than a more polymerized material because the silicate network consists of strong directional covalent bonds that resist deformation. Viscosity is a measure of the resistance of a liquid to flow, and flow requires the structure to deform and move. Since the crystallization is related to a diffusion of metal and oxygen atoms, the activation energy is strongly dependent on the viscosity behaviour of the silicate during the annealing. The increase of the concentration of OH$^-$ ions decreases the viscosity and therefore the annealing time that in turn will lower the activation energy. A similar effect is observed in the case of Fe incorporation, for example E_a/k = 26 300 K for Fe-rich olivine.[24]

Such a "simple" description may hide the intrinsic complexity of the annealing process that is strongly dependent on the experimental conditions and the starting materials used to produce amorphous materials. During the annealing, a "stall phase" may appear in which the heating of the material causes no further infrared detectable changes for some temperature-dependent time period. In the condensation experiments performed by Nuth and co-workers, described in the previous Section, there is initially a readjustment of the Si–O stretching and O–Si–O bending vibrations due to oxidation/reduction reactions to produce fully oxidized amorphous materials.[25] This reorganization phase is followed by grain growth through coagulation of condensed amorphous nanograins, giving rise to chemically homogenized amorphous materials that, in the case of magnesiosilica for example, are necessary for the nucleation and growth of forsterite, enstatite, and tridymite (a high temperature SiO_2 polymorph) nanocrystals, that eventually define the post-stall phase. Thus, condensed amorphous magnesiosilica grains need to coagulate into larger aggregates prior to crystallization of Mg-silicate subunits, but the processes and the individual grain distributions during a stall phase are chaotic in nature and hence unpredictable. Annealing for longer times and higher temperatures increases the abundance and sizes of forsterite grains, followed by those of enstatite, although an amorphous magnesiosilica phase still persists. Ultimately they form a composite aggregate consisting of forsterite, enstatite, and amorphous silica grains.

A stall phase, or its lack, is determined by various parameters during the condensation process, since the transition of submicron amorphous magnesiosilica to crystalline silicates (forsterite, enstatite) and SiO_2 (amorphous or crystalline) is a discontinuous process that can take different reaction pathways. A stall phase has also been observed in materials prepared with the sol–gel technique,[26] which is interesting because sol–gel samples are more highly ordered than condensates. The stall phenomenon is not then the result of some chemical rearrangement of the silicon oxidation state. Kaito *et al.* (2003)[27] were able to record the change in the physical structure of their condensates with transmission electron microscopy (TEM) as a function of temperature and time. They found that during the initial stage of the annealing a layer of polycrystalline materials was formed on the sample surface. During the stall phase such tiny surface crystals began to grow, but no crystallization was observed within the annealed materials. The driving force is the minimization of the surface energy that induces the growth of larger crystals from those of smaller sizes (Ostwald ripening). At the end of the stall phase, the crystallization began to penetrate also in the interiors of the material. In a different experiment, Fabian *et al.* (2000)[28] produced vapour-condensed smokes with Mg_2SiO_4 and $MgSiO_3$ compositions that were subsequently annealed at 1000 K. The annealed samples were mostly amorphous materials with pyroxene or olivine compositions that quickly crystallized to a mixture of forsterite and tridymite (but not enstatite) in amorphous silica. The already complete network of $(SiO_4)^{4-}$ tetrahedra as the basis of the molecular structure of the condensed material removed the need for a stall phase. A detailed account of the thermally supported transition from amorphous to crystalline is found in the article by Rietmeijer & Nuth (2013)[1] and references therein.

Low-temperature crystallization may occur under specific laboratory conditions. Kimura *et al.* (2008)[29] obtained the crystallization of laboratory-synthesized amorphous Mg-bearing silicate grains of 100 nm in diameter after exposure to CH_4 by 100 keV electron-beam irradiation in a transmission electron microscope at room temperature. Electrons induce the decomposition of methane molecules on the silicate surface, with the released carbon atoms free to react with oxygen and produce thermal energy locally. Such combustion energy can aid the growth of forsterite nuclei, without warming up the whole silicate particle. Forsterite of nanometre size has been produced at room temperature using laboratory-synthesized amorphous Mg-bearing silicate grains by covering them with a carbon layer produced in a methane gas atmosphere.[30] Such non-thermal crystallization was induced by the graphitization of the carbon layer due to oxidation in air. This kind of chemical-reaction-driven crystallization processes may explain how cometary silicates can crystallize, while still preserving volatile interstellar ices in their parent comets.

The inverse process, *i.e.* the amorphization of silicates is accomplished using irradiation with low-energy (keV) ions. Amorphization has been shown to occur in several experiments using 4–50 keV ions at fluences of 10^{15}–10^{17} ions cm^{-2}. In a series of experiments, K. Demyk[31] and collaborators irradiated olivine and pyroxene samples at the Centre de Spectrometrie Nucleaire et de Spectrometrie de Masse, Orsay, France with He^+ ions accelerated to energies of 4 and 10 keV. The implantation of helium ions into the mineral causes an increase (with the fluence) of the porosity and a change of the chemical composition, as evidenced by the modification of the material stoichiometry. The amorphization of the sample was revealed using TEM that evidenced a rim in the thinnest part of the sample not containing Bragg contours, whatever was the orientation of the sample. A perfect crystal should exhibit a typical contrast in the TEM micrograph with contours corresponding to the locus of the exact Bragg conditions. When an ion penetrates into a solid it loses energy mainly through ionization and elastic collisions with nuclei, until it is eventually stopped. Amorphization is driven by direct collisions of the incident He^+ ion with the atoms forming the lattice. This induces a displacement cascade at the end of each ion track, causing the local collapse of the crystalline structure. As shown in the experiments, significant amorphization of the solid occurs when the ion fluence is above a critical threshold as a result of many concurring individual events. Chemical modifications are induced by collisions between ions and the target. When the energy transferred to the target atom is sufficient for it to escape from the surface into the vacuum (*i.e.* sputtering), then fractionation may be induced by removing some species preferentially. For instance, in materials of olivine composition Mg and O are more sensitive to sputtering than Si and Fe. However, experiments performed in which submicrometre-sized enstatite ($MgSiO_3$) grains are irradiated with 50 keV He^+ ions[23] do not lead to significant sputtering of the grain surface or a change of chemical composition due to a selective sputtering of Mg, even in the case of high fluences. Apparently, such chemical changes are very sensitive to the adopted energy range. Finally, it should be noted that such disordered silicates produced by ion bombardment of initially crystalline materials may form at temperatures above the glass temperature. These materials are not true glasses, to which they tend when the extent of damage is high enough. Consequently, ion irradiation of initially crystalline material should lead to a continuum of structures ranging from perfect crystals to completely disordered materials.

The experiments described above aimed to simulate amorphization of silicates in low-velocity shocks by irradiating material samples with low-energy (keV) ions. However, this processing is inefficient, because in a 200 km s^{-1} shock, the range of the ions is much smaller than the size of most interstellar grains. Bringa *et al.* (2007)[32] point out that ions of much higher energy, which can penetrate even the largest interstellar grains, pervade the interstellar medium in the form of cosmic rays. To investigate the effects of cosmic rays on crystalline silicates, these authors irradiated samples of forsterite (Mg_2SiO_4) single crystals with triply ionized, 10 MeV Xe ions (lower than the real cosmic-ray energies). The Lawrence Livermore National Laboratory 4 MV ion accelerator was exploited to bombard a forsterite single crystal sample using a flux of $1-2 \times 10^{10}$ ions cm^{-2} s^{-1} with fluences up to 6×10^{13} ions cm^{-2}. At this energy, a large fraction of the total stopping power (energy loss per unit path length) is due to the inelastic collisions between bound electrons in the medium and the ion moving through it, and decreases with penetration depth as the ion slows down. TEM studies revealed that the sample was amorphized up to a depth about the half of the ion range. During an impact, a cylindrical pressure wave expands from the ion track ejecting molecules from the surface of the solid and creating a nanometre-sized track of dense electronic excitations that can be in part converted into atomic motion. The cylinder along the ion track is the region of amorphization of the material. Finally, chemical analysis of the amorphous region did not indicate any change in composition.

Ultraviolet radiation may generate silicate defects that can induce a series of chemical reactions: the non-bridging oxygen defects resulting from broken Si–O bonds are mobile and can migrate through the lattice.[33] Ultraviolet irradiation may be important on planetary surfaces (*e.g.* on Mars[34]), in which the dangling bonds resulting from radiation damage are highly reactive and can readily capture atmospheric oxygen to form superoxides. Low pressure can favour the stability of these phases by preventing rapid annealing of the broken bonds and destruction of the superoxides. Kimura & Nuth (2007)[35] synthesized refractory silicate grains from volatile starting materials such as silane or pentacarbonyl iron subjected to ultraviolet irradiation (or to an electrical discharge). The authors' purpose was to simulate silicate reassembling in the interstellar medium, and in general in low-density or cold regions, after the passage of a shock, in which dust particles could have been vaporized. For instance, silane, germane, phosphine, and other reduced metallic compounds are found in the upper atmospheres of giant planets in the Solar System, and might be present in giant extrasolar planets. Ultraviolet photolysis of such materials in the presence of an oxidant such as water vapour or OH radicals could result in the production of amorphous refractory grains.

4.3   Production of Carbon Dust Analogues

Gas-phase condensation techniques such as laser pyrolysis, laser ablation of graphite, resistive heating or arc discharge between graphite rods in quenching gas atmospheres, as well as plasma condensation methods, are the most exploited methods to produce particulate carbonaceous cosmic dust analogues. An alternative low temperature route is the production of hydrogenated carbon polymers through the photolysis of a series of organics.[36] Carbon materials show characteristics that are strictly related to the manufacturing process. Arc discharge and laser ablation methods, chemical vapour deposition (CVD), pyrolysis and flame synthesis are the most widely used processes for the production of special carbon materials such as fullerenes, nanotubes, and graphene, whose structures are relatively simple. By contrast, traditional high-temperature pyrolytic and combustion processes in very fuel-rich conditions of gaseous, liquid, and solid fuels also produce soot. This material exhibits a mixed ordered/disordered character, having a nanostructure mainly based on two-dimensional graphene layers grouped, cross-linked and/or layered on each other in a disordered way. Incomplete combustion leads to the formation of PAHs.

Exposure to ultraviolet photons, energetic particles (*e.g.* cosmic rays in space), shock waves and hot gas, hydrogen atoms, or thermal annealing alters the structure and properties of hydrocarbon dust. Thermal heating or ultraviolet irradiation darkens a-C:H to form hydrogen-poorer amorphous carbon. Thermal annealing of a-C:H materials in a vacuum produces modifications in the infrared spectra compatible with a progressive aromatization of the carbon structure, *i.e.* the carbon network tends to arrange in an increasing number of larger sp^2 clusters, due to hydrogen effusion. Thus, thermal treatment produces allotropic changes, with transformation of nanodiamonds into onion-like graphitic particles. Aromatization also occurs under ultraviolet irradiation as the energy of photons is high enough to break the C–H bond (~5 eV). X-rays induce a phase transition in a-C:H in terms of a volume expansion, graphitization, and change of local order of the irradiated sample area. Electron irradiation leads to structural alterations of the lattice. In the case of graphite, random displacements of the atoms and their subsequent rearrangements eventually lead to topological changes, such as the formation of carbon "onions". Transformation of diamond into a graphite-like carbon follows ion implantation. All kinds of carbon processes leading to a loss of hydrogen atoms from the structure induce an evolution from aliphatic-rich towards aromatic-rich materials, associated with a closing of the band gap of the material. Such transformations are reflected in variations of the optical properties of the material. We defer the discussion of this topic in the next Section.

In a laser ablation experiment, a laser beam is focused onto a metal–graphite composite target which is placed in a high-temperature furnace. At high laser flux, the material is typically converted to a plasma. The laser beam scans across the target surface under computer control to maintain a smooth and uniform face for vaporization. In an arc discharge experiment a vapour is created between two carbon electrodes, usually graphite. The anode is either pure graphite or contains metals. In the synthesis of carbon nanotubes in the soot, Iijima (1991)[37] used three components, argon, iron and methane. After the triggering of the arc between two electrodes, a plasma is formed consisting of the mixture of carbon vapour, the rare gas, and the vapours of catalysts. The plasma can be stabilized for a long reaction time by controlling the distance between the electrodes by means of the voltage control. Arc discharge and laser ablation methods for the growth of carbon nanotubes have been actively pursued in the past decades. In CVD precursors are transported in the vapour phase to decompose on a heated substrate to form a film. In this process a mixture of hydrocarbon gas (*e.g.* ethylene, methane or acetylene) and a process gas (*e.g.* ammonia, nitrogen, and hydrogen) are made to react in a reaction chamber on a heated metal substrate at high temperature (>1000 K), and atmospheric pressure. The mechanism involves both gas-phase reactions and heterogeneous processes that occur on the heated surface. The setup of a laser pyrolysis experiment for the production of carbonaceous materials is centred on a condensation chamber in which different temperatures can be realized by using either a pulsed or a continuous wave laser with different powers. The reaction gases flow into the reactor with concentrations and flow velocities that can be regulated independently by means of appropriate flow meters. The laser radiation induces the dissociation of the reaction gas and the subsequent condensation of carbon nanoparticles which can be extracted from the flow reactor and collected in a filter. Using this technique, Jäger *et al.* (2006)[38] condensed carbon nanoparticles produced by benzene, ethylene, and acetylene gas mixtures. The analysis revealed that the carbonaceous powder contained various PAHs.

Most of the primary cosmic carbonaceous material is thought to have formed as nano- and subnanometre-sized particles *via* gas-phase condensation in envelopes of carbon-rich evolved stars (see Chapter 5), through a poorly understood combustion-like process where small carbon chains (*e.g.* acetylene) form PAHs that nucleate into larger-size PAHs and, ultimately, into nanoparticles. Nucleation occurs above 2000 K followed by the growth of amorphous carbon materials on the condensation nuclei as the temperature decreases. As the temperature continues to fall, aromatic molecules form and condense onto the growing particles that eventually coagulate into larger structures such as those occurring in soot formation.[39] The temperature and the chemical composition of combustion flames and pyrolysis environments are very close to those in the photospheres of carbon-rich stars. Acetylene (C_2H_2) is the dominant carbon-bearing molecule after carbon monoxide in circumstellar shells characterized by temperatures of about 2000 K. Such similarity suggests that the chemistry responsible for the formation of PAHs and soot in hydrocarbon flames may be ongoing in astronomical sources.[40] The primary focus is on the formation of the first aromatic ring from small aliphatics, because this step is perceived as the rate-limiting step in the formation of larger aromatics. In combustion flames, the process starts with the breakdown of the fuel into smaller reactive molecules and radicals. Then, there are several possibilities, all involving radical chains such as CH_3, n-C_4H_3, n-C_4H_5, and propargyl C_3H_3. Some of them have been dismissed with arguments based on uncertain foundations.[41] The prevalent route should involve the recombination of two propargyl radicals to form cyclic and linear structures of benzene and the radical phenyl (C_6H_5). Frenklach and co-workers (*e.g.* Schuetz *et al.* 2003[42]) suggested that once benzene is formed the growth of aromatics proceeds through H abstraction (to produce the phenyl radical) followed by acetylene addition (the acronym for the whole process is HACA: H-abstraction-C_2H_2-addition[43]), in a repetitive reaction sequence to multiple cyclic structures. The transition of gas-phase species to solid particles is probably the least understood part of the soot formation process. The core of the process within the HACA context is the accumulation of particle mass *via* chemical reactions with gaseous precursors occurring simultaneously with the growth of particle size by collisions among PAH molecular species and clusters, until a critical nucleus of two PAHs is reached, each larger than the four-ring pyrene. Then PAH dimers, bound by van der Waals forces, collide with PAH molecules forming PAH trimers or with other dimers forming PAH tetramers, and so on, while individual PAH species keep increasing in size *via* molecular chemical growth reactions. While the nucleation kinetics control the number of nascent particles, coagulation and surface reactions involving acetylene control both growth and evolution of the particle number density.

Such a scenario was disputed by the results of an experimental and theoretical study of the dimerization and nucleation of pyrene at low and high temperatures performed by Sabbah *et al.* (2010).[44] These authors demonstrate that the equilibrium of this reaction strongly favours the dissociation of the pyrene dimer at high temperature and that physical dimerization (involving van der Waals forces) of pyrene cannot be a key step in carbon particle formation in hot environments. Therefore, it appears that a mechanism involving further chemical growth of PAHs to sizes well beyond that of pyrene is necessary before physical dimerization can play a major role in soot formation in combustion. Siegmann *et al.* (2002)[45] investigated the formation of carbonaceous particles in laminar, atmospheric pressure diffusion flames. On the basis of their experimental results, these authors put forward the hypothesis that soot synthesis can even precede PAH synthesis. In this picture, small hydrocarbon molecules, in particular acetylene, add to radical chains to produce unsaturated hydrocarbon radicals that condense and give rise to the first particles, consisting of chain-like aggregates held by covalent bonds (and not by van der Waals forces), *i.e.* aromatic compounds with few condensed rings connected by aliphatic bonds. The argument over formation routes still continues, with two unresolved issues concerning the mechanism for soot nucleation and how it is affected by individual PAHs, and the relative importance of nucleation *versus* surface growth. Possibly two reaction pathways are needed, one providing the high mass PAH stacking route (the small ones hardly condensing at high temperatures, see below), and another regarding the aliphatic aromatic chemical growth.

To (partially) address such questions Biennier *et al.* (2009)[46] produced carbonaceous particles by the pyrolysis of acetylene in a chamber of reduced dimension, which made it possible to use a heated porous graphite rod. The huge exchange surface granted by the porosity favours an efficient heat transfer from the rod to the gas flowing through it. The condensed particles were probed by off-apparatus techniques such as NMR, FT-infrared spectroscopy, and X-ray diffraction. The analysis suggests a chemical make-up of soluble low-mass aromatics surrounding small insoluble larger aromatic islands bridged by aliphatic groups. X-ray diffraction evidences a mostly amorphous network containing small aromatic islands generally composed of two graphene sheets connected by aliphatic bridges. The aromatic to aliphatic ratio is 3.2 for the whole sample. Such structures have already shown up in the observations (*e.g.* Dartois *et al.* 2007 [47]). These authors observed with the Spitzer satellite and UKIRT infrared telescope the extragalactic source IRAS 08572+ 3915, and detected the 3 to 7 μm carbonaceous absorption bands in the diffuse interstellar medium, with a high optical depth and good sensitivity. Dartois and co-workers placed an upper limit on the aromatic

C–H stretch, constrained the aromatic *versus* aliphatic C–H content of inter-stellar a-C:H, and set the lower limit $X_H \leq 0.2$ for the hydrogen content. Exploit-ing a ternary phase diagram where the hydrogen content and the two main bonding types (sp^2 and sp^3) for carbon constitute the poles, they were able to identify the carrier of the spectral features as an interstellar hydrocarbon belonging to the class of polymeric-like a-C:H, dominated by an aliphatic/olefinic backbone structure. The change from aliphatic to aromatic struc-tures may occur in environments that selectively dehydrogenate the a-C:H, providing an opportunity for aromatic molecules to form. These observa-tions, together with observations of very evolved stars (protoplanetary and planetary nebulae), suggest an evolution in which aliphatics are converted into aromatic structures. The condensates of Biennier *et al.* (2009)[46] may be placed at an intermediate level where most of aliphatic branches begin to be converted to aromatics, but the material is still far from being a fully devel-oped graphitic network. The absence of olefinic protons (that are discernible in young soot particles) supports the evolutionary stage of this phase of the carbon material.

An important point is to determine how the increase in temperature leads to the production of fullerene-containing particles and if chemical inter-mediates can be identified. A step towards this direction has been taken by Jäger *et al.* (2009)[48] who performed gas-phase condensation experiments to produce nanometre and sub-nanometre carbonaceous particles at both low (LT) and high temperature (HT) regimes. For HT condensations laser abla-tion of graphite was used, carried out with pulsed lasers characterized by high power densities able to induce temperatures of more than 3500 K in the condensation zone. Laser induced pyrolysis was exploited for both HT and LT experiments using pulsed (HT) and continuous-wave (LT) CO_2 lasers. The gas-phase hydrocarbon precursors were acetylene, ethylene (C_2H_4) and ben-zene. The chemical routes in the two cases are totally different, resulting in different structures and spectral properties. HT experiments provided very small fullerene-like soot condensates of a few nm in size, showing elongated or symmetric cage fragments, frequently linked by aliphatic bridges or by van der Waals forces. The formation pathway of the soot particles is char-acterized by the emergence of carbon chains, subsequently forming fuller-ene fragments of bowl shape with side chains. Such growth processes do not obey the laws of classical thermodynamics: far from equilibrium, a system may still evolve to some steady state, but in general this state can no lon-ger be characterized in terms of some suitably chosen potential. Irle *et al.* (2003)[49] present quantum chemical molecular dynamics simulations for the formation of fullerene molecules from ensembles of randomly positioned C_2 molecules in a periodic boundary box. Spontaneous self-assembly of fuller-ene molecules occurs through three distinct stages: nucleation of polycyclic structures, growth by ring condensation of attached carbon chains, and cage closure (see Figure 4.4).[49]

Similarly, long and branched carbon macromolecules may be the pre-cursors for cyclic structures with long carbon chains attached which can

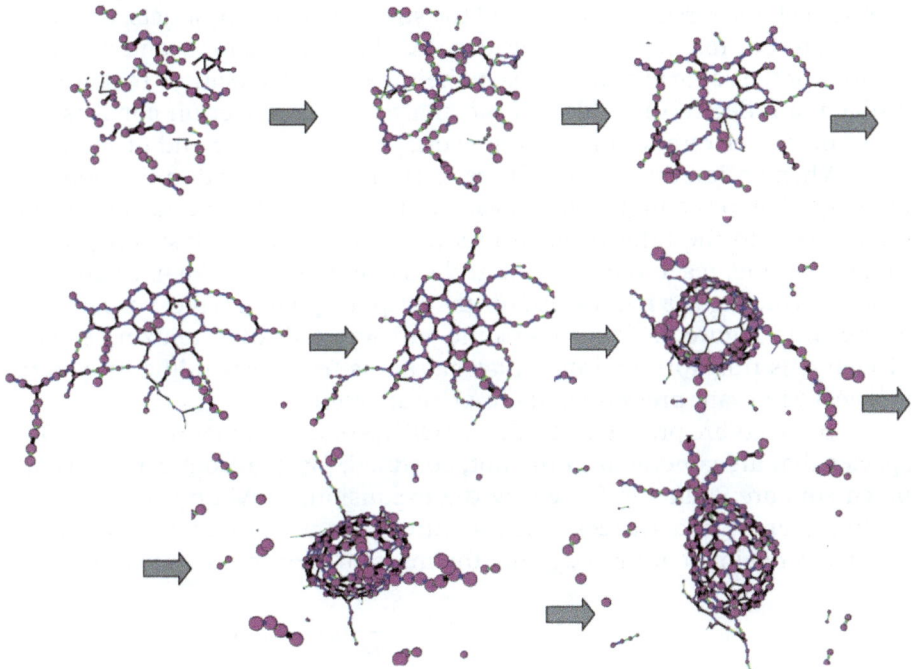

Figure 4.4 Snapshots of a trajectory leading to the formation of a fullerene molecule from ensembles of randomly positioned C_2 molecules. Single bonds are shown in black, double bonds in blue, and triple bonds in green. Adapted and reproduced with permission from Irle *et al.* (2003).[49]

condense and grow, for example, by the sticking of two linear chains attached to a nucleus. At low temperatures, the condensation by-products are mainly PAHs that are also the precursors or building blocks for the condensing soot grains. The low-temperature condensates contain PAH mixtures that are mainly composed of volatile three to five ring systems that condense on carbonaceous seeds *via* physi- and chemisorption. As a result, large carbonaceous grains are formed, revealing well-developed planar and slightly bent graphene layers in their interior. The HT soot grains consist of about 50% sp^2 hybridized carbon atoms, while in the LT case the content of sp^2 hybridized carbon atoms was found to range between 80% and 92%.

Contreras & Salama (2013)[50] studied the formation and destruction mechanisms of carbon dust analogues produced by means of collection of PAHs and hydrocarbon molecular precursors, using a newly developed facility, COSmIC (cosmic simulation chamber). COSmIC is composed of three major components: first, a pulsed discharge nozzle (PDN) where microscopic targets form in a supersonic free jet exposed to plasma discharge; second, high sensitivity (with an effective path length of many kilometres) cavity ring down spectroscopy (CRDS) that measures the spectral signature of the products; and lastly, an orthogonal reflectron time-of-flight (TOF) mass spectrometer

(MS) that characterizes the mass and the structure of the products. At higher masses, resolution is difficult because the flight time is long, and not all of the ions of the same m/z values reach their ideal TOF velocities. To solve this problem a reflectron is added to the analyser. The reflectron consists of a series of ring electrodes of very high voltage placed at the end of the flight tube. When an ion travels into the reflectron, it is reflected in the opposite direction due to the high voltage. Faster ions travel further and slower ions travel less into the reflectron. In this way both slow and fast ions, of the same m/z value, reach the detector at the same time rather than at different times, narrowing the bandwidth for the output signal. A typical experiment is shown in Figure 4.5:[50] a supersonic jet subjected to an ionizing electric discharge is transformed into a plasma that is freely expanding into astrophysically relevant pressure and temperature regimes.

The products are probed by CRDS or TOF-MS or a combination of both. The species that are generated in the hot, confined, plasma from carbonaceous precursors are suddenly frozen by the expansion, providing very efficient cooling over short distances. The measurements performed by Contreras & Salama were aimed at investigating the most efficient molecular precursors

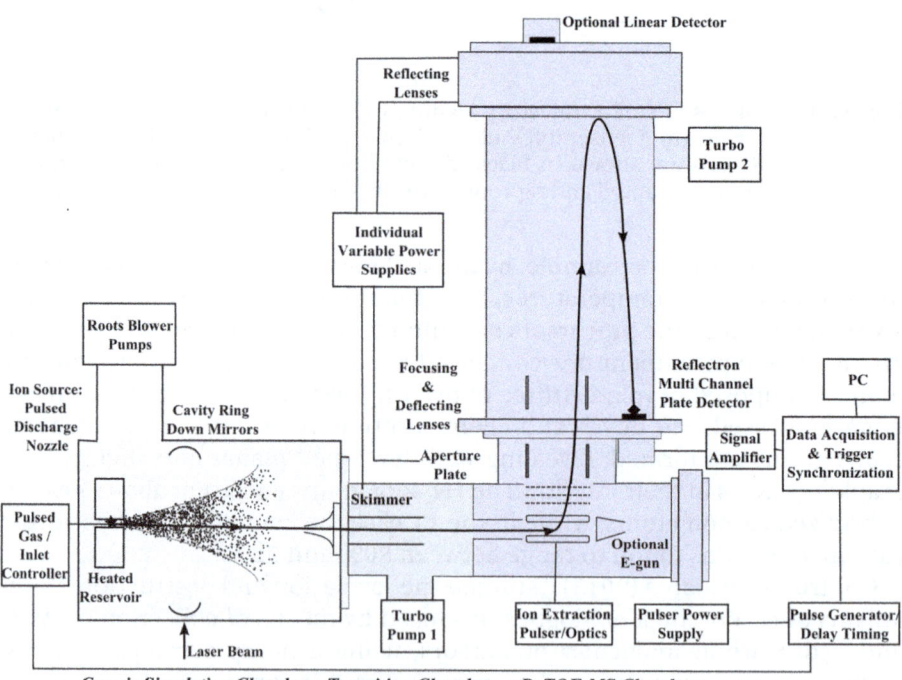

Figure 4.5 Schematic of the COSmIC apparatus. The black arrow indicates an ion trajectory originating from the pulsed discharge nozzle (PDN) into the chamber, followed by orthogonal acceleration and reversal of the flight path toward the detector. Reprinted with permission from Contreras & Salama (2013).[50]

in the chemical routes that eventually led to the formation of carbonaceous grains in the stellar envelopes of carbon stars. A variety of products is formed from each molecular precursor injected into the PDN. The products include neutrals, radicals, and ions of the molecular precursors, as well as neutrals, radicals, and ions of the fragments formed due to dissociation of the molecular precursors, pure mixtures of hydrocarbons seeded in an inert carrier gas (argon). Precursor species include saturated and unsaturated hydrocarbons, benzene, toluene (C_7H_8), pyridine (C_5H_5N, a benzene-like ring system where one of the carbon atoms has been substituted with a nitrogen atom), and various PAHs. The experiments clearly indicate that the reaction products are highly dependent on the molecular nature of the precursors. All precursors experience fragmentation, with losses of H, CH, and acetylene, among the plethora of parts and pieces specific to each hydrocarbon or PAH considered. Molecular growth through the recombination of the fragments is observed in the case of all unsaturated hydrocarbons and for small, mono-ring aromatic precursors. The increase in the molecular complexity is attributable to chain growth and ring formation. One of the most likely recombination pathways involves ion–molecule reactions, which tend to have minimal activation energies. No molecular growth is observed in the case of PAHs indicating that small PAH structures are not involved in the initiation of the chemical pathways that lead to the formation of solid grain particles. This may due to the stability of the PAH precursors, which only exhibit minimal fragmentation.

It is interesting to note that the molecular products observed in the COS-mIC experiments tend to support the assumptions made so far in the models that describe the chemical pathways to the formation of PAHs in circumstellar environments, involving the propargyl radical (see above). Such laboratory results point out that the most promising routes for molecular growth and for the formation of large molecular species in the outflows of carbon stars are indeed the routes that begin with small unsaturated carbon chains leading to the formation of benzene rings followed by the growth of larger aromatic structures through the HACA mechanism.

Addition of a phenyl radical with vinylacetylene may produce naphthalene (and possibly a step-wise growth of PAHs) *via* a neutral–neutral reaction that has been shown to proceed at very low interstellar temperatures.[51] In general, however, the addition of the phenyl radical to an olefinic double bond and an acetylenic triple bond is associated with entrance barriers. Such barriers are quite low, but still high enough to block the formation of naphthalene in cold molecular clouds. Combining crossed beam experiments with density functional modelling, Parker and co-workers identified a gas phase route to the formation of naphthalene that is barrier-less and exoergic. The major difference with the standard HACA system lies in the enhanced polarizability of vinylacetylene as compared to acetylene and ethylene; the enhanced polarizability triggers the formation of a weakly-bound van der Waals complex. The subsequent addition step is hindered by a barrier, but this barrier is located below the energy of the separated reactants (a so-called submerged barrier). Naphthalene can thus be produced through a single collision (neutral–neutral) reaction in the gas phase of the phenyl radical with vinyl-lacetylene. Although such a mechanism is very attractive because it challenges the conventional, high temperature theories of PAH synthesis, neither phenyl radicals nor vinylacetylene have been detected in interstellar clouds.

Dartois[36] and co-workers devised a method to produce a a-C:H polymer through the ultraviolet photolysis (using Lyman-α photons) of organic molecule precursors at low temperature. Laboratory ultraviolet photolysis is a "gentle" way to link together the basic units that build a-C:H. Plasma-produced a-C:H (*e.g.* quenched carbonaceous composite, see next Section), can be considered an extension of much higher energetic processing of a common material. The production is performed by slow deposition of an organic progenitor in the gas phase at ambient temperature onto a cold finger, with simultaneous ultraviolet irradiation. Selected precursors were saturated (*e.g.* methane) and unsaturated (*e.g. trans*-2-butene) aliphatic species, and xylene, representative of aromatic compounds (although xylene possesses an aliphatic character due to the methyl groups). The effect of nitrogen included in the aliphatic chain was tested by using diethylamine, $CH_3CH_2NHCH_2CH_3$, and triethylamine precursors. The decomposition of the a-C:H stretching mode absorption profile into individual characteristic group vibrations highlights the relative proportions of sp^3, sp^2 and sp^1 hybridized carbons not dominated either by diamond or graphene phases. These authors were able to assess an olefinic to aliphatic ratio of 0.05–0.1 in the produced samples, and a $CH_2:CH_3$ ratio of around 2. The aromatic contribution is more difficult to estimate because the only suitable band is the C=C stretch overlapped by the olefinic C=C. Dartois *et al.* (2005)[36] found from 5% to at most 20% aromatic C=C in the a-C:H network. Their inferred substructure unit for the photoproduced a-C:H presented is shown in Figure 4.6 (left panel[36]). Such

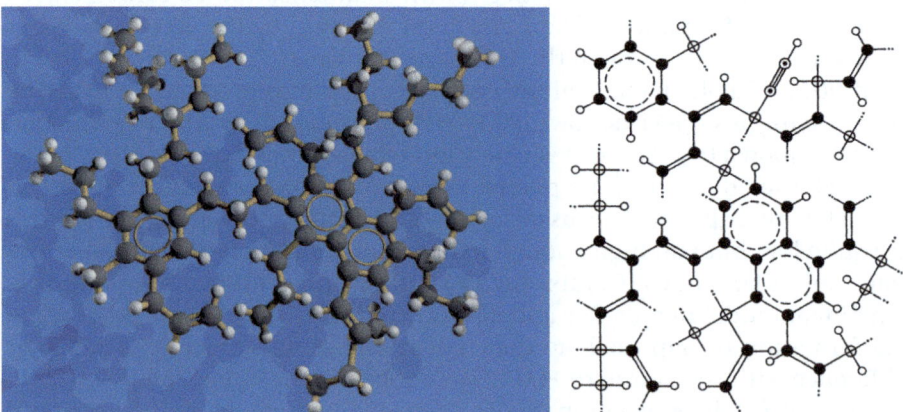

Figure 4.6   Left panel: schematic diagram of the expected typical substructure unit for the hydrogen-rich photoproduced a-C:H. Reproduced with permission from Dartois *et al.* (2005).[36] Right panel: an amorphous carbon material in the view of Jones *et al.* (1990).[3] Reprinted with permission.

structure is reminiscent of the interstellar carbon solid envisaged by Jones *et al.* (1990)[3] containing all the three types of carbon bonding (Figure 4.6, right panel[3]), and is in good agreement with the many experimental and theoretical works on plasma-produced a-C:H, but with a much higher hydrogen to carbon ratio.

Nanodiamonds have the most disparate origins. They are found in crude oil at concentrations up to thousands of parts per million, in meteorites and interstellar dust (see Chapter 7), and protoplanetary nebulae, as well as in sediment layers on Earth. They can be produced in the laboratory by CVD or by detonating high explosive materials. A typical CVD experiment would involve the injection of a mixture of methane and hydrogen gas in a plasma reactor. The diamond film, with variable morphologies, is deposited on a silicon substrate. Chemical vapour deposited diamond films are distinguished into three categories:[52] microcrystalline diamonds (0.5–10 mm), nanocrystalline diamonds (50–100 nm), and ultra-nanocrystalline diamonds (2–5 nm). Pure and mixed TNT detonations produce, among other carbon forms, nanocrystalline diamonds with diameters up to 4 nm in size. Structural studies have evidenced a diamond cluster core covered by a mixture of sp^2- and sp^3-bonded carbons in onion-like carbon shells and nanometre-sized graphite platelets. Thermal annealing and electron irradiation decrease the diamond content in favour of the formation of graphite flakes on the surface.

4.4 Optical Constants of Dust Analogues

Optical material properties enter into astronomical calculations through the response function which characterizes the reaction of a material to an external electromagnetic field. The ultimate goal of laboratory analogues of cosmic dust is to supply such basic data for the interpretation of ultraviolet, optical, and infrared spectroscopy of dusty regions, in the form of either complex refractive indexes or dielectric functions (an equivalent description for non-magnetic materials). However, the data used most commonly for modelling astronomical environments are synthetic optical functions such as those of Draine & Lee (1984).[53] This unsatisfactory position is forced on astronomers by the incomplete spectral coverage of individual laboratory datasets. The major drawback in using synthetic functions is their intrinsic weakness of being compilations of laboratory spectral data and astronomical observed dust opacities, sometimes modified specifically to match the astronomical observations. Thus, they do not represent the physics of real solids and any inference about the nature of dust in space, how it varies spatially or temporally, and should be taken very cautiously. Moreover, as confirmed in many experiments, dust analogue materials depend strongly on the experimental conditions, and quite frequently similar chemical compositions do not generate similar structural arrangements. Different structures, in turn, lead to different optical patterns. A discussion of the methods for comparing laboratory and astronomy properties of dust is given in for example Speck (2013).[54]

Such incompleteness in the laboratory data derives also from strictly technical problems, as most techniques of dust production are not suitable for optical studies in all spectral ranges. Transmission measurements in the regions of relative transparency (*e.g.* far-infrared) require amounts of material to detect the absorption that are a hundred times thicker than it would be actually needed in the range of electronic interband absorption, *i.e.* in the ultraviolet. Cosmic dust analogues should have their structure and composition accurately characterized. For amorphous materials, this is a rather complex procedure since chemical bonds vary from site to site. Moreover, for low frequency radiation the response to electromagnetic radiation is often significantly dependent on the sample temperature, and thus the determination of temperature-dependent absorption is required. The morphology of the sample should be also considered, as shape, size and agglomeration affect the results of optical measurements. For example, although porosity should not directly affect the spectra, voids give rise to extra reflections which increase extinction through increased scattering even though absorption is unchanged. In sufficiently small metallic particles the electron mean free path is limited by the particle boundary, with consequences for their optical properties. In general, the optical properties of small particles can considerably deviate from those of bulk materials because of the occurrence of surface modes.[55]

The infrared spectra of amorphous silicate materials have been investigated in the laboratory since the seventies (*e.g.* Day 1974[56]), shortly after the 10 μm feature was first observed in the spectra of several M-type giants and red supergiants.[57] In 2011 Speck[58] and collaborators presented a laboratory study of silicate glasses centred on olivines and pyroxenes. Such a choice is based on both cosmic and terrestrial hints suggesting a bulk cosmic silicate composition lying somewhere between pyroxene and olivine, close to $MgSiO_3$, and with an atomic $(Mg + Fe):Si$ ratio of slightly greater than 1. To complement the olivines and pyroxenes they also include samples from the melilite series, whose end-members are gehlenite ($Ca_2Al_2SiO_7$) and åkermanite ($Ca_2MgSi_2O_7$), and a "cosmic silicate" designed according to cosmic abundance ratios for Mg, Si, Al, Ca, and Na. This sample does not include volatile elements or iron (for chemical reasons: iron is expected to combine with other siderophile elements to form metal or metal sulfide grains rather than silicate), and is quite close to enstatite in composition. Finally, Speck *et al.* included obsidian, a naturally occurring glassy silicate, and silica (SiO_2) glass, whose structures are significantly different from olivines and pyroxenes. Samples were generated from mixtures of reagent-grade oxides and carbonates or by melting natural mineral samples. To compare different data sets it is convenient to convert the measured absorbance A to the imaginary part of the complex refractive index $k = 2.303A/4\pi dv$, where d is the thickness of the sample, and v the frequency. An example of the Speck and collaborators experiments, and their comparison with the existing literature is shown in Figure 4.7 (ref. 17, 58–61) where it is evidenced that variations in the spectral patterns exist even for compositions that are apparently the same.

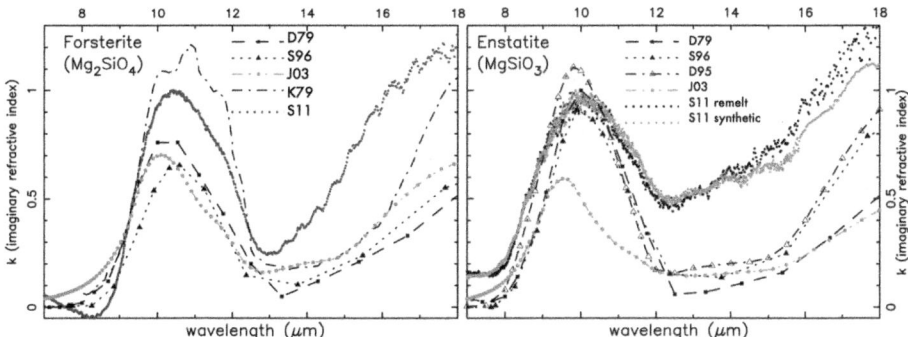

Figure 4.7   Comparison of laboratory optical data for two "amorphous" silicates. D79 = Day (1979)[59]; D95 = Dorschner *et al.* (1995)[60]; J03 = Jäger *et al.* (2003b)[17]; K79 = Krätschmer & Huffman (1979)[61]; S96 = Scott & Duley (1996)[62]; S11 = Speck *et al.* (2011)[58]; re-melt = material produced by melting natural mineral samples; synthetic = produced by from mixtures of reagent-grade oxides and carbonate. Adapted and reproduced with permission from Speck *et al.* (2011).[58]

The case of forsterite composition highlights the differences arising from producing samples by means of different techniques: Krätschmer & Huffman (1979)[61] exploited ion irradiation of crystalline forsterite, Day (1979)[59] CVD, Scott & Duley (1996)[62] laser ablation of crystalline samples, and Jäger *et al.* (2003) the sol–gel method. The spectra arising from materials produced by the last three techniques are similar. They likely represent chaotic silicates rather than glassy silicates as produced by Speck *et al.* (2011).[58] In the case of enstatite composition, samples prepared by the laser ablation technique, CVD synthesis, and "quenching of melt to glass" method by Dorschner *et al.* (1995),[60] provide spectra showing (again) similar trends, while glassy silicates show broader features. The sol–gel prepared sample gives a rather different response. The origin for such discrepancy should be traced back to the material structure of the sample, which is much affected by the incorporation of water, mainly dissolved as network-modifying hydroxyl ions that act on the polymerization degree (see Section 4.2.2). Also, the incorporation of cations (except for Na) reduces the polymerization of the SiO_4 tetrahedra, thus increasing the number of non-bridging oxygen atoms, and results in a shift of the Si–O stretching band towards longer wavelengths. The major effect of replacing some of the magnesium with iron is a broadening of the 10 and 20 μm bands. This behaviour is affected by the oxidation state of the incorporated iron. While Fe^{2+} simply replaces magnesium as a "network modifier" cation (Mg^{2+} and Fe^{2+} have approximately the same size) Fe^{3+} is a network former, and thus enhances the degree of polymerization in the network, modifying the spectrum. Aluminium is another ion that affects the state of polymerization, because it is preferentially a network former.[58,63] Figure 4.8 (ref. 58 and 64) shows a comparison between the "cosmic silicate", $(Na_{0.11}Ca_{0.12}Mg_{1.86})(Al_{0.18}Si_{1.85})O_6$, synthesized

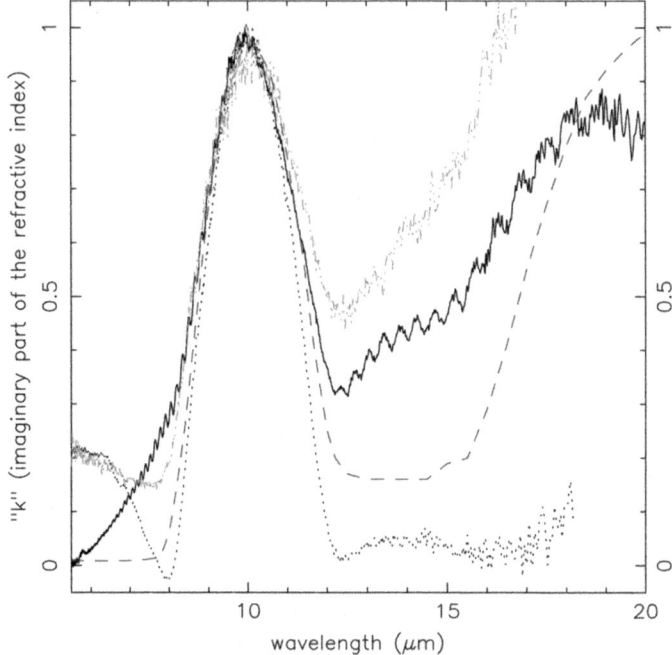

Figure 4.8 Comparison of "dirty" silicates. Speck *et al.* (2011)[58]: astronomical
silicate (solid black line); basalt (dotted black line); enstatite re-melt
(dotted-dashed light grey line). The dashed dark grey line is the "dirty"
silicate from Jäger *et al.* (1994).[64] Reprinted with permission from Speck
et al. (2011).[58]

by Speck *et al.* (2011)[58] and the "dirty silicate", $Mg_{0.50}Fe_{0.43}Ca_{0.03}Al_{0.04}SiO_3$,
produced by Jäger *et al.* (1994).[64]

The 10 μm features are very similar in both peak position and line
width, but the ratio of the strength of Si–O stretching and bending modes
varies. This is unfortunate, since it shows that the coexistence of several elements in a material tends to wash out relevant different factors affecting
for example composition, oxidation state, and the polymerization degree
(*i.e.* the amount of non-bridging oxygen atoms, see Section 4.2.2). The
chemical state of the iron can also influence the absorption of silicates in
the visible-near infrared. Ferric iron causes a strong charge transfer transition absorption, while ferrous iron only produces relatively weak crystal
field bands in the near-infrared.[65] In general, spectral parameters of amorphous silicate materials are sensitive to the degree of disorder, porosity, and
sample chemistry, including the oxidation state, polymerization, Mg:Fe
ratio, and water content. This in addition means that morphological information can only be obtained from the spectra if the influence of the other
parameters is sufficiently well understood and can be disentangled. A com-
pilation of optical properties for astronomically relevant materials is found

in the Heidelberg-Jena-St.Petersburg-Database of Optical Constants. The database also contains an extended reference list for olivines, pyroxenes and other silicates (see Henning & Mutschke 2010[66] for an excellent review on the optical properties of cosmic dust analogues). Relevant data have also been published by various groups in Japan, especially the group of C. Koike at Kyoto Pharmaceutical University, who systematically investigated the variation in the infrared bands with composition for a series of synthetic silicate minerals (*e.g.* Chihara *et al.* 2007[67]). Koike and co-workers also considered evolutionary changes of silicates in shocks, occurring for example during the formation of dust by planetesimal collisions (see Chapter 7). Impacts affect the mineralogical properties of the planetesimal components, changing their infrared spectroscopic properties. Morlok, Koike *et al.* (2010)[68] found that infrared spectra of shocked (in the laboratory) meteoritic samples (taken from the Murchison carbonaceous chondrites, see Chapter 7) show similarities to the astronomical spectra of dust in various young stellar objects and debris disks, and also with those of cometary dust (*e.g.* Hale Bopp dust).

So far, few sets of data have been derived from measurements on carbon dust analogues over a wide wavelength range. These synthetic materials attempt to simulate the highly variable optical properties of carbon by different hydrogen concentrations in condensation experiments or by different annealing temperatures in pyrolysis experiments. The mid-infrared spectra of gas-phase condensed carbonaceous solids are characterized by aromatic and aliphatic bands, partly superimposed on broad plateau features around 8 and 12 μm. In the HT fullerene-like condensates produced by Jäger *et al.* (2009)[48] (Section 4.3) the infrared spectrum shows strong saturated and aliphatic $-CH_x$ absorptions at 3.4, 6.8, and 7.25 μm. Such groups are mainly responsible for the links between the fullerene fragments. The aromatic C–H stretching at 3.3 μm has not been observed, but aromatic $-C=C-$ groups are identified between 6.2 and 6.25 μm, together with out-of-plane bending vibrations of aromatic $=C-H$ groups between 11 and 14 μm. $-C\equiv C-$ triple bonds are evidenced by features at 3.03 and 4.7 μm, suggesting the possible presence of polyynes, supposed to be intermediates in the formation of fullerene-like soot grains. The carbonaceous material produced by Jäger *et al.* (2009)[48] through LT condensation of hydrocarbons in laser pyrolysis (Section 4.3)—basically a mixture of soot and PAHs—shows a series of aromatic bands including 3.3, 6.3, 8.6, 11.3, 12.3, and 13.3 μm as well as aliphatic infrared bands at 3.4, 6.8, and 7.25 μm ($-C-H$ stretching and deformation bands). Other bands at 6.8, 7.25, 7.93, 8.47, 9.28, 9.72, 10.5 μm mainly arise from $-C-C-$ stretching and C–H deformation. Other minor vibration modes are due to $-C-H$ twisting, wagging, rocking, and $-C-C-$ deformation bands. The condensate produced by Biennier *et al.* (2009),[46] discussed in Section 4.3, is composed of small crystalline aromatic aliphatic linked units surrounded by low-mass aromatics. The strongest features in the infrared spectrum of the sample are aromatic features at 3.29 ($-C-H$ stretch), 6.30 (C=C stretch), and 11.36, 11.98 and 13.31 μm out-of-plane bending vibrations of $=C-H$ groups.

The presence of aliphatic groups, *i.e.* CH_3-, $-CH_2$-, and $=CH-$ bonded to or bridging aromatic rings in the material is evidenced by the 3.38, 3.42 and 3.50 μm peaks, identified as aliphatic asymmetric and symmetric stretches. The spectral positions of the infrared features of the extract and the residue are very close to the ones obtained by Jäger *et al.* (2009).[48] Evidently, the condensates produced by Biennier *et al.* (2009)[46] and Jäger *et al.* (2009)[48] are "relatives". Similar prominent bands have been identified by Dartois and collaborators for their hydrogenated amorphous carbon polymer produced *via* photolysis of organic molecules at low temperature (see Section 4.3). These authors started from different precursors and ended up with similar spectra, although not so similar to be indiscernible. They analysed in detail the structure of C–H stretching feature, decomposing this band into individual characteristic group vibrations to gain information to the relative proportions of sp^3, sp^2 and sp^1 hybridized carbons (with the exception of course of the dehydrogenated phases).

These findings are related to the carriers of the so-called UIBs (see Chapter 2). A popular hypothesis identified these bands with the AIBs arising from the vibrational modes of gas-phase, free-flying PAH molecules. Amorphous solids with a mixed aromatic/aliphatic structure, such as a-C:H have also been proposed (Jones *et al.* 1990), although bulk materials challenge this proposal because their heat capacities are so large that equilibrium temperatures are very low. To overcome this difficulty, Duley & Williams (2011)[69] proposed that the heating of a-C:H dust might occur *via* the release of stored chemical energy, while Kwok & Zhang (2013)[70] argued that the UIBs may arise from coal- or kerogen-like organic nanoparticles, consisting of chain-like aliphatic hydrocarbon material linking small units of aromatic rings. Such particles are nanometres in size. The heat capacity of one of them is so small that the absorption of a single UV photon causes a significant rise in the particle's temperature. These small entities are called MAONs: mixed aromatic/aliphatic organic nanoparticles).

This discussion leads to a new major problem regarding the structure of the UIB carriers: are they predominantly aromatic, like PAHs, or largely aliphatic but mixed with small aromatic units (like for example MAONs)? Pino *et al.* (2008)[71] suggested that the spectral position of the C=C aromatic stretch (at 6.30 μm) may be an indicator of the carbon skeleton, *i.e.* this feature might trace the evolution of the environment of the aromatic units embedded in an aliphatic network. They set up an experiment at the Laboratoire de Photophysique Moléculaire in Orsay for the production and spectroscopic analysis of laboratory samples of carbonaceous dust analogues. The (fairly standard) apparatus combines a reactor chamber and a sampling chamber at much lower pressure where a molecular jet seeded with the reactor products is formed. The reactor products are deposited onto a substrate placed in the molecular jet for *in* and *ex situ* analysis. The outcomes were a wide range of soot material samples, whose infrared absorption spectra were recorded. The spectral analyses, including band shape and position variations, were used to interpret the diversity and evolution of the features in the astronomical spectra. Such spectra may be classified into three major groups, A, B, and C, characterized by the frequency of the C=C stretching mode.[72] The carriers of class C infrared emission spectra were proposed to be dominated by aliphatic branches linking aromatic units embedded within the network. At the end of the evolution, class A spectra correspond to more mature structures in which aliphatics have been converted into aromatic cycles (see Figure 4.9).[71,72]

A variety of astrophysical objects fall in the class A category, while class C sources are less frequent. On the basis of the band wavelength locations and the $sp^2:sp^3$ ratios, the laboratory condensates of Biennier and co-workers can be placed at the border between class C and class B, whereas Jäger and co-workers' (2009)[48] products belong to class B (see Figure 4.9). Yang *et al.* (2013)[73] analysed a wide range of astronomical environments, including reflection nebulae, HII regions, molecular clouds, photodissociation regions, planetary nebulae, and protoplanetary nebulae. They determined the ratio of the power emitted from the 3.4 μm feature to that from the 3.3 μm feature to be approximately 0.1, and the fraction of C atoms in aliphatic form to be 2% of the total carbonaceous matter. As we learned earlier, a similar analysis performed by Dartois *et al.* (2007)[47] along a line of sight to an extragalactic source showed that the chemical environment of the diffuse interstellar medium is highly aliphatic, so this is a clear indication that carbon materials respond differently to different environments. The evolution of hydrocarbon materials is therefore of key interest for the study of interstellar dust properties and should be taken into proper account in modelling. The current efforts to develop a comprehensive

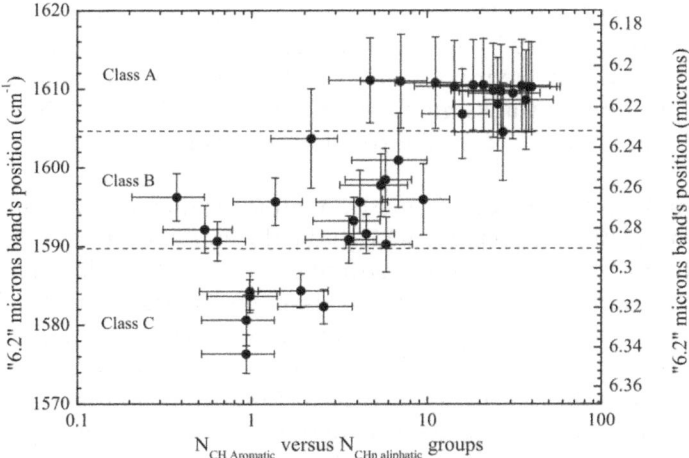

Figure 4.9 Carbon–carbon stretch position as a function of the aromatic C–H bond to aliphatic group ratio. The wavelength ranges of each AIB class (A, B, and C, see text) as given by Peeters *et al.* (2002)[72] are shown. Reprinted with permission from Pino *et al.* (2008).[71]

model of interstellar dust (see Chapter 3) are still rather limited, as they are based on a single physical parameter, *i.e.* the optical gap energy, a quantity depending primarily on the medium-range order, including the kind of subunits, mean lengths of subunits and distances between subunits. Nevertheless, it is important to recall that interstellar extinction curves can be phenomenologically described by means of a very limited number of empirical parameters,[74,75] so that grain models are not required to be too complex.

Another outstanding question regarding the composition of the interstellar medium concerns the origin of the strong absorption bump at 217.5 nm, which is widely believed to be due to $\pi* \leftarrow \pi$ transitions in carbonaceous species but has not yet been produced in the laboratory. This failure may possibly be due to the effect of particle shape or particle clustering that modifies significantly the intimate extinction properties of (hydrogenated) carbon nanoparticles. The only way to overcome this problem is to measure the absorption characteristics of isolated non-agglomerated carbon grains, for example exploiting a matrix-isolation technique.[76] The samples produced by Th. Henning and collaborators appeared to possess a feature of sufficient strength and sharpness at the right wavelength position. Using laser ablation of a graphite surface, Gadallah *et al.* (2011)[77] produced nano-sized a-C:H materials which were subsequently irradiated by strong ultraviolet doses in a high vacuum. A decrease in $sp^3 : sp^2$ hybridization ratio with increasing ultraviolet dose was found (in agreement with many preceding experiments), corresponding to a heavy structural modification in the processed materials. The variation in the internal structure leads to dramatic changes in the spectral properties in the far-ultraviolet to visible range, and eventually to the appearance of a new band centred at approximately 217.5 nm. Unfortunately, the band strength was not enough to constrain the required amount of carbon needed to produce the interstellar feature within the modest interstellar elemental budget. Moreover, the band profile was Gaussian rather than Lorentzian and the carrier of that band could not be firmly identified in the small amounts of processed material.

Other studies have suggested that the organic residue produced in the laboratory from a hydrocarbon plasma might be the carrier of the 217.5 nm ultraviolet absorption bump, and the origin of some interstellar infrared spectral features otherwise attributed to PAHs have even assigned to this residue. One apparatus employed to produce such materials was originally designed for the study of the formation of molecules by the chemist A. Sakata and his colleagues at the University of Electro-Communications in Tokyo. The hydrocarbon plasma is produced in a microwave discharge tube by molecular components (*e.g.* CH_4), and injected into a vacuum chamber. Experimental outcomes from the plasma were mainly polyynes and PAHs, but also abundant solids. It was quickly discovered that such solids had bands similar to the UIBs and the 217.5 nm bump.[78] The solid material was termed quenched carbonaceous composite (QCC) because was synthesized from carbon (and hydrogen), formed through rapid cooling, and its composition was a mixture of chemical compounds. Various types of QCCs may be produced, such as organic materials (filmy-QCCs) whose major components are compact PAHs with masses in the range 200–500 amu, and a carbonaceous material with onion-like structure (dark-QCCs). The variety of structures in QCCs correlates with the peak position of the ultraviolet spectrum. Dark-QCC (the material peaking at 220 nm) contains multi-layer, carbonaceous onion-like particles. These materials represent a step in the evolution of condensates; they eventually develop a carbon network with a bent, ribbon-like structure, whose precursors are probably the onion-like particles. The filmy-QCC is (relatively) rich in hydrogen compared to dark-QCC. When this form of QCC is heated in vacuum (thermally altered filmy-QCC), it becomes carbonized and shows strong infrared bands and an underlying weak continuum. Filmy-QCCs show photoluminescence consistent with the extended red emission observed in the Red Rectangle.[79]

Except for few dissenting voices (see Kwok 2013 [80]), the PAH hypothesis has become extremely popular in the astronomical community. Two groups independently proposed the existence in space of free-flying PAHs.[81,82] Such molecules have many appealing characteristics: they absorb ultraviolet radiation effectively, convert it with high efficiency to vibrational excitation, and release it as infrared emission, in bands qualitatively matching the observed UIBs. They are also very resistant to photodissociation (provided they are large enough), and can possibly form in C-rich outflows of evolved stars, with chemical pathways (as we have seen above) inspired from combustion chemistry. Shortly thereafter, it was also realized that interstellar PAHs, depending on size and charge state, would also have discrete bands in the visible, making them very plausible candidate carriers for the diffuse interstellar bands, the DIBs.[83,84] Purely energetic considerations, based on the overall integrated emission in the UIB bands, assuming unit conversion efficiency and ultraviolet absorption cross-sections,[11] led to the estimate that about 15–20% of the available interstellar carbon must be locked in PAHs to produce the observed infrared emission fluxes. This implied that PAHs must be taken into account as a necessary ingredient of interstellar extinction models. From this point on, all interstellar extinction models were updated to include some sort of "astronomical PAH" component. Remarkably, PAHs were assumed to contribute only to the non-linear far-ultraviolet rise in interstellar extinction, while the bump was still entirely attributed to a population of classical small "astronomical graphite" particles. This was very convenient, as it provided an attractive physical explanation of the observed behaviour of these two components of the extinction curve, which were known to vary independently in observations. All interstellar extinction models thus adopted an average cross-section of the neutral PAHs from which the $\pi* \leftarrow \pi$ transitions were simply removed, leaving only the non-linear rise in the far-ultraviolet due to the onset of $\sigma* \leftarrow \sigma$ transitions. Probably, the reason for this misconception is the fact that PAHs were believed to be largely ionized in the diffuse interstellar medium, so that the variation of the position of π∗ ← π transitions in different molecules would smooth them out over such a wide wavelength range to make them irrelevant. Moreover, direct measurements of the cross-sections of PAH cations were (and still are) rather difficult, since PAH cations are very reactive species. Several years later, Duley & Seahra (1998),[85] using the discrete dipole approximation to study the spectra of PAHs, showed that the π∗ ← π transitions appeared to be almost unaffected by ionization. This was soon confirmed experimentally.[86] This result, together with the realization that the interstellar carbon budget might be insufficient for a separate population of carbonaceous particles to produce the bump, caused a general update of the photoabsorption cross-sections that define the present contribution of PAHs to interstellar extinction models (see Chapter 3). Currently, the situation is again fluid, and the role of the aliphatic component must be considered as an important component of the chemical make-up of carbonaceous particles. A thorough review on the role of PAHs in the extinction is given in Mulas *et al.* (2011),[87] while a general description of PAHs can be found in the book of A. Tielens (2007).[88]

To study PAHs in the laboratory, using spectroscopy or intramolecular processes for example, gas-phase isolation is required. Over the past two decades much laboratory effort has been based on matrix-isolation (MI) spectroscopy, in which guest molecules or atoms are trapped in rigid host materials. Because the species are embedded in a host material, the diffusion process is prevented and bimolecular reactions cannot take place, except with the host material. Several kinds of host materials are used, such as crystalline solids, polymers or glasses formed during the freezing of the liquid or solidification of the gas phase. Although MI data may deviate from gas-phase spectra, various gas-phase spectra revealed band positions that are in very good agreement with those reported in MI studies. Other studies use laser technology that permits the study of ultrafast dynamics with femtosecond lasers or high-resolution spectroscopy with continuous lasers. In the PAH hypothesis, the UIBs are generally attributed to emission from highly vibrationally excited molecules in the neutral and, particularly, cationic states. It is then necessary to record the infrared spectra of cationic PAHs in the laboratory. In the mid- and far-infrared spectral ranges, the absorption properties of strongly bound ions can be studied with infrared multiple-photon dissociation spectroscopy. At a vibrational resonance, the molecules absorb multiple infrared photons in a non-coherent fashion. After absorption, intramolecular vibrational redistribution randomizes the absorbed energy over the manifold of vibrational states heating the complex until the energy is sufficient for statistical dissociation to occur. The intensity of photofragment ions, as well as the remaining parent ion, is recorded using a TOF-MS. The fragment ion yield as a function of infrared wavelength is the infrared multiple-photon dissociation spectrum, which has been shown in many examples to be a good surrogate for the infrared absorption spectrum. A detailed review of the state of the art of the spectroscopy of PAHs is given in Pino *et al.* (2011).[89]

Further Reading

1. F. J. M. Rietmeijer and J. A. Nuth, *Astrophys. J.*, 2013, **771**, 34.
2. T. I. Mori, I. Sakon, T. Onaka, H. Kaneda, H. Umehata and R. Ohsawa, *Astrophys. J.*, 2012, **744**, 68.
3. A. P. Jones, W. W. Duley and D. A. Williams, *Q. J. R. Astron. Soc.*, 1990, **31**, 567.
4. O. Guillois, I. Nenner, R. Papoular and C. Reynaud, *Astrophys. J.*, 1996, **464**, 810.
5. J. B. Pollack, O. B. Toon and B. N. Khare, *Icarus*, 1973, **19**, 372.
6. N. J. Woolf and E. P. Ney, *Astrophys. J.*, 1969, **155**, L181.
7. J. A. Nuth, III and J. H. Hecht, *Astrophys. Space Sci.*, 1990, **163**, 79.
8. T. P. Stecher and B. Donn, *Astrophys. J.*, 1965, **142**, 1681.
9. J. S. Mathis and G. Whiffen, *Astrophys. J.*, 1989, **341**, 808.
10. E. L. Wright, *Astrophys. J.*, 1987, **320**, 818.
11. C. Joblin, A. Léger and P. Martin, *Astrophys. J.*, 1992, **393**, L79.
12. R. Papoular, J. Conard, O. Guillois, I. Nenner, C. Reynaud and J.-N. Rouzaud, *Astron. Astrophys.*, 1996, **315**, 222.
13. T. M. Steel and W. W. Duley, *Astrophys. J.*, 1988, **315**, 33.
14. A. Higa, T. Oshiro, Y. Saida, M. Yamazato and M. Toguchi, *New Diamond Front. Carbon Technol.*, 2006, **16**, 245.
15. P. Mohr, H. Mutschke and F. Lewen, in *The Life Cycle of Dust in the Universe: Observations, Theory, and Laboratory Experiments*, ed. A. Andersen, M. Baes, H. Gomez and D. Watson, 2013, p. 140.
16. J. M. Burlitch, M. L. Beeman, B. Riley and L. Kohlstedt, *Mater. Sci.*, 1991, **3**, 692.
17. C. Jäger, J. Dorschner, H. Mutschke, Th. Posh and Th. Henning, *Astron. Astrophys.*, 2003, **408**, 193.
18. J. A. Nuth, III, F. J. M. Rietmeijer and H. G. M. Hill, *Meteorit. Planet. Sci.*, 2002, **37**, 1579.
19. F. J. M. Rietmeijer, A. Pun, Y. Kimura and J. A. Nuth, *Icarus*, 2008, **195**, 493.
20. F. J. M. Rietmeijer and J. A. Nuth, *EOS, Trans., Am. Geophys. Union*, 2000, **81**, 409.
21. I. Prigogine, *Astrophys. Space Sci.*, 1979, **65**, 371.
22. T. Sabri, L. Gavilan, C. Jäger, J. L. Lemaire, G. Vidali, H. Mutschke and Th. Henning, *Astrophys. J.*, 2014, **780**, 180.
23. C. Jäger, D. Fabian, F. Schrempel, J. Dorschner, Th. Henning and W. Wesch, *Astron. Astrophys.*, 2003, **401**, 57.
24. J. R. Brucato, V. Mennella, L. Colangeli, A. Rotundi and P. Palumbo, *Planet. Space Sci.*, 2002, **50**, 829.
25. S. L. Hallenbeck, J. A. Nuth, III and R. A. Nelson, *Astrophys. J.*, 2000, **535**, 247.
26. S. P. Thompson and C. C. Tang, *Astron. Astrophys.*, 2001, **368**, 721.
27. C. Kaito, Y. Ojima, K. Kamitsuji, *et al.*, *Meteorit. Planet. Sci.*, 2003, **38**, 49.
28. D. Fabian, C. Jäger, Th. Henning, J. Dorschner and H. Mutschke, *Astron. Astrophys.*, 2000, **364**, 282.

29. Y. Kimura, Y. Miyazaki, A. Kumamoto, M. Saito and C. Kaito, *Astrophys. J.*, 2008, **680**, L89.
30. C. Kaito, Y. Miyazaki, A. Kumamoto and Y. Kimura, *Astrophys. J.*, 2007, **666**, L57.
31. K. Demyk, Ph. Carrez, H. Leroux, P. Cordier, A. P. Jones, J. Borg, E. Quirico, P. I. Raynal and L. d'Hendecourt, *Astron. Astrophys.*, 2001, **368**, L38.
32. E. M. Bringa, *et al.*, *Astrophys. J.*, 2007, **662**, 372.
33. F. J. Grunthaner and P. J. Grunthaner, *Mater. Sci. Rep.*, 1986, **1**, 65.
34. A. S. Yen, F. J. Grunthaner, S. S. Kim and M. H. Hecht, *Lunar Planet. Sci.*, 1999, **30**, 1924.
35. Y. Kimura and J. A. Nuth, III, *Astrophys. J.*, 2007, **664**, 1253.
36. E. Dartois, G. M. Muñoz Caro, D. Deboffle, G. Montagnac and L. D'Hendecourt, *Astron. Astrophys.*, 2005, **432**, 895.
37. S. Iijima, *Nature*, 1991, **354**, 56.
38. C. Jäger, S. Krasnokutski, A. Staicu, F. Huisken, H. Mutschke, Th. Henning, W. Poppitz and I. Voicu, *Astrophys. J., Suppl. Ser.*, 2006, **166**, 557.
39. I. Cherchneff, in *PAHs and the Universe*, ed. C. Joblin and A. G. G. M. Tielens, 2011, p. 177.
40. M. Frenklach and E. D. Feigelson, *Astrophys. J.*, 1989, **341**, 372.
41. M. Frenklach, *Phys. Chem. Chem. Phys.*, 2002, **4**, 2028.
42. C. A. Schuetz, M. Frenklach, A. C. Kollias and W. A. Lester, *J. Chem. Phys.*, 2003, **119**, 9386.
43. M. Frenklach and H. Wang, *Proc. Combust. Inst.*, 1991, **23**, 1559.
44. H. Sabbah, L. Biennier, S. J. Klippenstein, I. R. Sims and B. R. Rowe, *Phys. Chem. Lett.*, 2010, **1**, 2962.
45. K. Siegmann, K. Sattler and H. C. Siegmann, *J. Electron Spectrosc. Relat. Phenom.*, 2002, **126**, 191.
46. L. Biennier, R. Georges, V. Chandrasekaran, B. Rowe, T. Bataille, V. Jayaram, K. P. J. Reddy and E. Arunan, *Carbon*, 2009, **47**, 3295.
47. E. Dartois, T. R. Geballe, T. Pino, A.-T. Cao, A. Jones, D. Deboffle, V. Guerrini, Ph. Bréchignac and L. d'Hendecourt, *Astron. Astrophys.*, 2007, **463**, 635.
48. C. Jäger, F. Huisken, H. Mutschke, I. L. Jansa and Th. Henning, *Astrophys. J.*, 2009, **696**, 706.
49. S. Irle, G. Zheng, M. Elstner and K. Morokuma, *Nano Lett.*, 2003, **3**, 1657.
50. C. S. Contreras and F. Salama, *Astrophys. J., Suppl. Ser.*, 2013, **208**, 6.
51. D. S. N. Parker, F. Zhang, Y. S. Kim, R. I. Kaiser, A. Landera, V. V. Kislov, A. M. Menel and A. G. G. M. Tielens, *Proc. Natl. Acad. Sci. U. S. A.*, 2012, **104**, 5274.
52. L. Fayette, B. Marcus, M. Mermoux, G. Tourillon, K. Laffon, P. Parent and F. Normand, *Phys. Rev. B*, 1998, **57**, 14123.
53. B. T. Draine and H. M. Lee, *Astrophys. J.*, 1984, **285**, 89.
54. A. K. Speck, in *The Life Cycle of Dust in the Universe: Observations, Theory, and Laboratory Experiments*, ed. A. Andersen, M. Baes, H. Gomez and D. Watson, 2013, p. 2.

55. F. Borghese, P. Denti and R. Saija, *Scattering by Model Nonspherical Particles*, Springer, 2007.
56. K. L. Day, *Astrophys. J.*, 1974, **192**, L15.
57. F. C. Gillett, F. J. Low and W. A. Stein, *Astrophys. J.*, 1968, **154**, 667.
58. A. K. Speck, A. G. Whittington and A. M. Hofmeister, *Astrophys. J.*, 2011, **740**, 93.
59. K. L. Day, *Astrophys. J.*, 1979, **234**, 158.
60. J. Dorschner, B. Begemann, T. Henning, C. Jaeger and H. Mutschke, *Astron. Astrophys.*, 1995, **300**, 503.
61. W. Krätschmer and D. R. Huffman, *Astrophys. Space Sci.*, 1979, **61**, 195.
62. A. Scott and W. W. Duley, *Astrophys. J., Suppl. Ser.*, 1996, **105**, 401.
63. H. Mutschke, B. Begemann, J. Dorschner, J. Guertler, B. Gustafson, Th. Henning and R. Stognienko, *Astron. Astrophys.*, 1998, **333**, 188.
64. C. Jäger, H. Mutschke, B. Begemann, J. Dorschner and Th. Henning, *Astron. Astrophys.*, 1994, **292**, 641.
65. H. Mutschke, in *The Life Cycle of Dust in the Universe: Observations, Theory, and Laboratory Experiments*, ed. A. Andersen, M. Baes, H. Gomez and D. Watson, 2013, p. 42.
66. Th. Henning and H. J. Mutschke, *Nano.*, 2010, **4**, 041580.
67. H. Chihara, C. Koike and A. Tsuchiyama, *Astron. Astrophys.*, 2007, **464**, 229.
68. A. Morlok, C. Koike, N. Tomioka, I. Mann and K. Tomeoka, *Icarus*, 2010, **207**, 45.
69. W. W. Duley and D. A. Williams, *Astrophys. J.*, 2011, **737**, L44.
70. S. Kwok and Y. Zhang, *Astrophys. J.*, 2013, **771**, 5.
71. T. Pino, E. Dartois, A.-T. Cao, Y. Carpentier, T. Chamaillé, R. Vasquez, *et al.*, *Astron. Astrophys.*, 2008, **490**, 665.
72. E. Peeters, S. Hony, C. Van Kerckhoven, A. Tielens, L. J. Allamandola, D. M. Hudgins, *et al.*, *Astron. Astrophys.*, 2002, **390**, 1089.
73. X. J. Yang, R. Glaser, A. Li and J. X. Zhong, *Astrophys. J.*, 2013, **776**, 110.
74. J. A. Cardelli, G. C. Clayton and J. S. Mathis, *Astrophys. J.*, 1989, **345**, 245.
75. E. L. Fitzpatrick and D. Massa, *Astrophys. J.*, 2007, **663**, 320.
76. M. Schnaiter, H. Mutschke, J. Dorschner, Th. Henning and F. Salama, *Astrophys. J.*, 1998, **498**, 486.
77. K. A. K. Gadallah, H. Mutschke and C. Jäger, *Astron. Astrophys.*, 2011, **528**, A56.
78. A. Sakata, S. Wada, Y. Okutsu, H. Shintani and Y. Nakada, *Nature*, 1983, **301**, 493.
79. S. Wada, Y. Mizutani, T. Narisawa and A. T. Tokunaga, in *Organic Matter in Space* ed. S. Kwok and S. Sandford, 2008, p. 417.
80. S. Kwok, *Stardust: The Cosmic Seeds of Life*, Springer, 2013.
81. A. Léger and J.-L. Puget, *Astron. Astrophys.*, 1984, **137**, 5.
82. L. J. Allamandola, A. G. G. M. Tielens and J. R. Barker, *Astrophys. J.*, 1985, **290**, 25.
83. G. P. van der Zwet and L. J. Allamandola, *Astron. Astrophys.*, 1985, **146**, 76.

84. A. Léger and L. D'Hendecourt, *Astron. Astrophys.*, 1985, **146**, 81.
85. W. W. Duley and S. Seahra, *Astrophys. J.*, 1998, **507**, 874.
86. X. D. F. Chillier, B. M. Stone, F. Salama and L. J. Allamandola, *J. Chem. Phys.*, 1999, **111**, 449.
87. G. Mulas, G. Malloci, C. Joblin and C. Cecchi-Pestellini, in *PAHs and the Universe*, ed. C. Joblin and A. G. G. M. Tielens, 2011, p. 327.
88. A. G. G. M. Tielens, *The Physics and Chemistry of Interstellar Medium*, Cambridge University Press, 2007.
89. T. Pino, Y. Carpentier, G. Féraud, H. Friha, D. L. Kokkin, T. P. Troy, N. Chalyavi., Ph. Bréchignac and T. W. Schmidt, in *PAHs and the Universe*, ed. C. Joblin and A. G. G. M. Tielens, 2011, p. 355.

Section II

The Formation of Dust and Its Evolution in the Interstellar Media of Galaxies

CHAPTER 5

Dust Formation in Stellar Environments

5.1 The Origins of Interstellar Dust: Introduction

Interstellar dust grains do not originate in the interstellar medium. The gas densities in the interstellar medium are far too low for the nucleation of grains to occur from interstellar atoms and molecules in the available timescale. However, astronomical observations indicate that certain local-ised regions of interstellar space—specifically, some types of circumstel-lar regions—are the birthplaces of dust. From these special regions of relatively high density, dust grains can apparently be formed and ejected into the general interstellar medium; there, they may evolve in response to the interstellar radiation and particle fields and to chemical reactions with interstellar atoms and molecules. Therefore, after newly-formed dust is injected into the interstellar medium, it may change its chemical struc-ture and physical nature. We'll discuss some aspects of that evolution in the following chapter (Chapter 6). In interstellar clouds, dust provides surfaces on which heterogeneous chemistry occurs (Chapter 8) and also provides a substrate for the deposition of icy mantles of simple molecules; a fairly complex chemistry may be promoted in these mantles, as we will describe in Chapter 9. Dust that is newly-formed in circumstellar regions is not the same as interstellar dust, but by trying to understand the dust formation processes and the chemical and physical nature of newly-formed dust, we can also hope to understand better the chemical and physical nature of interstellar dust. Some newly-formed dust and some evolved (interstellar) dust can be found in collected meteoritic material; we shall discuss this *presolar* dust in Chapter 7.

The Chemistry of Cosmic Dust
By David A. Williams and Cesare Cecchi-Pestellini
© David A. Williams and Cesare Cecchi-Pestellini 2016
Published by the Royal Society of Chemistry, www.rsc.org

An excellent, very comprehensive and high level research text on circumstellar dust shells has recently been published by Gail & Sedlmayr (2014).[1]
Those authors provide much more detail on dust formation than can be
included in this chapter. A most important book on astromineralogy, edited
by Henning (2003),[2] includes not only dust formation in stellar outflows but
also considers Solar System solids and laboratory analogues.

Variable stars have long been favourite targets of observation by astronomers. For example, Edward Pigott[3] reported in 1797 that the star now known
as R Coronae Borealis had disappeared from his view in 1783 but then recov-
ered and remained fairly constant in brightness until it disappeared again
some ten years later. This star is now classified as an irregular variable. The
accepted interpretation that is placed on these and similar observations was
not developed until the 20th century: we now accept that dust forms from
time to time in the extended stellar envelope, just as a candle flame generates
soot. The dusty atmosphere blocks out the light of the star, but then drifts
away from the star. This drift is possibly driven by stellar radiation pressure
on the dust grains, which then drag the gas with it; or the drift may be driven
by underlying shocks at the stellar surface. The dusty gas becomes diluted in
the expansion as the envelope drifts into interstellar space, so that its optical
extinction is reduced and the star apparently brightens once again. The event
in R Coronae Borealis and in some other stars evidently repeats irregularly.

The process of dust formation is a delicate one. The gas needs to enter a
zone in which the pressure and temperature are appropriate for dust formation, and to remain there long enough for the formation to be achieved. Dust
formation involves the creation of a nucleation centre on which condensation
may occur. Thus, there are constraints on the physical properties of the gas
and also on the gas dynamics. Evidently, the nature of the dust that is formed
depends on the elemental chemical abundances in the atmosphere. The most
important elements for dust formation are carbon, oxygen, magnesium, silicon, sulfur, and iron. In general, if there is an excess of carbon over oxygen,
then some kind of carbon dust tends to form, whereas an excess of oxygen
over carbon leads to the formation of various kinds of oxides and silicates.

While visual detections of variable extinction suppressing the light of a
star are taken as indicators of the presence of dust, observations in the infrared are particularly useful to confirm those indications and to identify the
chemical nature of the dust through absorption in vibrational modes of the
material. For example, absorptions at 9.7 and 18 μm are associated with Si–O
stretch and bend modes of amorphous olivine, while crystalline forsterite
has a set of features at 9.8, 11.4, 16.2, 19.4, 23.4, and 33.9 μm. Infrared observations have revealed that a variety of astronomical objects are associated
with dust, and are indeed sites of dust formation.

Many of these sources of dust are stars with cool extended atmospheres,
such as red giants and supergiants. In particular, these sources include stars
known as asymptotic giant branch (AGB) stars; these are stars of low and
intermediate mass (up to ~10 solar masses) that are near the end of their

lives, having consumed much of the hydrogen and helium fuels in their cores. They develop extended cool envelopes in which the elemental abundances of potential dust components (such as carbon, oxygen, and metals) are strongly enhanced above the values typical in the Milky Way (see Table 1.1 (ref. 4–6)). The envelopes are observed to drift away from the star, eventually becoming detached as the AGB stars evolve into well-known short-lived objects: the *protoplanetary nebulae* and *planetary nebulae*. The stars become white dwarfs (very hot stars powered by gravitational collapse). Protoplanetary and planetary nebulae contain (and possibly modify) the dust from the preceding AGB stage of their existence. Stars with cool stellar envelopes are of two main types, carbon-rich and oxygen-rich, forming carbon dust or silicate and other dust, respectively. These stars are believed to be major contributors to the interstellar dust population.

Brown dwarfs (stars of exceptionally low mass and temperature in which nuclear fusion does not occur) are observed to be rich in dust; however, this is probably interstellar dust from the pre-existing dense core of gas from which the brown dwarf formed, rather than truly new dust. Similarly, dust is also found in great abundance associated with prestellar and young stellar objects; this is also believed to be interstellar dust from the prestellar core in which star formation occurred.

In contrast to the relatively low temperature situations associated with stellar giants and supergiants, dust is also associated with winds from very high temperature stars. These include massive stars with very high mass loss rates: the Wolf–Rayet (WR) stars; red super giant (RSG) cool and large stars, and the outflows associated with luminous blue variable (LBV) stars. However, while these objects are important members of the stellar zoo, they are rare and unlikely to be major contributors to the interstellar dust population.

Massive stars at the ends of their lives may explode in supernovae. Conditions in the ejecta from the stellar explosion would appear to be very hostile to dust formation, but these ejecta are observed in some cases to be sites of copious dust formation, beginning within a year or so of the outburst, and possibly (at least in the case of supernova SN1987A) continuing for several decades. Dust was detected in the ejecta of SN1987A a couple of years after the outburst in 1987, and the mass of that dust was estimated then to be $\sim 10^{-4}$ solar masses ($M_\odot$); that mass was estimated to have increased enormously by 2011 to ~ 0.4–$0.7\ M_\odot$ of dust.[7] Supernovae may be significant contributors to interstellar dust.

Towards the end of their lives, binary stars of modest mass may generate explosions known as novae ("new stars") that are optically as bright as stars, but fade within about a year. Some of these novae may be periodic. The formation of dust does not occur in all novae, but when it does it may be detected by an abrupt extinction in the optical spectrum of a nova accompanied by a rise in infrared emission from warm dust. Dust formation may occur within a few months after outburst, implying a very rapid formation process under apparently rather hostile conditions.

Table 5.1   A list of the main contributors to gas and dust in the Milky Way, with estimated injection rates[a] ($M_\odot$ pc^{-2} Myr^{-1}).

Source	Gas	Carbon dust	Silicate dust
AGB (C-rich)	750	3	
AGB (O-rich)	750		5
OB stars	30		
Wolf–Rayet	100	0.1	
Red supergiants	20		0.03
Novae	6	0.3	0.03
SN type Ia		0.3	2
SN type II	100	2	10

[a]Data are taken from Tielens *et al.* (2005),[4] Massey *et al.* (2005),[5] and Ferrarotti & Gail (2006).[6]

The main contributors to interstellar dust in the interstellar space of the Milky Way are listed in Table 5.1, and images of some typical dust-forming objects are shown in Figure 5.1.[8] The dominant contributors of dust to the Milky Way galaxy are believed to be AGB stars, and these objects have received considerable observational and theoretical attention. We shall discuss the process of dust formation in these objects in the following section. Ideas concerning the formation of dust in novae and supernovae are continuing to be evaluated, and we shall discuss those topics in subsequent sections.

5.2   Dust Formation in AGB Stars

There are several distinct stages that may for convenience be considered separately, in the sequence of events leading to the formation of dust in AGB stars; see Figure 5.2 for a schematic diagram summarizing these stages. The first important stage is the determination of the elemental abundances appropriate for the stellar atmospheres of these stars. Elemental abundances in AGB stars differ from the mean interstellar values because of major structural rearrangements occurring inside the star as it approaches the end of its life. The atmospheres of the stars become enriched in elements from deep in the stellar interiors; these elements (especially carbon, nitrogen, oxygen, and some metals) are the result of the thermonuclear processes energizing the star. These elements may promote dust formation. In particular, the abundances of carbon and oxygen are crucial, and there are distinct differences in the chemistry and potential for dust formation between carbon-rich and oxygen-rich stars.

The second stage is the determination of the nature of the stellar atmosphere, prior to any dust formation. The gaseous stellar atmosphere is, by interstellar standards, very dense, and the temperature sufficiently high to allow rapid conversion to thermodynamic equilibrium in the form of the set of stable molecules for the appropriate temperature and pressure. It is from this molecular gas that dust may form, if conditions permit.

The final stages take place in the envelope of gas as it drifts away from the star, cooling and becoming more tenuous as it does so. If the physical

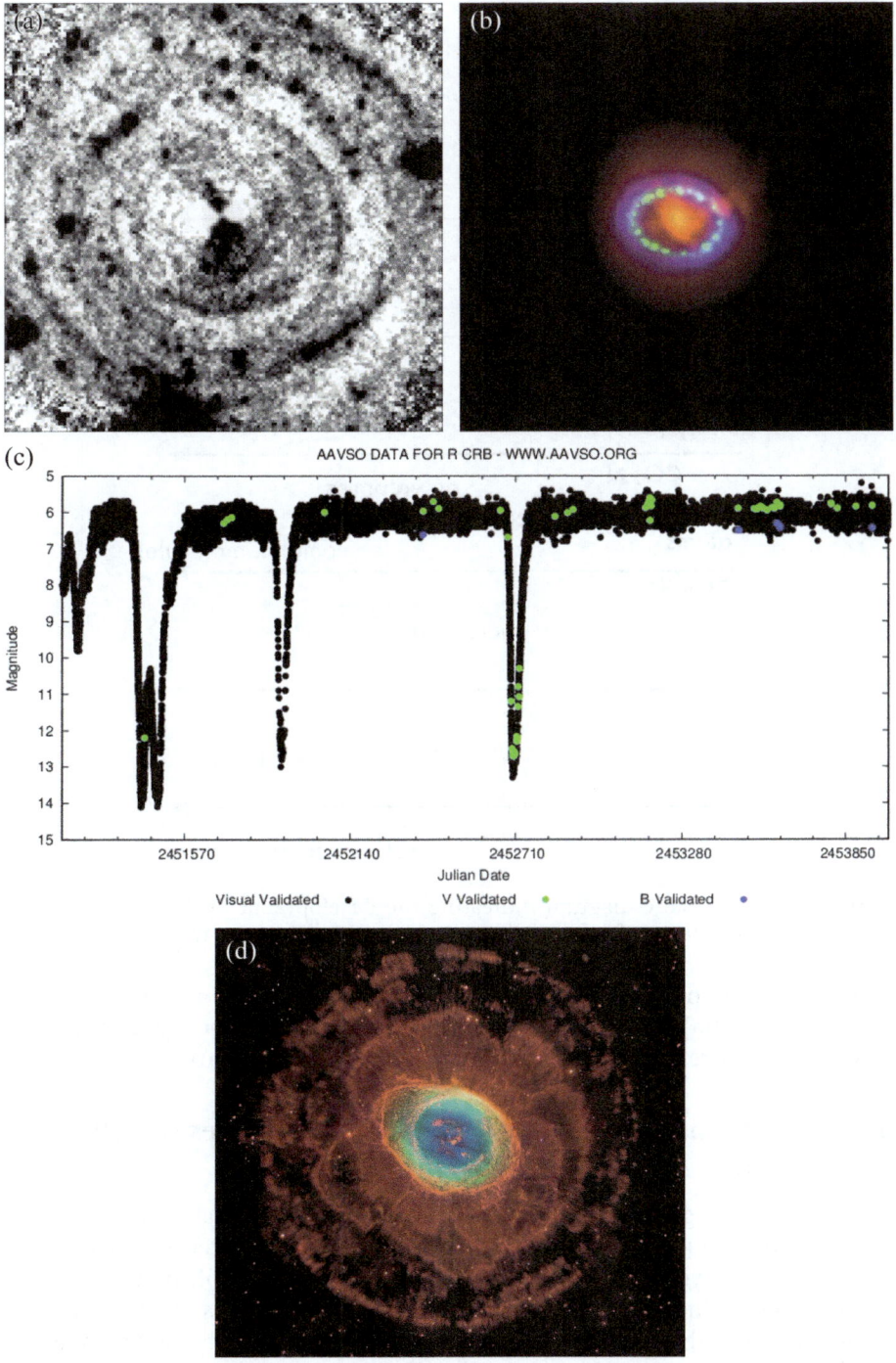

Figure 5.1   Some dust-forming objects; (a) IRC +10216; reproduced with permission from Mauron & Huggins (2000);[8] (b) SN1987A; courtesy ESO; (c) R Corona Borealis light curve, courtesy AAVSO; (d) Ring nebula (credit: Hubble, Large Binocular Telescope, Subaru; composition and copyright: Robert Gendler).

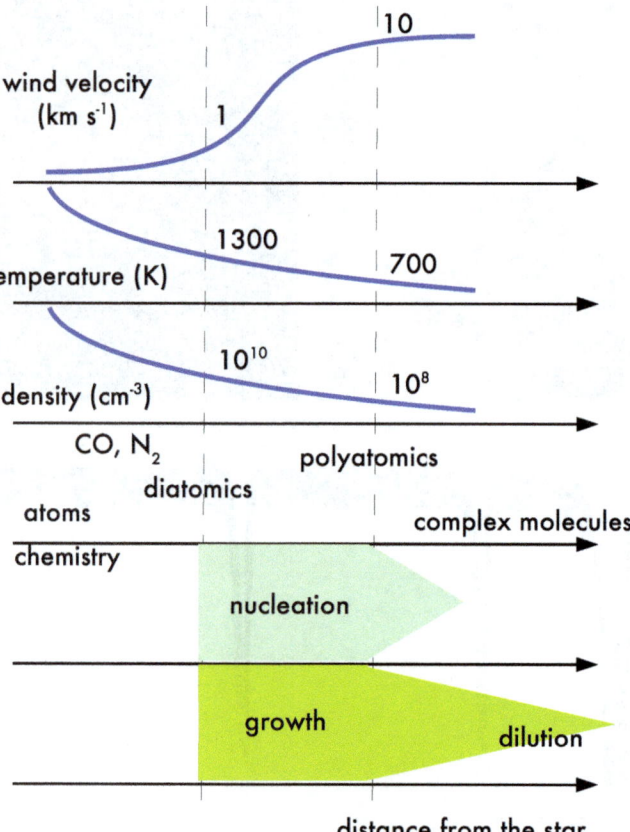

Figure 5.2   Schematic diagram indicating the development of dust formation in the outflow from a cool star. Based on Gail & Sedlmayr (2014).[1]

conditions are appropriate during this expansion, then dust may nucleate and grow. We shall describe all these processes rather cursorily, and readers who require more detail should consult Gail & Sedlmayr (2014).[1]

5.2.1   Elemental Abundances in the Atmospheres of AGB Stars

The nuclear energy sources for stars that become AGB stars are hydrogen-burning (to produce helium nuclei) and helium-burning (to produce carbon, nitrogen, and oxygen nuclei, in particular). The precise evolutionary path of a star is determined by the stellar mass. For stellar masses greater than 0.5 solar masses both these energy sources are involved. If the stellar mass is smaller than this limit, then no helium-burning occurs and so there is no enrichment of the elements that may be involved in dust formation (and so no dust formation is possible). For these very low stellar masses the stellar evolution in any case is very slow.

Hydrogen-burning proceeds most rapidly at the centre of the star where the temperature is highest. Therefore, a helium core develops and grows as the hydrogen is converted to helium, and the helium core is surrounded by a hydrogen-burning shell. The helium core contracts and becomes denser and hotter while the envelope expands. Eventually, the helium core ignites, the details depending sensitively on the stellar mass. The important point for present considerations is that a carbon-oxygen core develops inside the helium core through reactions in which three ^{4}He nuclei (α-particles) are converted to a ^{12}C nucleus, and a ^{12}C nucleus and one α-particle create a ^{16}O nucleus. Reactions of these nuclei with protons, α-particles, electrons and positrons, γ-rays and neutrinos in the CNO cycle generate other isotopes of carbon (^{13}C), and oxygen (^{15}O, ^{17}O, ^{18}O), isotopes of nitrogen (^{13}N, ^{14}N, ^{15}N), and other elements. The important balance between oxygen and carbon in the core is controlled by the stellar mass which determines the temperature (see below).

Although this carbon-oxygen core is deep inside the star, a convection zone in which turbulent mixing is occurring ensures that abundance gradients are removed. Convection is particularly important in stars of low and intermediate mass; in low mass stars the convective zone may occupy the entire bulk of the star, while in a star like the Sun convection affects the material lying in about the outer third of a radius. In AGB stars however, the convection zone can be deep, so that material (enriched in C, O, and N) from the central core can reach the stellar atmosphere. This process is called "dredge-up", and the evidence supporting it lies in the isotopic anomalies arising in the helium core chemistry that are detected in presolar grains in meteorites (see Chapter 7). Dredge-up may occur a number of times during the life of an AGB star.

After the helium shell has ignited and the helium fuel begins to be consumed, helium-burning declines. Hydrogen-burning in a shell surrounding the helium shell then resumes, introducing some restructuring in which the core becomes smaller and hotter. Eventually, helium-burning may be able to resume, until—again—much of the helium is consumed. Then hydrogen-burning resumes once again and the cycle may repeat. These switches of energy source and consequent restructuring affect the stellar energy output, and such a star is called a *thermally pulsing AGB star*. The pulsation stimulates mass loss through a stellar wind, although radiation pressure on dust grains produced in the envelope of the star may eventually dominate as the driver of the wind. The maximum mass loss in the wind may occur at a very high rate, as high as 10^{-4} $M_\odot$ per year. Evidently, such a high mass loss rate cannot be sustained for very long.

Detailed models show that the likelihood of an AGB star becoming a carbon-rich star (as opposed to remaining an oxygen-rich star) is strongly dependent on stellar mass. Detailed computations show that for stellar masses less than 1.55 solar masses, stars remain oxygen-rich, but at that mass, a short-lived carbon-rich phase (3.7×10^4 year) can arise following the oxygen-rich phase whose duration is almost ten times longer. However, for a stellar mass of 2 solar masses, the oxygen-rich and carbon-rich phases have similar durations (at around 3.5×10^5 year), while for stellar masses of 3 solar masses

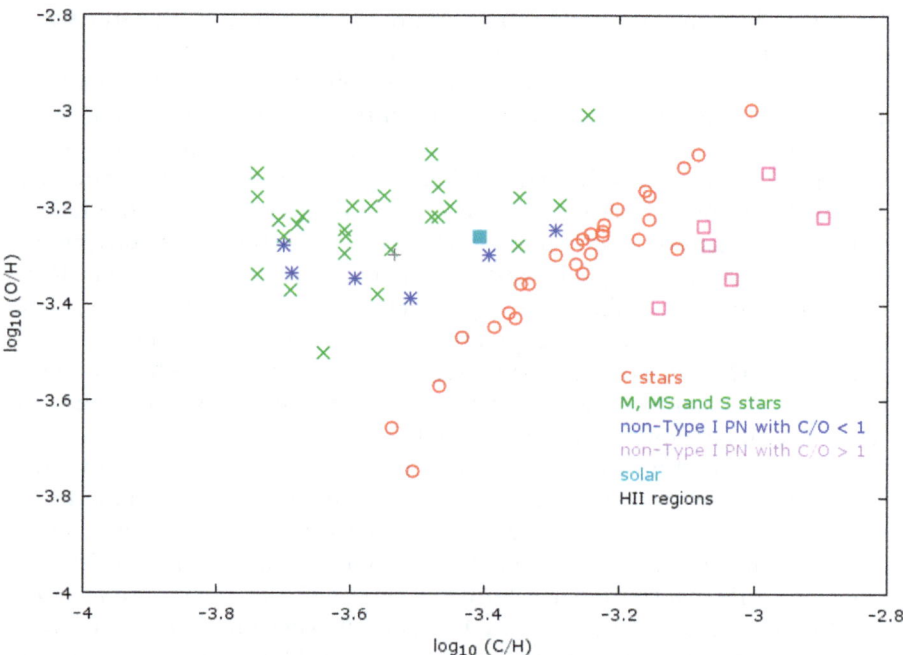

Figure 5.3 Typical ratios O:H and C:H measured in AGB stars. Data from Herwig (2005).[10]

(which corresponds to the duration of a star's AGB phase), the carbon-rich phase duration (~1.6 My) is about four times longer than the oxygen-rich phase. At five solar masses and above, AGB stars do not have a carbon-rich phase. These model results have been broadly confirmed by observational studies (*cf.* Girardi & Marigo 2007[9] for stars in the Large Magellanic Cloud). However, it should be noted that the fate of an AGB star depends on its metallicity (*i.e.* the relative abundances of elements other than hydrogen and helium).

Relative to the Solar System values, carbon and oxygen fractional abundances are slightly reduced in abundance in O-rich stars while that of nitrogen is enhanced. In C-rich stars, carbon is strongly enhanced by dredge-up, so that the abundance of carbon exceeds that of oxygen. Typical ratios O:H and C:H measured in AGB stars are shown in Figure 5.3.[10] The chemical compositions of C-rich star atmospheres are rather different from those of O-rich stars, with the latter presenting more erratic C:O ratios.

5.2.2 Chemical Composition of the Stellar Atmosphere

The physical conditions in the atmospheres of AGB and other cool stars are such that the gas is almost entirely molecular. Given the elemental abundances (discussed in subsection 5.2.1), and with assumptions about the temperature and pressure in the atmosphere, it is possible to determine the chemical nature of the stellar atmosphere. When this material is set in motion by pulsations or shocks to form the stellar wind, it becomes the environment in which dust formation may occur. Thus, the chemical nature of the atmosphere is the starting point for dust formation.

The problem is a standard one for thermodynamics: given the elemental abundances in a gas with specified temperature and pressure, what is the chemical composition of the gas in equilibrium? As is well known, the equilibrium state corresponds to the minimum of the Gibbs function at the specified pressure and temperature for all of the species involved in the system. With the use of conservation equations for each element and tabulated thermodynamical data for each species, a set of algebraic equations for the partial pressure of each species can be solved iteratively, and their abundances evaluated for any pressure and temperature. Many such calculations have been made over the last half century. Some sample results are presented in Figure 5.4 (ref. 11) for O-rich and C-rich environments.

In the O-rich case, almost all carbon is locked in CO. Some of this CO, about 1%, may be converted to CO_2 at sufficiently low temperatures. Any other carbon-bearing species are very low in abundance. The excess oxygen (*i.e.* the oxygen not bound in CO) is—for temperatures below about 1600 K—mainly bound in H_2O, while SiO takes up most of the silicon. The metals Mg and Fe are almost entirely atomic, while the metals Al and Ca form a variety of compounds that may survive to quite high temperatures. Sulfur is a relatively high abundance element; it is present mainly in atomic form for temperatures above ~1100 K, and as H_2S (~1000 K) and S_2 (for temperatures below 1000 K).

However, taking account of the relative elemental abundances, it is clear that the species most likely to be involved in dust formation in O-rich conditions are water, atomic iron and atomic magnesium, and silicon monoxide. Thus, the potentially very complicated situation for abundant dust formation in circumstellar envelopes becomes tractable, with (ignoring CO as an unlikely partner) only four main possible contributors (H_2O, Fe, Mg, and SiO). However, minor components of dust may involve other species.

In the C-rich case, almost all the oxygen is locked in CO, while the excess carbon feeds a very rich chemistry. However, this chemistry is dominated (for the stated pressure) by atomic carbon (above 1700 K), by C_2H (~1600 K) (C_2H is not included in the results shown in Figure 5.4), and C_2H_2 for lower temperatures. The metals Fe and Mg are atomic, while Si is atomic at higher temperatures and present as SiS at temperatures below ~1100 K. Thus, carbon dust is likely to be an amorphous hydrogenated solid, while metals may also form dust.

There is a critical case in which the oxygen is completely locked in CO and SiO, the carbon completely locked in CO, and silicon is completely locked in SiO and SiS. This rather special situation would promote dust formation from the species SiO and SiS and from the metals.

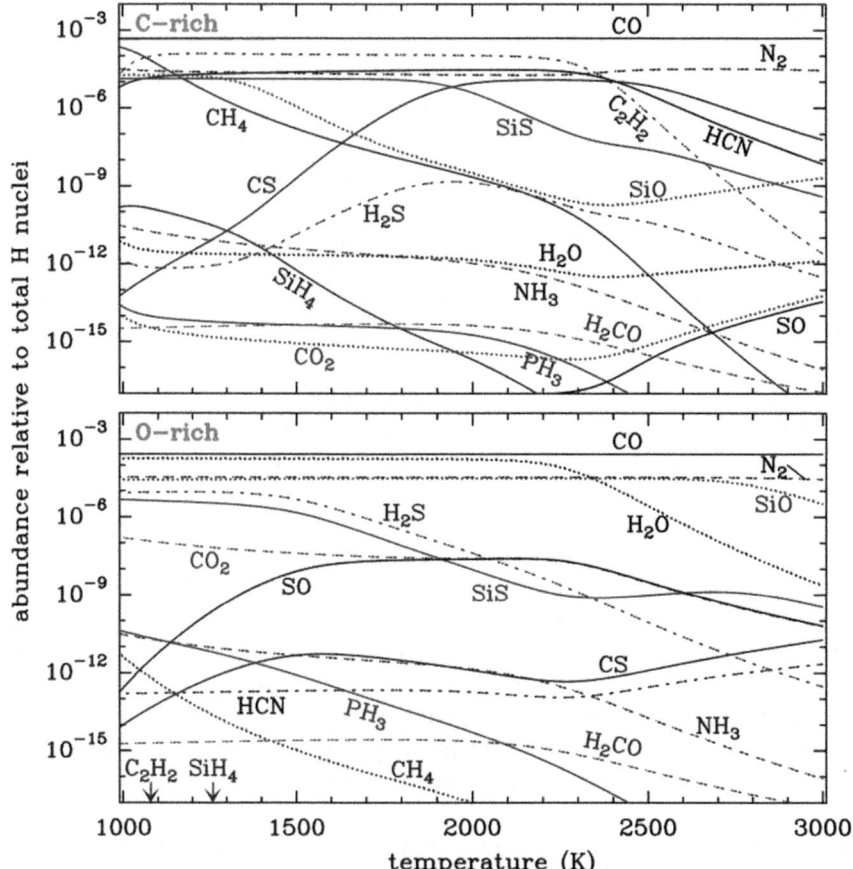

Figure 5.4   Thermodynamic equilibrium abundances as a function of temperature
for C- and O-rich gas (with C:O abundance ratios of 1.5 and 0.5, respectively) and with a constant total atomic hydrogen number density of
$10^{14}\,\mathrm{cm}^{-3}$. Reproduced with permission from Agúndez *et al.* (2010).[11]

5.2.3   Dust Formation in the Stellar Outflow

5.2.3.1   Nucleation

The picture of dust formation in stellar outflows that is summarized in
Figure 5.2 envisages the formation of nuclei which form centres of growth
for dust grains. Once the dust grains have formed, they may become an
important driving force in maintaining the outward flow of the stellar envelope
through the action of stellar radiation pressure.

These nuclei do not necessarily have the same composition as the material
that condenses on them during grain growth, but are formed from the first
materials to condense out from the gas. It is necessary, therefore, to consider
the chemistry that occurs in the outflow, as described briefly in the previous
section, but to include also the possibility that dust grains form in the gas.

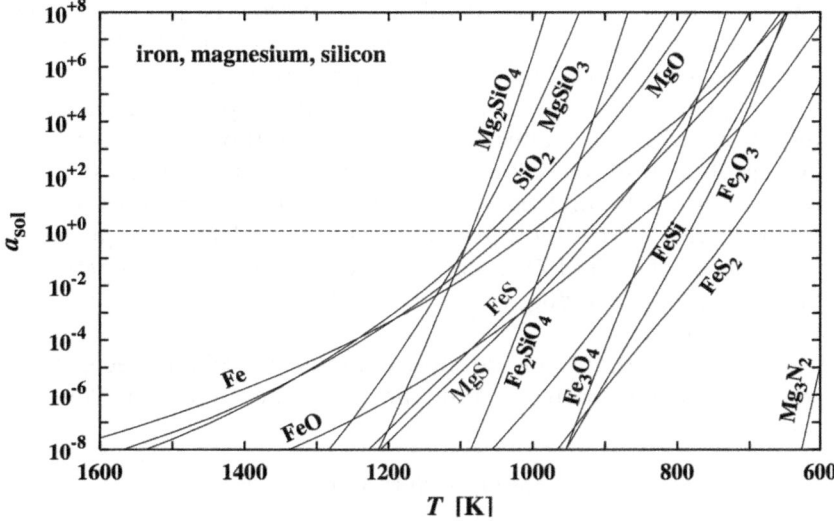

Figure 5.5 Pseudo-activities for a number of solid silicon, magnesium and iron compounds, as a function of temperature, in a gas with pressure 5×10^{-9} bar. Reproduced with permission from Gail (2003).[12]

Such solids may be in equilibrium with the gas, if accretion and evaporation are in balance. If accretion is dominant, then the dust grains grow, and if evaporation is dominant, then the dust grains tend to disappear.

Evidently, this is a complex situation, but one can approach it *via* the pure gas-phase calculations reported in the previous section. As shown by Gail & Sedlmayr (2014),[1] for a given pressure and temperature one can calculate from the partial pressures in the gas and the free energy, quantities called the *pseudoactivities*, a^c, which define equilibrium ($a^c = 1$), or growth, ($a^c > 1$) or decline ($a^c < 1$) in the nucleation cluster. When growth occurs, then partial pressures in the gas must change, but such calculations indicate the nucleation clusters most likely to form in the cooling gas; these are the materials with largest a^c at the highest temperatures. Figure 5.5 shows some pseudoactivities computed by Gail (2003)[12] for an O-rich atmosphere.

In this computation, of the elements Fe, Mg, and Si, forsterite (Mg_2SiO_4) and enstatite ($MgSiO_3$) show the highest values of a^c. However, other more minor species have much larger pseudoactivities at high temperatures, suggesting that these minor species—with much smaller abundances than the main constituent elements of the dust—may form first in the cooling outflow, and these may be the nucleation centres on which other more abundant materials condense as the temperature falls. The minor solids include corundum (Al_2O_3) which forms at high temperature, and gehlenite ($Ca_2Al_2SiO_7$), also formed at a fairly high temperature. Other low abundance elements may form the solids titanium oxide (TiO_2), perovskite ($CaTiO_3$), and zirconium oxide (ZrO_2). Results similar to those in Figure 5.5 suggest that the first solid materials to form in the cooling outflow involve these minor elements which in themselves are incapable of providing the entire dust population. However, they are able to act as nucleation centres for the condensation of more abundant dust grain constituents, as the temperature declines.

A similar picture emerges from computations of pseudoactivities for a carbon-rich gas with abundances corresponding to those in the envelope of an AGB star. Here, however, carbon—the dominant dust-forming material—may form solids at the highest temperatures in the cooling outflow. Other more minor species with high pseudoactivities at relatively high temperatures include SiC, TiC, and ZrC. Many species have high pseudoactivities at lower temperatures, including the relatively abundant Fe_3C at temperatures below about 1000 K.

It is worth noting that CO is possibly the most abundant species in both oxygen-rich and carbon-rich environments. It is believed that the Boudouard reaction:

$$2CO \rightarrow C(s) + CO_2$$

is energetically feasible at a temperature of ~1000 K. It is thought to be responsible for terrestrial soot formation. However, the reaction must be more complicated than indicated. It may contribute to the formation of nuclei in stellar outflows.

The growth of homogeneous nuclei can be addressed either by a kinetic theory, or—more usually in the astrophysical context—by a classical nucleation theory. The latter theory is used because of a lack of cluster data for use in rate equations, even though it cannot correctly account for the properties of small clusters. Both are beyond the scope of this book. Further information may be found in Gail & Sedlmayr (2014).[1]

5.2.3.2 Growth of Dust Grains

The growth mechanism is constrained by abundances of species that may contribute to growth. In the case of oxygen-rich outflows in which forsterite (Mg_2SiO_4) dust is present, growth must occur from the most abundant species relevant to forsterite. These species are SiO, Mg, and H_2O and a net reaction

$$SiO + 2Mg + 3H_2O \rightarrow Mg_2SiO_4(s) + 3H_2$$

must proceed if growth is to occur. The precise details of this process are unclear.

Similarly, carbon dust grains must grow from the most abundant hydrocarbon, acetylene (C_2H_2), in the gas in a set of reactions summarized as

$$C_2H_2 \rightarrow 2C(s) + H_2$$

where the solid is some form of soot. However, soot is composed largely of sheets of carbon with a graphene-like structure, in hexagons of carbon atoms. The bonding in these carbon sheets is aromatic, whereas the acetylene has triple bond structure. A complex sequence of reactions is required to generate this overall reaction (see for example Cherchneff 1998;[13] see also the discussion in Section 4.3 on the HACA mechanism).

Grain growth may be opposed by evaporation or by chemical attack. However, if these processes are ignored, then a simple calculation shows that in the entire outflow, from the point at which dust growth begins to the point at which the dust grains emerge into the interstellar medium the maximum radii of grains of any likely type are in the order of 0.1 μm. This value is consistent with the typical maximum radius found for interstellar grains, deduced from measurements of interstellar extinction (see Chapter 2). It is likely that a range of grain sizes will emerge from stellar envelopes. However, as we shall describe in Chapter 6, dust populations from cool stellar envelopes and other sources will—when ejected into the interstellar medium—be subject to a variety of modifying processes. The chemical structure may be modified, and the physical nature of the grains, including the size distribution, may be affected in situations of grain growth *via* grain–grain collisions or shattering or other erosion processes in shocks.

5.3   Dust Formation in Supernovae

5.3.1   Supernovae

Table 5.1 indicates that supernovae are estimated to be important sources of interstellar dust in the Milky Way galaxy. Indeed, in the early Universe (up to ~one billion years after the Big Bang), supernovae may be the only sources of dust and heavy elements, because low mass stars have not had time to evolve and become AGB stars. The conclusion that supernovae are required to be important dust producers—implied directly by observational evidence—may at first sight seem surprising given the extremely harsh physical environments that exist in supernovae ejecta. A supernova is a stellar explosion in which the star may briefly become about as optically bright as the host galaxy of which it is a member. The rise to peak brightness is abrupt and fades (typically) in a few weeks. In this time the supernova emits about as much energy as a star such as the Sun emits in its entire lifetime of more than 10 billion years. The stellar explosion leaves a long-lived signature in the interstellar medium, a supernova remnant (see Figure 5.6). The number of these remnants that have been detected in the Milky Way galaxy suggests that supernovae occur in our galaxy at a rate of about three per century. However, the last supernova actually detected in the Milky Way was in 1604, and in the intervening period we might have expected 12 detections. Evidently, supernovae in the Milky Way are hard to detect because they tend to lie in the plane of the galaxy where extinction due to interstellar dust is high. Supernovae can be more easily detected in external galaxies, and the peak brightnesses of certain types of supernovae are even used as "standard candles" to infer distances to high redshift galaxies of which they

Figure 5.6   False colour image of the supernova remnant Cassiopeia A, using data from observations made using the Hubble (visible data, orange) and Spitzer (IR data, red) telescopes and the Chandra X-ray Observatory (data are blue and green). Courtesy: NASA/JPL-Caltech.

are members. As remarked in the introduction to this chapter, SN1987A—a supernova detected in 1987 in a neighbouring galaxy to the Milky Way, the Large Magellanic Cloud—formed some dust within a year after outburst, and a quarter century later appears to have formed a very large amount of dust (see Section 5.3.2, below).

Supernovae are Nature's dramatic signatures of the end phase of stellar existence. The explosions can occur in two main ways. We will describe here very crudely some basic ideas about these explosions, so that we have some conception of the extreme conditions under which dust formation must occur.

In the first type of supernova (Type I), the process may begin with a stellar binary, two stars of modest but rather unequal mass in mutual orbit around their centre of mass. The more massive star of the two evolves more rapidly, passes through its giant phase and is in on track to become a white dwarf – a star in which thermonuclear reactions have converted hydrogen and helium in the core to carbon and oxygen nuclei, but which lacks the mass to drive the central temperatures high enough for further thermonuclear conversions to occur. Without a thermonuclear energy source, the white dwarf star will slowly cool and fade. However, when the lower mass companion star passes into its giant phase it may be able to transfer large amounts of mass quickly to the white dwarf. If so, the abrupt increase in mass and pressure may drive an increase in the white dwarf temperature to such an extent that reactions synthesizing heavy elements occur throughout most of the star. The resulting injection of energy is the cause of the stellar explosion, and material is expelled with velocities on the order of ten thousand km s^{-1}; these high speed ejecta drive a fast shock into the surrounding interstellar medium, raising the temperature of that interstellar gas to very high values, on the order of millions of Kelvin. This hot gas is the supernova remnant; it radiates mainly in X-rays, and cools slowly. The high luminosity of the supernova is created by energetic photons released in decays of radioactive isotopes produced in the explosion, in particular, the sequence $^{56}Ni \rightarrow \, ^{56}Co \rightarrow \, ^{56}Fe$.

The second main route to a supernova (Type II) involves a single, massive star, rather than a binary star of modest masses. Thermonuclear reactions occur most rapidly at the centre of the star, so here hydrogen is first converted to helium, then subsequently to carbon, neon, oxygen, silicon, and iron which—in the simplest picture of the core of a star—form layers, or shells, with iron at the centre and hydrogen at the outside (the "onion layer" model). All these reactions are exothermic, so each reaction generates energy, drives up the temperature and powers the next stage of reaction. However, iron is as far as this sequence can go, because iron represents the most stable nucleus. To go beyond iron to more massive nuclei, however, reactions would absorb energy, lowering the temperature and switching off the energy sources. Thus, eventually a massive star runs abruptly out of energy. Without internal thermal support, the star implodes, and material rebounds from the core, generating huge quantities of neutrinos, and sending shock waves into the outer layers of the star, ejecting them into the surrounding space. Depending on the stellar mass, the stellar core becomes either a neutron star or a black hole. The main radioactive isotopes in core collapse supernovae are ^{56}Ni, ^{57}Ni, and ^{44}Ti; the first of these produces ^{56}Co which itself is radioactive and may power the light curve (*i.e.* the variation of brightness with time).

Core collapse supernovae are generally fainter than white dwarf supernovae, because the yields of ^{56}Ni are lower. But the total energy released—arising from the gravitational potential energy released in the core collapse—is greater. In general, supernovae are important sources of heavy elements in the interstellar medium, by fusion for elements up to iron and by nucleosynthesis for elements more massive than iron. Somehow, they are also important sources of dust. One of the key questions relating to dust formation is the extent to which the various shells of H, He, C, ..., Fe in the "onion layer" model remain separate during the explosion, and the extent to which they can mix. Obviously, if hydrogen can mix to some extent, then the chemistry that may be possible in a carbon/oxygen layer is potentially much richer. Many current models of core collapse supernovae adopt the idea that the shells are fragmented into macroscopic clumps—what might be called the "bag of marbles" concept—in which each marble is a macroscopic piece of the former "onion shell" (see Figure 5.7 (ref. 14)). Each type of "marble" in the "bag" has a unique filling factor in the region of interest. This arrangement would seem to promote mixing between what were originally distinct elemental layers. In fact, the extent of mixing that might occur is poorly known.

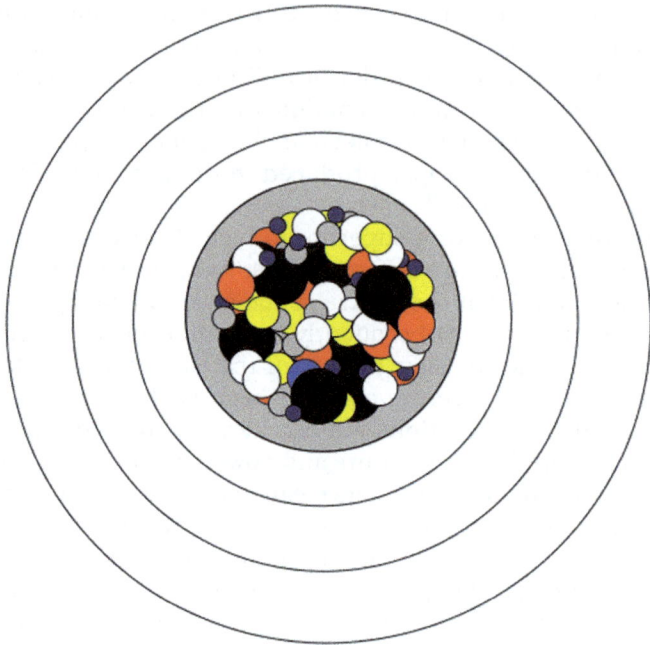

Figure 5.7   Schematic diagram showing the structure of the "bag of marbles" concept, in which "onion layer" shells of the interior of a pre-supernova
star have fragmented into clumps. Reproduced with permission from
Jerkstrand *et al.* (2011).[14]

5.3.2   The Formation of Dust in Supernovae

Observations of dust formation in supernovae tend to show that dust appears
after the detection of some simple chemical tracers. For example, the
well-studied recent supernova SN1987A showed emission in rovibrational
bands in the mid-infrared from CO and SiO molecules around one hundred
days after outburst, while the infrared continuum emission from solid dust
grains was first detected a year or so after outburst. Evidently, chemistry pre-
cedes dust formation in SN1987A and in many other supernovae. The chemistry in SN1987A was not a transient phenomenon, but has continued to the
present; Kamenetzky *et al.* (2013)[15] reported detections of the same species
at millimetre wavelengths in cool gas. Many supernovae are observed to contain rather small amounts of dust, less than about one tenth of one percent
of a solar mass. Such low dust contributions would seem to imply that supernovae could not be major dust contributors, as recognized by Cherchneff &
Dwek (2010).[16] While a low dust mass seemed to be initially the case also
for SN1987A, observations using the Hershel Space Observatory[7,17] and the
Atacama Large Millimeter Array[18] (ALMA) confirm that an amount of dust on
the order of one solar mass of material is present in SN1987A. Further, the
ALMA observations show that this dust is concentrated around the explosion

site, and has not been affected by the impact of the ejecta on surrounding interstellar or circumstellar material, see Figure 5.8.[18] For the ejecta from SN1987A, that impact will come later (see Section 5.3.3 below).

The chemistry in the supernova ejecta is very different from that in AGB stellar envelopes, as the environment is much more hostile to chemistry. Firstly, the ejecta contain radioactive elements such as ^{56}CO that generate γ-rays which, by a variety of processes, create fast particles and a harsh UV radiation field, both intrinsic to the gas. Therefore, no shielding for nascent molecules can be provided by, for example, ionization continua. The other major difference from the chemistry that occurs in AGB envelopes is that the gas may be hydrogen-poor or even contain no hydrogen at all, if no mixing with the outer shells occurs. Without hydrogen, or with hydrogen severely reduced in abundance, the main entry routes to chemistry in either the interstellar medium or in AGB envelopes are impeded or even suppressed.

However, early theoretical work (*e.g.* Lepp *et al.* 1990;[19] Rawlings & Williams 1990[20]) showed that, in spite of the hostility of the environment and the possible lack of hydrogen, a significant chemistry could be established. In the absence of hydrogen, entry routes to the chemical network of a zone in the ejecta containing carbon and oxygen could occur at early times through three-body reactions such as

$$C + O + He \rightarrow CO + He$$

or more generally by slow radiative associations such as

$$C + O \rightarrow CO + h\nu$$

A detailed model has been explored by Sarangi & Cherchneff (2013),[21] who follow the chemistry and the evolution of dust formed in the ejecta of supernovae (with solar metallicity) of various progenitor masses after outburst. They consider the various zones separately, and present results for the entire ejecta. The dust is assumed to form from the precursor molecules and atoms, and the process is largely complete after about five years because of the removal of the molecular components into the dust, and because the timescale of dust formation becomes too long. The dust mass builds from a very low value to a value of a few percent of a solar mass, and contains a wide variety of materials. While the prediction of maximum dust mass from this model is still somewhat low compared to that required by some observations, these results show that very significant amounts of dust are likely to arise in this way; see Figure 5.9.[22] Total masses (all zones) of molecules and of the various dust components predicted by the Sarangi & Cherchneff model for a progenitor mass of 25 solar masses and ^{56}Ni mass of 0.075 solar masses are shown in Table 5.2.

While the Sarangi and Cherchneff model does not predict dust masses quite as large as those detected by Matsuura *et al.* (2011)[7,17] and Indebetouw *et al.* (2014)[18] in SN1987A, their results confirm that dust masses should grow

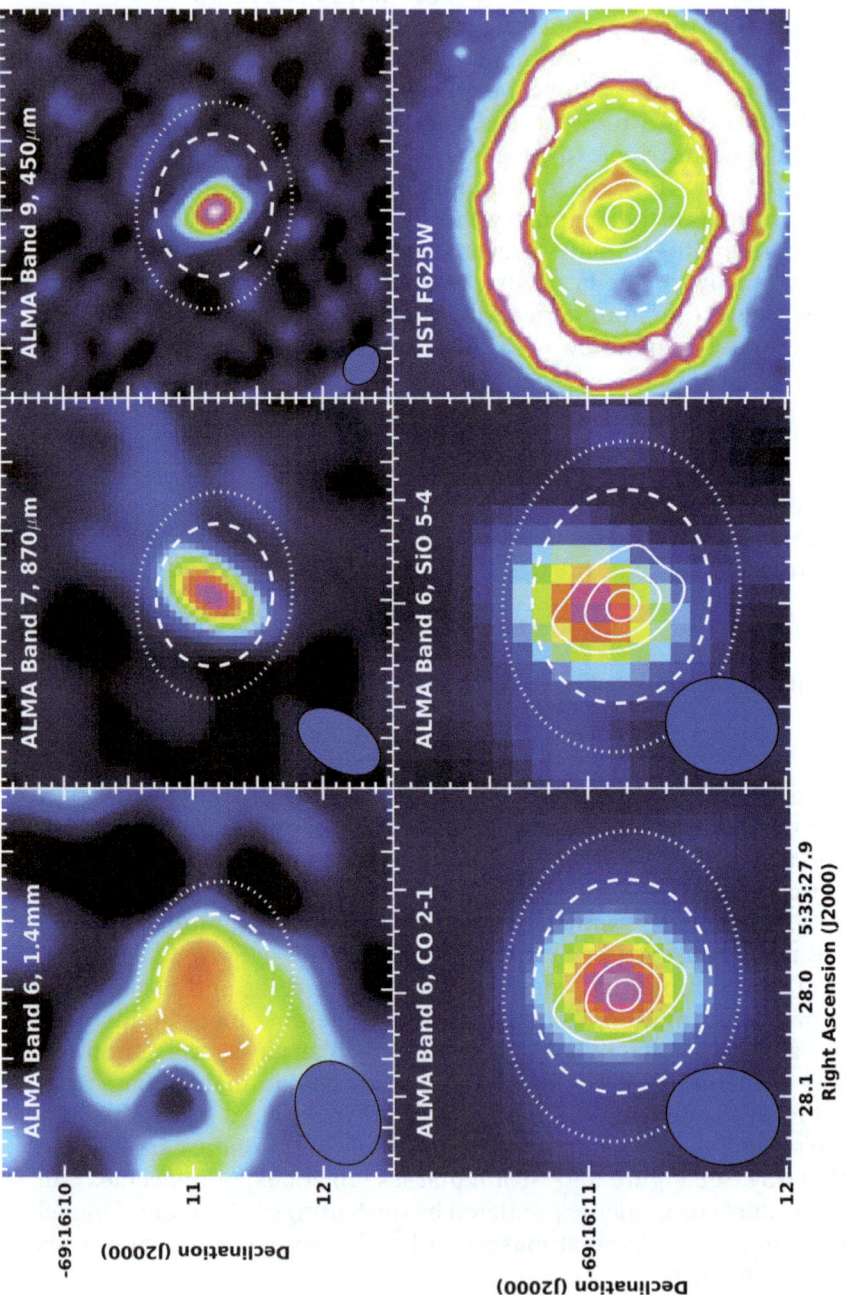

Figure 5.8   ALMA images of SN1987A, showing dust emission at three infrared wave-bands. The dust is concentrated around the explosion site. The dotted ellipse marks the position of the torus of synchrotron radiation at centimetre radiation, and the dashed line marks the reverse shock where the ejecta impacts on surrounding interstellar gas. Emission from the shock has been artificially removed. Reproduced with permission from Indebetouw *et al.* (2014).[18]

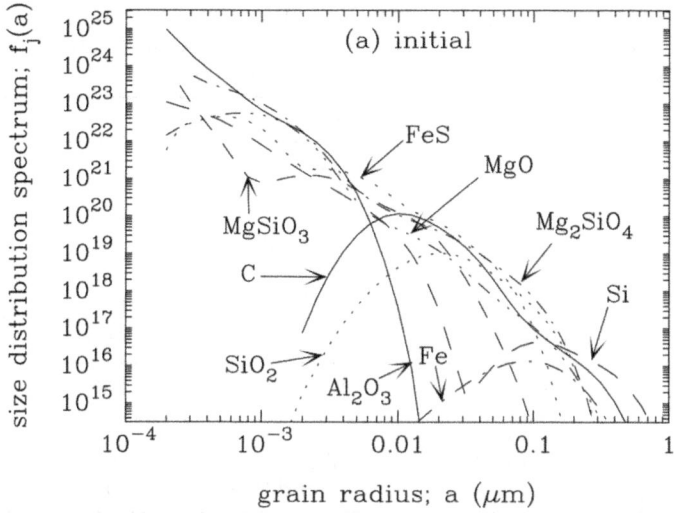

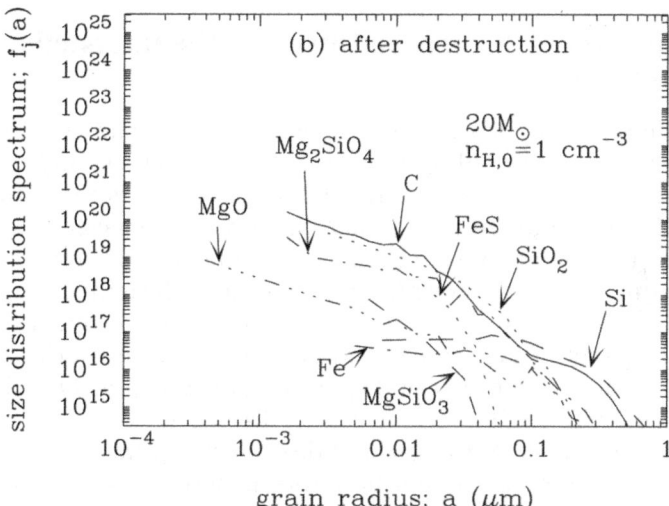

Figure 5.9   Size distribution of each dust species before (a) and after (b) destruction by shocks in a supernova remnant. The expansion is assumed to occur into an interstellar medium with H atom number density of 1 cm^{-3}. Reproduced with permission from Nozawa *et al.* (2007).

rapidly after outburst, from small values to something on the order of one tenth of a solar mass. Further, the dust is expected to contain a wide variety of materials. Other modellers[23-25] find similar results but with somewhat larger dust masses, especially for very large progenitor masses (*i.e.* more than 100 solar masses). These newly-formed dust grains are expected to have sizes in the approximate range from 1 nm to 1 μm.

Table 5.2 Masses of molecules and upper limits on dust masses (all in solar masses) at 2000 days post-outburst for a supernova with a 25 solar mass progenitor star that has 0.075 solar masses of ^{56}Ni.

Molecule	Total mass
SiO	2.2×10^{-6}
O_2	1.76
CO	0.86
SO	6.8×10^{-2}
SiS	0.1
Grand total	2.82

Dust	Total	Dust	Total
Mg_2SiO_4	3.2×10^{-2}	C	1.0×10^{-2}
SiO_2	9.1×10^{-4}	Si	2.5×10^{-3}
Al_2O_3	4.1×10^{-2}	Mg	2.2×10^{-3}
Fe	5.7×10^{-4}	SiC	8.2×10^{-4}
		Grand total	0.09

5.3.3 Injection of Supernova Dust into the Interstellar Medium

However, the formation of dust in the ejecta is not the end of the story. The supernova ejecta are travelling with very high velocities, and—sooner or later—must impact on the surrounding material which is either material in the nearby interstellar medium or material previously ejected from the progenitor star. The impact of very high velocity material on this ambient gas sets up a strong reverse shock which abruptly raises the temperature to more than one million K in the post-shock gas (evident in the optical emission, bottom-right panel of Figure 5.8) and creates a reverse shock that travels inward, dissociating molecules and sputtering dust. Figure 5.8 indicates the position of this reverse shock in the case of SN1987A.

In the case of SN1987A, the gas and dust of the ejecta have not yet been affected by the reverse shock, nor have they penetrated into the post-shock high temperature gas where dissociation of molecules and erosion of dust will also take place. However, supernovae that are somewhat older than SN1987A do show evidence of this reverse shock and its effect on the ejecta. Wallström *et al.* (2013)[26] presented observations, using the Herschel Observatory, of high-J rotational levels of CO from a dense knot of gas in the supernova remnant Cassiopeia A (SNR Cas A; see Figure 5.6). This object is about ten thousand light years away from Earth. Its outburst was not detected on Earth but its current appearance is believed to correspond to an evolutionary age of about three centuries (*i.e.* it could have been detected in the 17th century). Thus, Cas A is about ten times older than SN1987A. In this interval, the reverse shock is beginning to affect the ejecta. The high-J emission from CO shows large line-widths and is in fairly high density gas. It is believed to arise from CO that has re-formed in the post-shock gas. Thus, although the reverse shock is dissociating molecules, some—at least—of the molecular material may be re-constituted.

From the point of view of understanding the nature and amount of dust that is ultimately injected into the interstellar medium, it is important to consider the effect of this shock system on the dust and the efficiency of dust survival. Nozawa *et al.* (2007)[22] have investigated the evolution of dust formed in supernovae explosions in the early Universe, and their results are also instructive for understanding the effect on dust formed in more recent events. Nozawa *et al.* find that both the radius and chemical composition of the dust grains are important in determining whether the grains survive and reach the interstellar medium. For a supernova in an interstellar gas of one H-atom per cm^3 the smallest grains (radius < ~50 nm) are destroyed by sputtering. Somewhat larger grains (up to ~200 nm) are trapped in the dense shell of gas behind the forward shock, while grains larger than this limit are injected into the interstellar medium without much decrease in size. The effects on an ensemble of grains of a wide range of sizes and with different compositions are summarized in Figure 5.9.

The prediction of these investigations is that the overall composition of the dust is changed considerably as a result of this processing. Grains formed in supernovae do not necessarily survive into the interstellar medium, with small grains being particularly vulnerable; the average grain composition after the initial formation is heavily modified as a result of this processing before the material reaches the interstellar medium – where it may undergo further changes, see Chapter 6.

5.4 Dust Formation in Novae

5.4.1 Novae

Novae come in a variety of shapes, sizes and properties. But they share a common description and origin. The basic common property of novae is that they are stars whose optical brightness is observed to rise abruptly by a factor of about 10^4 or more and then to fade over a period of weeks or months. In pre-telescope times, the pre-nova star would have been unobserved and therefore the newly bright star would have been regarded as a new star, *i.e.* *nova stella*. The light curve (the plot of brightness against time) generally shows a steep early decline and then a more gentle reduction. An excellent and comprehensive text on novae has been edited by Bode & Evans (2008).[27]

After the early decline (by a factor of a few tens in optical brightness) the light curves of some novae enter a temporary "transition" in which the brightness falls almost to the pre-nova level, but then recovers after some months and joins the extrapolated light curve. When infrared observations became possible, this optical "transition" was seen to be accompanied by a rise in infrared emission. It was inferred that these novae were forming dust in a cloud that was initially so dense that the underlying star was optically obscured. Later, the dust cloud dispersed into the surrounding space and so this extinction was removed, and the light curve rejoined the extrapolated optical light curve. Novae that enter such a transition—the dust-forming novae—are the ones that concern us in this chapter. However, it should be noted that other novae, especially those showing a fast decay, do not enter a "transition" and their light curves decay fairly smoothly from the maximum to the post-nova stage. Yet other novae show a series of brightness oscillations in the transition region, before rejoining the extrapolated light curve.

As is the case with supernovae, observations in the Milky Way do not detect all the novae occurring. There is a strong concentration towards the plane of the Milky Way where interstellar dust is concentrated and extinction caused by this dust is strongest. In any case, novae are easy to miss, unless the region of their location happens to be currently under study. An estimate of the total rate of novae in the Milky Way is 34^{+15}_{-12} per year.[28]

The common origin of novae is this: they all arise in close binary stellar systems in which one of the two stars is a white dwarf, *i.e.* a star of modest mass that has used up all its hydrogen and helium fuel but whose self-gravity is insufficient to promote a further stage of evolution. The gravity is, however, strong enough to drag from the atmosphere of its partner star, a red giant, hydrogen-rich material which accumulates first in an accretion disk from which it falls on to the surface of the white dwarf. It is compressed and heated at the surface of the white dwarf, eventually reaching such conditions that hydrogen fusion occurs in a thermonuclear runaway explosion. In the resulting high temperatures, heavy atoms capable of accepting a proton do so, and distinctly non-solar CNO abundances arise and are ejected into the interstellar medium. In particular, novae are important sources of ^{13}C, ^{15}N, and ^{17}O and other isotopes in the interstellar medium. A nova explosion typically ejects ~10^{-4} solar masses of material at velocities ~10^3 km s^{-1} into the interstellar medium. This is material of high metallicity, *i.e.* it is enriched in heavy elements relative to solar abundances. The mass injection rate from novae into the interstellar medium of the Milky Way galaxy is, however, distinctly less than that of supernovae.

The ejecta are observed from infrared observations to contain molecules and dust. Molecular spectra (especially of CO, CN, and H_2) appear in the spectra of the ejecta first, typically a few days or weeks after the optical maximum. They are best detected through their rovibrational spectra in the near and mid-infrared, but molecules may sometimes also be seen in their optical spectra. In a dust-forming nova, the dust appears later, and the time of maximum infrared emission from the dust-forming shell is typically a few months after outburst. Dust can be detected in the optical and ultraviolet through the extinction it causes, though this requires knowledge of underlying nova spectrum. The infrared spectra are more informative; the continuum can be featureless, suggesting that it arises from small carbon grains. Specific features between 3.3 and 11.3 µm commonly attributed to hydrocarbons are seen in some novae, and others at 9.7 and 18 µm from silicates are also seen. Both types of grain can be seen in the same nova.

Table 5.3 Selected data on some novae ejecta. The table gives year of nova outburst, dust condensation temperature (K), time after outburst of maximum IR emission from dust (days), dust type (C = carbon, S = silicate, SiC = silicon carbide, HC = hydrocarbon), dust grain maximum radius (μm), and dust and gas masses in solar masses. Data extracted from Gehrz (2008).[29]

Parameter	NQ Vul	LW Ser	QU Vul	QV Vul	V838 Her	V705 Cas
Year	1976	1978	1984	1987	1991	1993
$T_{cond}(K)$	1100	1000			1266	1150
$t_d, IR_{max}(d)$	80	75	240	83	18	38
Dust type	C	C	S	C, S, SiC, HC	C	C, S, HC
Dust radius	0.71 μm	0.50 μm		0.63 μm	0.18 μm	0.7 μm
Dust mass	3.5×10^{-7}	1.6×10^{-6}	$\sim 1 \times 10^{-8}$	1.0×10^{-6}	4.8×10^{-9}	8.2×10^{-7}
Gas mass	1×10^{-4}	2×10^{-5}	3×10^{-4}	3×10^{-5}	3.5×10^{-4}	1×10^{-5}

Indeed, Gehrz (2008)[29] notes that: "Novae can produce every known type of astrophysical grain, in some cases showing the signatures of three or four different grain components during the temporal development of a single outburst." From the observations, one may deduce that dust grains in novae ejecta grow to 0.2–2 μm in radius. There are some similarities between comet dust and nova dust (see Chapter 7), suggesting that nova dust may be incorporated in star-forming regions. We show in Table 5.3 (ref. 29) information about some recent novae, including data relevant to dust and gas in the ejecta.

5.4.2 Dust Formation

As indicated by the observations, the formation of molecules appears to be a precursor to the formation of dust in novae ejecta. At first sight, it may appear that chemistry in novae ejecta is similar to that in supernovae ejecta, since both types of ejecta originate in a stellar explosion, start at high density and elevated temperature, expand at high velocity, and become diluted as they move into circumstellar space. In fact, there are significant differences in the chemistry. Novae ejecta are rich in hydrogen, whereas supernovae ejecta may be hydrogen-poor or hydrogen-free. Novae are affected by a strong radiation field that moves to shorter wavelengths and becomes more intense as the photosphere contracts, while supernovae have an unusual radiation field generated from γ-rays by radioactive elements. Novae ejecta have high metallicities, promoting molecule formation.

Early work on nova chemistry[30] demonstrated the importance of the photophysics and of the shielding against photolysis that molecular hydrogen provides not only for itself but also—fortuitously—for carbon. Thus, where hydrogen is atomic, carbon is ionized, and where hydrogen is molecular carbon atoms may be neutral. A crucial part of the chemistry is therefore associated with the formation of molecular hydrogen in these dust-free regions. Three-body reactions and reactions with H⁻ ions are found able to provide the necessary H_2. The overlapping Lyman and Werner bands of H_2 (that lead to photodissociation) become saturated and provide a *quasi*-continuum absorption over wavelengths from 91–110 nm that protects carbon atoms from photoionization in the radiation field. In such a shielded zone a chemistry initiated by C atoms with H_2 and O atoms can generate a rich but simple chemistry on timescales of days. As expected, CO is the dominant species.

Dust nucleation may occur on large molecules, such as PAHs. Flame chemistry has been widely studied but it seems unlikely that the chemical routes determined in the laboratory apply in novae, since flame chemistry is generally not subject to an intense radiation field as is the case in nova chemistry. For example, Dhanoa & Rawlings (2014)[31] have shown that the abundance of acetylene, C_2H_2, is suppressed in novae and is unable to contribute to the formation of large molecules that may act as dust nucleation centres. Pontefract & Rawlings (2004)[32] extended and improved earlier work,[20] and have explored the chemistry in novae ejecta to estimate the abundances of large molecules, taking account of the high radiation fields present in the gas, and allowing growth of molecules by reactions with C atoms and H_2 molecules. Results of computations reported by Evans & Rawlings (2008)[33] are in Figure 5.10. Evidently, large abundances of substantial hydrocarbons can be attained, even in the presence of intense radiation fields and allowing for the restricting effects of reactions with O atoms.

The computed high abundances of large hydrocarbons arise in high density gas in which the shielding by optically thick H_2 suppresses the UV field of the nova, in which—by our previous discussion—the carbon is mainly in the form of neutral atoms. Rawlings suggests that the nucleation may occur in clumpy regions of higher than average density. If the carbon ionization front driven by the nova radiation advances through the potential dust-forming shell *before* condensation occurs, then dust will not be formed. However, if the advance of the carbon ionization front reaches the potential dust-forming shell *after* condensation occurs, then dust will have formed and may survive. Evidently, the conditions for dust formation are complex. Condensation can only occur for temperatures lower than a "condensation temperature", which depends on gas pressure, metallicity and relative elemental abundances. For typical conditions, condensation temperatures lie in the range between one and two thousand K (see Table 5.4 (ref. 33 and 34)).

Observational evidence suggests that nova dust grains are larger than interstellar grains by about a factor of 3 in radius. Presumably, once condensation is possible, it occurs rapidly and goes to completion. It may be that nova dust, like supernova dust, must pass through shocked regions to enter the interstellar medium (see 5.3.3), and there suffer some erosion or shattering.

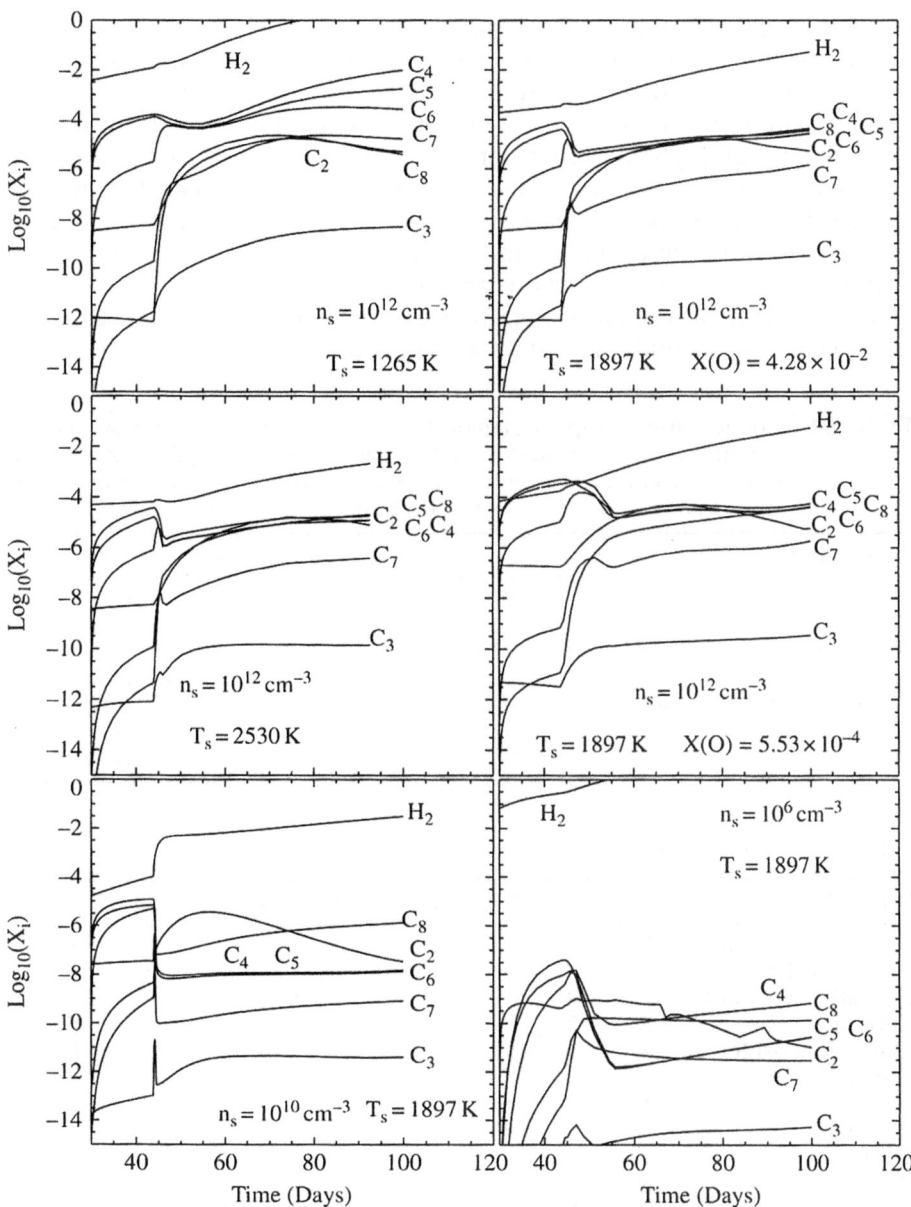

Figure 5.10 Time-dependent chemistry leading to the formation of hydrocarbons in the ejecta of novae, for several combinations of density and temperature and oxygen abundances. Reproduced with permission from Evans & Rawlings (2008);[33] the figure is based on unpublished work by M. Pontefract and J. M. C. Rawlings.

5.5   Conclusions

We have seen in this chapter that dust grains are injected into the interstellar medium from several different kinds of astronomical objects. These are all locations where the densities are much higher than in interstellar clouds, and also with higher temperatures. The physical process depends on gas initially at a suitably high pressure entering a cooling and dissipative phase, and thereby passing through a set of physical conditions in which dust formation is encouraged to occur. We have emphasized that AGB stars and supernovae are the most important sources, and that the dust grains coming from these sources are predominantly carbons and silicates.

However, it is important to note that a much wider variety of astronomical objects than simply AGB stars and supernovae can contribute to the

Table 5.4   Condensation temperatures (K) for various condensates in nova winds, for three different values of the C:O ratio in the ejecta; the gas pressure is 10^3 dyne cm^{-2}. A null entry indicates that condensation is unlikely for those parameters. Data from Evans & Rawlings (2008);[33] Ebel (2000).[34]

Condensate		$T_{cond}(K)$ for C:O		
		0.85	1.00	1.15
Olivine	Mg_2SiO_4	1390	1190	1150
Ortho-pyroxene	$MgSiO_3$	1260	—	—
Graphitic carbon	C	—	1150	1690
Silicon carbide	SiC	—	1630	1720
Corundum	Al_2O_3	1690	—	1240
Cohenite	Fe_3C	—	1160	1470

Table 5.5   Chemical inventory in dust factories.

Material	AGB	Post-AGB	PN	Nova	RSG	WR	LBV	SN
Amorphous silicates	X	X	X	X	X		X	X
Crystalline forsterite	X	X	X		X		X	
Crystalline enstatite	X	X	X		X		X	
Chromite	X							X
Aluminium oxide	X			X				X
Spinel	X							X
TiO_2	X				X			
Hibonite	X							
MgO	X							
Fe	X							X
PAHs	X	X	X	X	X	X	X	
a-C:H	X	X	X	X		X		
Graphite	X	X		X				X
Diamond		X						X
SiC	X	X	X	X				X
Other carbides	X							X
Si_3N_4								X
MgS	X	X	X				X	

population of interstellar dust grains, and that the range of materials being injected into the interstellar medium is much wider than the silicates and carbons that dominate the chemical abundances from the main dust sources. To emphasize this point, we show in Table 5.5 an observational inventory of dust formed in circumstellar sources, and which will be injected into the interstellar medium. The table is an updated version of a table provided by Tielens *et al.* (2005).[4] It demonstrates the chemical richness to be found in the solid phase of the interstellar medium. However, in abundances, the solid component of the interstellar medium is mainly silicates and carbons. Of course, once ejected into the interstellar medium, dust grains are subjected to a variety of processes that may convert them into other physical and chemical forms. We shall discuss some of those modifying processes in the following Chapter.

Further Reading

1. H.-P. Gail and E. Sedlmayr, *Physics and Chemistry of Circumstellar Dust Shells*, Cambridge University Press, 2014.
2. *Astromineralogy*, ed. Th. K. Henning, Springer-Verlag, Berlin, Heidelberg, New York, 2003.
3. E. Pigott, *Philos. Trans. R. Soc. London*, 1797, **87**, 133.
4. A. G. G. M. Tielens, L. B. F. M. Waters and B. J. Bernatowicz, in *Chondrites and the Protoplanetary Disk*, ed. A. N. Krot, E. R. D. Scott and B. Reipurth, 2005, p. 625.
5. P. Massey, B. Pletz, E. M. Levesque, K. A. G. Olsen, G. C. Clayton and E. Josselin, *Astrophys. J.*, 2005, **634**, 1286.
6. A. S. Ferrarotti and H.-P. Gail, *Astron. Astrophys.*, 2006, **447**, 553.
7. M. Matsuura, E. Dwek, *et al.*, *Science*, 2011, **333**, 1258.
8. N. Mauron and P. J. Huggins, *Astron. Astrophys.*, 2000, **359**, 707.
9. L. Girardi and P. Marigo, *Astron. Astrophys.*, 2007, **462**, 237.
10. F. Herwig, *Annu. Rev. Astron. Astrophys.*, 2005, **43**, 435.
11. M. Agúndez, J. Cernicharo and M. Guélin, *Astrophys. J.*, 2010, **724**, L133.
12. H.-P. Gail, in *Astrochemistry*, ed. Th. K. Henning, Springer-Verlag, Berlin, Heidelberg, New York, 2003, p. 55.
13. I. Cherchneff, *The Molecular Astrophysics of stars and Galaxies*, ed. T. W. Hartquist and D. A. Williams, Clarendon Press, Oxford, 1998, p. 265.
14. A. Jerkstrand, C. Frannson and C. Kozma, *Astron. Astrophys.*, 2011, **530**, 45.
15. J. Kamenetzky, R. McCray, *et al.*, *Astrophys. J.*, 2013, **773**, L34.
16. I. Cherchneff and E. Dwek, *Astrophys. J.*, 2010, **713**, 1.
17. M. Matsuura, E. Dwek, *et al.*, *Astrophys. J.*, 2015, **800**, 50.
18. R. Indebetouw, M. Matsuura, *et al.*, *Astrophys. J.*, 2014, **782**, L2.
19. S. Lepp, A. Dalgarno and R. McCray, *Astrophys. J.*, 1990, **358**, 262.
20. J. M. C. Rawlings and D. A. Williams, *Mon. Not. R. Astron. Soc.*, 1990, **246**, 208.

21. A. Sarangi and I. Cherchneff, *Astrophys. J.*, 2013, **776**, 107.

22. T. Nozawa, T. Kozasa, A. Habe, E. Dwek, H. Umeda, N. Tominaga, K. Maeda and K. Nomoto, *Astrophys. J.*, 2007, **666**, 955.

23. P. Todini and A. Ferrara, *Mon. Not. R. Astron. Soc.*, 2001, **325**, 726.

24. T. Nozawa, T. Kosaza, H. Umeda, K. Maeda and K. Nomoto, *Astrophys. J.*, 2003, **598**, 785.

25. R. Schneider, A. Ferrara and R. Salvaterra, *Mon. Not. R. Astron. Soc.*, 2004, **351**, 1379.

26. S. H. J. Wallström, C. Biscaro, F. Salgado, J. H. Black, I. Cherchneff, S. Muller, O. Berné, J. Rho and A. G. G. M. Tielens, *Astron. Astrophys.*, 2013, **558**, L2.

27. *Classical Novae*, ed. M. F. Bode and A. Evans, Cambridge University Press, Cambridge, 2008.

28. M. J. Darnley, M. F. Bode, *et al.*, *Mon. Not Roy. Astron. Soc.*, 2006, **369**, 257.

29. R. D. Gehrz, in *Classical Novae*, ed. M. F. Bode and A. Evans, 2008, p. 167.

30. J. M. C. Rawlings, *Mon. Not. R. Astron. Soc.*, 1988, **232**, 507.

31. H. Dhanoa and J. M. C. Rawlings, *Mon. Not. R. Astron. Soc.*, 2014, **440**, 1786.

32. M. Pontefract and J. M. C. Rawlings, *Mon. Not. R. Astron. Soc.*, 2004, **347**, 1294.

33. A. Evans and J. M. C. Rawlings, in *Classical Novae*, ed. M. Bode and A. Evans, 2008, p. 308.

34. D. S. Ebel, *J. Geophys. Res.*, 2000, **105**, 10363.

CHAPTER 6

Dust Evolution in the Interstellar Medium

6.1 Introduction

We have seen in Chapter 5 that dust grains may be formed in several types of near-stellar environment. However, once they enter the interstellar medium these grains will be modified in size, shape and even composition by a variety of physical processes. Evidently, we must regard grains from *near-stellar* regions as the raw material from which true *interstellar* grains are formed. Since the primary object of this book is to discuss the chemistry occurring on interstellar grains (see Section III) and its consequences (see Section IV), we need to discuss the processes that modify the grains injected into the interstellar medium from near-stellar sources.

Of course, these modifying processes may continue to act throughout the lifetime of a grain (which is probably some hundreds of millions of years for a dust grain in the Milky Way galaxy), and so to some extent the nature and properties of interstellar dust grains may be time-dependent. Therefore, we should really consider the possibility that grains *evolve* in the interstellar medium. If so, the properties of dust inferred by astronomers from their observations on a particular line of sight may in fact be an average over many different evolutionary states achieved in a variety of physical conditions. This time- and space-averaging is, however, only rarely attempted in the modelling of interstellar grains. In this chapter, we shall briefly discuss several kinds of modifying processes that are likely to occur in the interstellar medium, and consider the implications for chemistry on the surfaces of interstellar grains.

What are these modifying processes? Grains may grow in the interstellar medium by the accretion of atoms and molecules from the gas phase, the most

The Chemistry of Cosmic Dust
By David A. Williams and Cesare Cecchi-Pestellini
© David A. Williams and Cesare Cecchi-Pestellini 2016
Published by the Royal Society of Chemistry, www.rsc.org

obvious example of this being the accumulation of ice mantles on the surfaces of dust grains (see Chapter 9 for details). Such ice mantles are generally detected in cooler and denser regions of interstellar medium. Grains may also grow by coagulation of smaller grains to make larger ones; since the number density of grains in the interstellar space of the Milky Way is very low on average (typically, one large grain in a million cubic metres of the averaged interstellar medium) this process requires huge enhancements in the number density of grains for the process to be effective on reasonable timescales. But it is clear that this process does happen. The accumulation of dust grains (initially of a size up to a few tenths of a micrometre) into planetesimals (kilometre-sized solid objects which have appreciable self-gravity) and then the growth of planetesimals into megametre-sized planets in star- and planet-forming regions is obviously just an extreme example of this coagulation process (see Chapter 10).

Opposing these growth mechanisms are processes that may reduce or entirely remove the solid dust component. One such process is the *shattering* of grains in *grain–grain* collisions at high energy; such collisions redistribute grain mass into units of smaller size, generating a size distribution that favours small dust grains over large ones. Shattering may also cause *vaporization* of some of the dust material as atoms or molecules into the gas phase, and may remove smaller grains entirely. *Gas–grain* collisions at sufficiently high energy may lead to *sputtering*; this is a process in which the grain surface is eroded and atomic or molecular material is ejected to the gas phase. The collision energy required to drive all these processes may be thermal or non-thermal. Interstellar shocks provide high temperature post-shock zones in which a range of high gas–grain and grain–grain streaming velocities occur. The details of these dynamical processes are beyond the scope of this book, but it is necessary to understand the implications of these collisional processes for the nature of grains, their size distributions and their structures, and we discuss these consequences in the following Sections.

In diffuse regions of interstellar space, starlight may induce chemical changes within some dust materials. The obvious example is the photodarkening of HAC induced by stellar ultraviolet radiation (discussed in Chapters 3 and 4). In this process, polymeric H-rich carbonaceous material may be converted to an aromatic H-poor form of carbon. The polymeric-to-aromatic chemical change implies that corresponding changes in the optical and chemical properties of the material will occur. The photodarkening process can be reversed by H-insertion into the aromatic carbon lattice, in regions of higher temperature than interstellar clouds.

The passage of cosmic rays through dust grains may induce some structural or chemical changes in the material of interstellar dust. An example of this process is the conversion of crystalline silicates to an amorphous form. In the interstellar medium the crystalline form is present, but rare. Crystalline silicates have been detected in protoplanetary disks and some other circumstellar regions. Such crystalline to amorphous structural changes may affect the chemical reactivity of the grain surface. Amorphization can be effectively induced by low energy helium ions (He^+), while heavy cosmic rays are predicted to be less effective.

Various other processes are likely to occur in interstellar and circumstellar regions of space. Thermal annealing may give rise to the formation of crystalline silicates in circumstellar disks. High pressures achieved in shocks have been suggested as a mechanism to create the nanodiamonds that are found in collected carbonaceous grains embedded in collected interplanetary particles (see Chapter 7). Chemisputtering can affect the chemical nature of the surface of silicate grains formed in the outflows of AGB stars (see Chapter 5).

We shall describe in more detail some of these processes in the following sections. Different models of interstellar grains emphasize different aspects of the formation and evolution of interstellar grains. For example, Asano *et al.* (2014)[1] emphasize grain formation, grain growth, shattering in shocks and coagulation. We shall concentrate mainly on the dust grain models discussed in Chapter 3.

6.2 Disruption and Destruction of Dust Grains in Interstellar Shocks

6.2.1 Model Predictions

As we have seen in Chapters 2 and 3, the size distribution of interstellar grains is conventionally inferred from fits to observations of interstellar extinction (and other phenomena) along a particular line of sight, and the results are often interpreted as implying a dust grain size distribution, with the number of grains (assumed spherical and of radius a) in the range $a \rightarrow a + da$ proportional to $a^{-m}da$ within a specified range of radii, with m often taken to be 3.5, or thereabouts. Such a distribution implies that there are many more small grains (say, $a \sim 10$ nm) than large (say, $a \sim 100$ nm). This size distribution is of considerable interest, as it must arise from the balance between the growth of grains in competition with grain destruction mechanisms. Therefore, destruction mechanisms have been the subject of many detailed theoretical studies.

Results from a comprehensive model have been presented by Jones *et al.* (1996;[2] see also Jones *et al.* 1994;[3] Serra Díaz-Cano & Jones 2008;[4] Bocchio *et al.* 2014[5]). These studies include grain destruction in which grain material is returned entirely to the gas phase through gas–grain thermal and non-thermal sputtering and by vaporization of grains in grain–grain collisions; it also includes a new study of grain disruption in which grain–grain collisions lead to the shattering of grains so that the grain size distribution is changed but the mass in grains may remain unchanged. The relative motions between grains and between gas and grains are supposed to occur in magnetically-influenced interstellar shocks; these shocks cause significant grain destruction and disruption to occur in one particular phase of the interstellar medium – a low density ($n_H \sim 0.25$ cm^{-3}) and warm ($T_K \sim 10^4$ K) partly ionized gas. Because the gas is threaded by a magnetic field, the motions of ions and charged grains are affected by the field, and differential velocities arise between different grains and between gas and grains. Shock velocities in the interstellar medium may lie in the range ~ 1–10^3 km s^{-1} and streaming velocities are of similar magnitude.

These shocks may arise from explosions in the interstellar medium from supernovae and other energetic events. Supernovae deposit up to 10^{44} Joules of energy, more or less instantaneously and effectively at a single point in the interstellar medium, and the shock waves generated by this explosion pervade and heat much of the low-density gas in the interstellar medium.

Jones *et al.* (1996)[2] developed a new theory of shattering of grains in interstellar shocks, and confirmed its accuracy by comparing results from it with data from laboratory experiments. Their theory is able to predict the shattered mass and the shattered fragment size distribution arising from a target-projectile collision at a specified velocity and for specific materials. Jones *et al.* then coupled this shattering theory with a detailed shock model in which gas–grain impacts are also included. Typical results are shown in Figure 6.1.[4]

The figure shows the computed grain size distribution of spherical grains, $n(a)$, as a function of radius, a, assuming that the grains are composed of graphite and initially had the MRN distribution, $n(a)da \propto a^{-3.5}da$, between

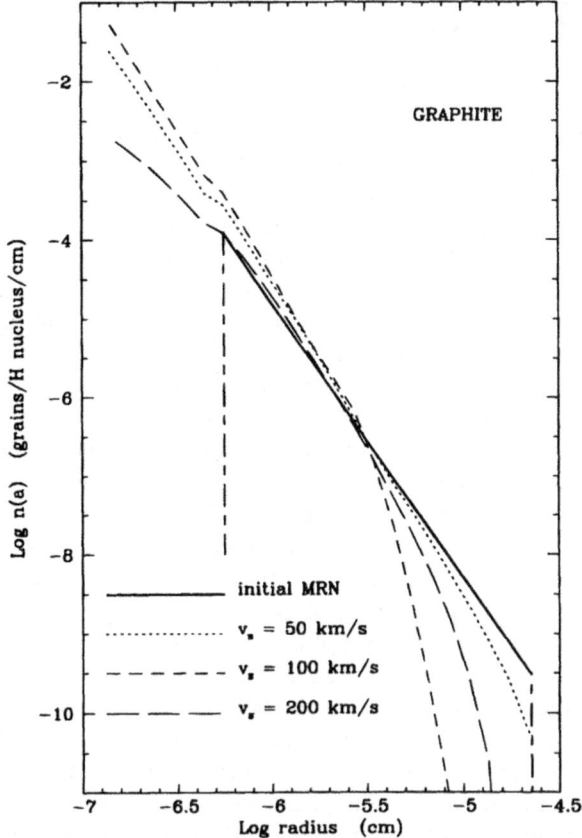

Figure 6.1   The change in the size distribution of a population of graphite grains. The initial population has an MRN size distribution (solid line), and the effects of shocks with three different speeds are shown. Reproduced with permission from Jones *et al.* (1996).[2]

the limits indicated by the vertical lines (see Chapter 3 for a discussion of the MRN distribution). Results are shown for three different shock velocities. The figure shows the effect of a single passage through a shock. Evidently, there is a significant loss of grains of larger sizes in all three cases. The highest velocity shock (at 200 km s^{-1}) is less effective in destroying grains than the slower shocks (50 and 100 km s^{-1}) because the fastest grains are less influenced by the magnetic field than slower grains. Serra Díaz-Cano & Jones (2008)[4] made a similar calculation for hydrogenated amorphous carbon grains, and found that the erosion of these grains is about three times greater than for graphite (an additional factor of 3 is found by Bocchio *et al.* 2014 [5]).

Of course, the initial size distribution may differ from the canonical MRN distribution adopted by Jones *et al.* Therefore, Jones *et al.* also consider the consequences of passage through a shock for a test particle of radius 100 nm, for various materials, see Figure 6.2. Apart from ice, which is rapidly eroded by sputtering for all shock velocities above a low threshold velocity of 1.8 km s^{-1}, the destruction of grains of other materials is generally dominated by shattering, rather than sputtering.

The Jones *et al.* study reaches a number of conclusions, including the following points. The threshold velocities for grain–grain shattering are on the order of 1 km s^{-1}, and the effects of shattering generally dominate those of vaporization. The shattered fragment size distribution is slightly less steep than the canonical (and assumed initial) MRN distribution. The maximum fragment size increases with increasing shock velocity; eventually, as the velocity increases the target grain is completely disrupted, and then for even larger shock velocities the maximum fragment size decreases. Shattering of large grains is generally very effective for shock velocities larger than 100 km s^{-1}, in passage through a single shock. The critical velocity for *catastrophic shattering* (*i.e.* the complete shattering of a large target grain by a small projectile grain) may be ~10^3 km s^{-1} for robust materials such as diamond and silicon carbide. In comparison to shattering, sputtering is effective mechanism for returning grain material to the gas phase, and both thermal and non-thermal sputtering may be important. Vaporization during shattering is relatively unimportant compared to the vaporization caused by sputtering.

6.2.2 Implications for Surface Chemistry of the Jones *et al.* Model of Grain Destruction in Shocks

Given the conclusions made by Jones *et al.*, we may infer some significant consequences for chemistry on the surfaces of interstellar dust grains:

(1) It is often suggested that grains from near-stellar sources (principally AGB stars) are relatively large; however, the size distribution of near-stellar dust is poorly known. As a result of shattering in interstellar shocks, a new size distribution of interstellar dust grains will emerge in which there are many more small grains than large, as indicated and supported by fitting of observational data of interstellar extinction (see Chapter 3). Therefore, the surface area for potential

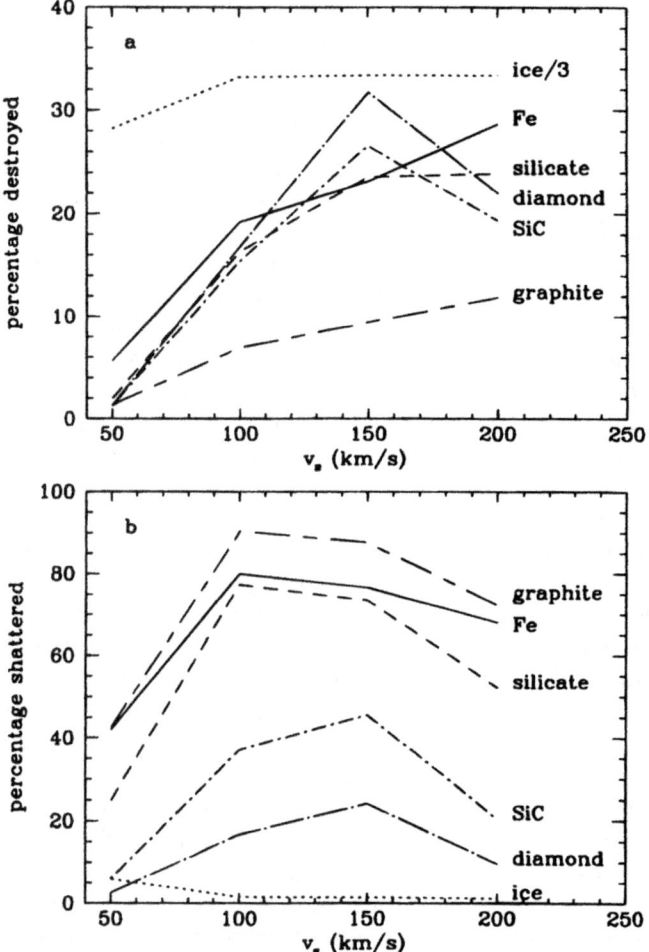

Figure 6.2   Shock destruction of a test grain of radius 100 nm, for various materials, as a function of shock speed. (a) total destruction (shattering, vaporization, and sputtering); (b) shattering only. Reproduced with permission from Jones *et al.* (1996).[2]

heterogeneous chemistry in the interstellar medium is much larger than would be the case for a population of large grains.

(2) Shattering affects different materials in different ways. The computed shattering thresholds for ice (in the very unlikely event that ice is present in the warm low-density interstellar environments considered), silicates and carbons correspond to shock speeds of just a few km s^{-1}, and catastrophic shattering occurs for shock speeds on the order of a hundred km s^{-1}. However, other potential grain materials such as silicon carbide or diamond are much more robust, suggesting that if grains of these robust materials are injected from near-stellar regions and exist in the interstellar medium, they may be present in relatively large

sizes. If so, they provide less surface area for chemistry. Therefore, the materials most likely to provide the largest areas for surface chemistry are carbons and silicates.

(3) Shattering of silicates and of carbons may lead to morphologically different systems. Silicates retain a 3D structure, and thus provide small grains of a range of sizes. Under shattering, hydrogenated amorphous carbons are likely to lead to a population of PAHs containing on the order of a hundred carbon atoms. While PAHs—like all interstellar molecules—are subject to chemical destruction, evidently they can be replaced since shocks continue to affect carbonaceous grains. Thus, a carbon surface involving sp^2 bonding is likely to be important in surface chemistry, on both large and small grains and (obviously) on PAHs.

(4) Grain surfaces are intermittently cleaned by sputtering in interstellar shocks. However, this process probably occurs on average approximately once in a hundred million years. The interval between cleaning events is very long, and therefore the precise nature of the surfaces of dust grains is determined by the local physical conditions, wherever the grains are located.

(5) Computed values of the critical velocities for the shattering threshold are 2.7 and 1.2 km s^{-1} for silicates and HAC, respectively, and critical velocities for catastrophic shattering are 175 and 75 km s^{-1}, according to Jones *et al.* (1996).[2] Evidently, silicates are more robust than carbons. Hence, in a model in which carbon is deposited as a thin layer on the surface of silicate grains, the passage through frequent relatively low-velocity shocks will remove carbon first, and, if the shock velocities are no larger than a few tens of km s^{-1}, leave much of the silicate untouched. Carbon deposition then resumes.

6.3 Response of Dust to Electromagnetic Radiation

6.3.1 Carbonaceous Dust

Conclusion (5) of the previous section, along with the detection in many appropriate laboratory experiments of the deposition of solid carbon under a wide range of conditions, led to the development of the model described in Section 3.3.1: the so-called *unified* model. The consequences of such a model have been discussed in Chapter 3 and the ability of the model to match in detail the observed ISECs for various lines of sight in the Milky Way and in relatively nearby galaxies has been established, assuming that the grains have evolved over a long period. Here, we discuss in particular the evolutionary aspects of the model and describe the implications for the nature of the surface of dust grains.

As described in Chapter 3, the time-dependence explicit in the unified model depends on two factors. Firstly, carbon is deposited from the gas phase on to the surfaces of bare silicate (or, indeed, other) grains and converted in the hydrogen-rich environment to a hydrogenated polymeric carbon. The timescale associated with this deposition is simply determined by the rate of collision of gas phase carbon atoms and ions with grain surfaces, *i.e.* by the gas number density, the carbon relative abundance in the gas, the gas temperature, and the total surface provided by dust grains per unit volume of space (implying assumptions concerning the grain size distribution and the dust-to-gas mass ratio). Depending on the precise details adopted, the deposition timescale is on the order of a million years in typical diffuse inter-stellar clouds, and may be shorter in higher density regions.

Secondly, this solid carbon is subjected to the local radiation field, of which the most important component is ultraviolet from hot stars. As many exper-iments have demonstrated, irradiation by ultraviolet tends to drive out the hydrogen and convert the polymeric (sp^3 bonding) carbon to graphitic (sp^2) form, a process known as photodarkening (see Chapter 3). The optical and chemical properties of the solid carbon evidently change during the photodark-ening. Using laboratory data and estimates of the intensity of the unshielded interstellar radiation field, one can estimate the photodarkening timescale in interstellar space. This timescale is on the order of 10^5 years, and will be some-what longer inside clouds where the radiation field is partially extinguished.

The lifetimes of diffuse clouds (through which starlight can penetrate fairly readily) are believed to be in the order of a few million years, up to ten million years, before they are disrupted by various dynamical effects in the interstel-lar medium. Therefore, the effects of carbon deposition and photodarkening should be apparent in dust-related signatures such as extinction. Computa-tions of the time-dependent structure of dust grains according to the uni-fied model, and of the consequent time-dependent extinction, have been made by Cecchi-Pestellini *et al.* (2010)[6] using the accurate procedures for the optics of grains of stratified layers with different compositions—in this case vacuum, silicate, carbon (sp^2), and carbon (sp^3)—as developed by Iatì *et al.* (2008).[7] For these computations, the grains were for convenience assumed to be spherical, but this is not a necessary assumption. We show in Figure 6.3 instantaneous normalized interstellar extinction curves computed by these methods at an evolutionary time of 2 million years, for several clouds of dif-ferent assumed hydrogen number density.

Figure 6.3 shows how the computed normalized ISEC for a cloud of spec-ified hydrogen atom number density (300 cm^{-3}) evolves in time. Note that the conventional normalization adopted is in the visible, so that the varia-tions are most apparent in the ultraviolet. At time zero, the grains are bare silicates, but as time elapses carbon is deposited as sp^3 (polymeric) and is converted to sp^2 (graphitic form). At an evolutionary age of about 0.5 million years, the computed curve is close to the average ISEC for the Milky Way, while further evolution causes the curve to lie well below the average ISEC. The carbon deposition rate is the same for all curves in Figure 6.3, because the same gas number density is adopted in these calculations. In Figure 6.4 we show the effect of changing the density (and the carbon deposition rate) on the instantaneous ISECs at an evolutionary timescale of 2 million years.

Evidently, the density of the cloud has an important effect on the computed normalized ISEC. The lower density cases generate ISECs that are similar to

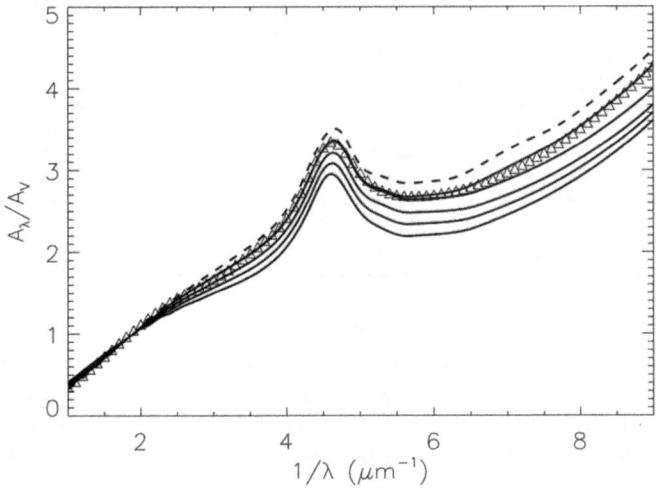

Figure 6.3   Evolution of the instantaneous normalized extinction curves for Milky
Way dust:gas ratio, with H atom gas number density of 300 cm^{-3}, gas
kinetic temperature of 100 K and a photodarkening time of 10^5 year.
Dashed line: initial conditions (*i.e.* bare silicates); solid lines: extinc-
tion curves at 0.5, 2, 5, 7 and 10 million years. Triangles: average ISEC.
Reproduced with permission from Cecchi-Pestellini *et al.* (2010).[6]

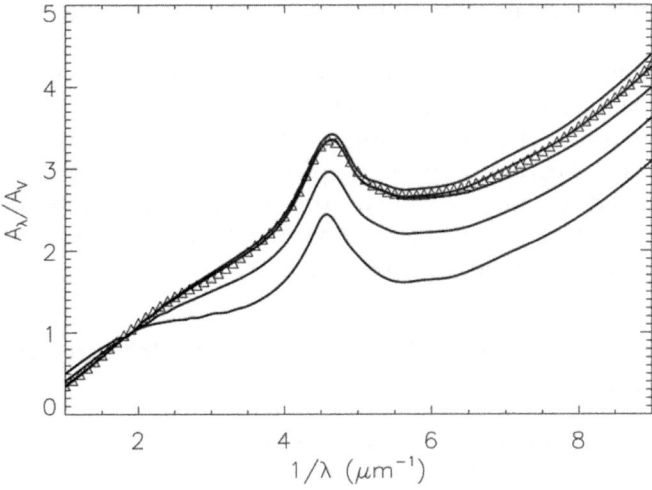

Figure 6.4   Instantaneous normalized extinction curves at 2 million years of evolution. The kinetic temperature is 100 K and the photodarkening time is
10^5 year. Solid lines, H atom number density of 30, 100, 300, 1000, and
3000 cm^{-3} (top to bottom). Triangles: average ISEC. Reproduced with
permission from Cecchi-Pestellini *et al.* (2010).[6]

Table 6.1   Time interval (in millions of years, My) in which the computed normal-ized extinction, A_λ/A_V, lies within ±5% of the observed average ISEC at $\lambda^{-1} = 6\ \mu m^{-1}$, for several values of the photodissociation timescale, τ_{pd}, and for several values of the interstellar cloud number density of H atoms in all forms, n_H. Data from Cecchi-Pestellini *et al.* (2010).[5]

τ_{pd} (years)	$n_H = 30\ cm^{-3}$ (My)	$n_H = 100\ cm^{-3}$ (My)	$n_H = 300\ cm^{-3}$ (My)	$n_H = 1000\ cm^{-3}$ (My)
10^4	0.50–34.3	0.18–22.3	0.09–4.7	0.07–1.3
10^5	0.45–36.3	0.16–15.2	0.08–4.0	0.07–0.9
10^6	0.77–46.2	0.33–9.2	0.16–2.3	0.07–0.5

the average ISEC for the Milky Way. At a number density of 100 cm^{-3} the computed ISEC closely approaches the average ISEC at an evolutionary time of 2 million years, rather later than the higher density case shown in Figure 6.3. The sensitivity to the photodarkening timescale has also been explored. This timescale is not of crucial importance for values of this timescale lying between 10^4 and 10^6 years.

Lines of sight in the interstellar medium may pass through regions in vari-ous stages of evolution. However, it is worth noting that the computed ISECs evolve through the observed average ISEC shape, and are in the vicinity of the average ISEC of the Milky Way galaxy for significant periods. Table 6.1 gives the time interval for which the computed ISEC lies within 5% of the observed Milky Way average. For clouds of 30 and 100 H atoms cm^{-3}, (*i.e.* diffuse clouds), this interval is a very substantial fraction of the expected life-time of a cloud (typically ~10 million years). For denser clouds, this interval is shorter.

Observational evidence in support of this evolutionary model is shown in Figure 6.5.[8] Here, data for a large number of Milky Way ISECs are interpreted *via* the unified model to give points on a plot of carbon mantle thickness *versus* fraction of carbon mantle in sp^2 form. Superimposed are lines cor-responding to different evolutionary tracks defined by the photodarkening rate. Almost all the data points are contained between tracks corresponding to a small range of photodarkening rate.

6.3.2   Consequences of the Unified Model for Surface Chemistry

Assuming that the unified model is a realistic description of interstellar dust, what—if any—are the implications for surface chemistry? The model makes clear predictions for the nature of the surface on assumed spherical grains:

(1) For ISECs for the Milky Way and similar environments, the outer sur-face of the stratified grain is predicted to be sp^3 carbon, *i.e.* a form of HAC.

(2) However, the total sp^2 + sp^3 outer layer is very thin, ~1 nm, *i.e.* just a few monolayers. Almost all of this (much more than 90%) is sp^2 carbon.

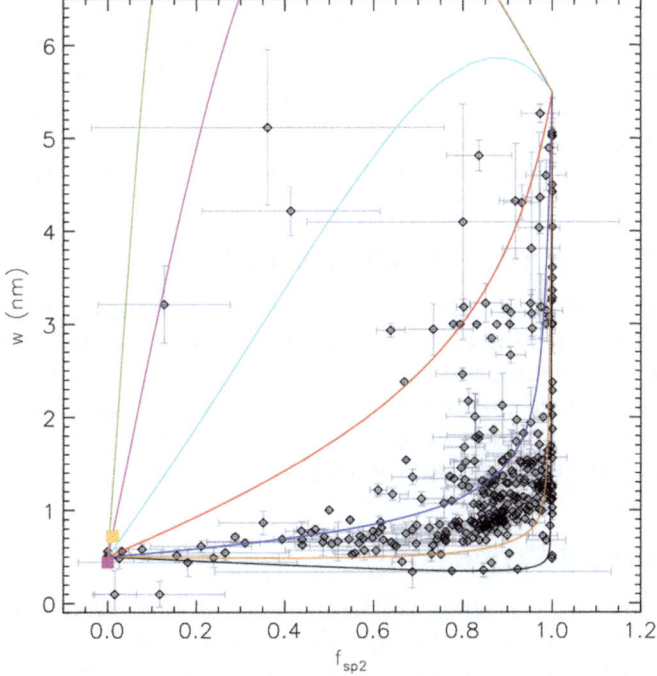

Figure 6.5 Comparison of observationally inferred data and modelled data for the carbon mantle thickness, w, and the fraction of sp^2 carbon in the mantle, f_{sp2}. The theoretical solid lines are computed for different values of a scaled photodarkening rate. The orange and blue lines encompassing the bulk of the observational data refer to scaled photodarkening rates of 3×10^{-9} and 1×10^{-9} cm^3 K$^{-1/2}$ per year, respectively. Reproduced with permission from Cecchi-Pestellini *et al.* (2014).[8]

Therefore, there is insufficient sp^3 carbon to form a monolayer. The implication of these results is that much of the surface is sp^2, with "islands" of sp^3 carbon on the surface.

(3) The underlying silicate grains are unlikely to be spherical. Therefore, at edges and corners of the grains, the inner layers of the carbon mantles, *e.g.* sp^2 carbon, will also be exposed, probably with reactive edges.

(4) The model predicts that an amount of carbon similar to that in mantles is present in the form of PAHs. These molecules must be of substantial size (~50 or more carbon atoms) to withstand destruction by the radiation field; they will provide suitable surfaces for chemistry.

(5) Bare silicate surfaces are predicted to be rare in the Milky Way galaxy. In a more active galaxy, such as the Small Magellanic Cloud, where shocks are more frequent, bare silicates may be more prevalent.

(6) The silicate cores are predicted to be porous, with a high degree of porosity. If so, they may be aggregations of smaller particles. The interstices between such small particles offer a degree of trapping within which a complex cavity chemistry may occur (see Chapter 12).

(7) The conclusions above are obtained for physical conditions that are similar to those in the Milky Way. Other situations may provide different conclusions. For example, if the local ultraviolet radiation field is much smaller than in the Milky Way, then photodarkening will be much slower, and the proportion of sp^3 carbon may be much larger. If so, the favoured surface for chemistry may be polymeric carbon.

6.4 Response of Dust to Fast Particles

Olivines irradiated by H^+, He^+ and Ar^+ ions with energies in the keV range produce some interesting effects.[9] Irradiation causes the surface to become enriched in magnesium; this is interpreted as magnesium atom migration driven by the injected positive ions. The iron valence state is also affected by the irradiation. Surface roughness increases with irradiation, and this enhances the surface chemical reactivity.

6.5 Thermal Processing

While silicates in the interstellar medium are almost entirely amorphous, collected cometary particles are crystalline (see Chapter 7). Silicates in protoplanetary disks also show crystallinity in their spectra. Thermal annealing of amorphous silicates to form crystalline structures requires temperatures in excess of 1000 K; however, comet-forming regions are cold, as indicated by the presence in comets of volatile molecules such as carbon monoxide. How can crystalline silicates, requiring a hot origin in which amorphous interstellar dust is converted to crystalline cometary dust, be compatible with high volatility molecules (see Li 2009[10])?

Observations by Ábrahám *et al.* (2009)[11] of an eruptive star with repetitive outbursts showed that features attributed to crystalline forsterite were formed in the outburst but were not present in the quiescent spectrum. The crystalline signature was attributed to thermal annealing of dust from amorphous to crystalline in the inner part of the circumstellar disk. Models of mixing of large grains in protoplanetary disks[12] showed how grains in the inner disk could be driven by radiation pressure and transported to the cooler comet-forming regions in the outer part of the disks. This scenario gives support to the view that thermal annealing of silicates from amorphous to crystalline may occur in specific regions of near-stellar space, where temperatures and densities may be abnormally high compared to the cool and diffuse interstellar medium.

6.6 Conclusions

As we have seen in Chapter 3, there are various models of interstellar dust, all capable of accounting for observations of interstellar extinction and other phenomena. The reason for the existence of multiple solutions is that the fitting procedures have abundant free parameters. Nevertheless, there is little disagreement among predictions concerning the nature of surfaces of the dust grains. For physical conditions that do not differ enormously from those of the diffuse matter in the Milky Way, surfaces of silicates, of polymeric and of graphitic carbons are expected, and PAHs are probably present. There is some disagreement as to the proportions of these surfaces that may be available. Interstellar silicates are inferred from extinction observations to be almost entirely amorphous. These, then, are the surfaces that will be emphasized in Chapter 8, in which reactions on bare grain surfaces will be considered.

There are, of course, other potential grain materials that have been discussed (see Chapters 3 and 4, and especially Chapter 7). Certainly, the evidence discussed there suggests that a wide variety of surfaces must exist in interstellar space. However, a consideration of elemental abundances suggests that the surfaces mentioned above are likely to be the most important for surface chemistry in the interstellar medium.

Further Reading

1. R. Asano, T. Takeuchi, H. Hirashita and T. Nozawa, *Mon. Not. R. Astron. Soc.*, 2014, **440**, 134.
2. A. P. Jones, A. G. G. M. Tielens and D. J. Hollenbach, *Astrophys. J.*, 1996, **469**, 740.
3. A. P. Jones, A. G. G. M. Tielens, D. J. Hollenbach and C. F. McKee, *Astrophys. J.*, 1994, **433**, 797.
4. L. Serra Díaz-Cano and A. P. Jones, *Astron. Astrophys.*, 2008, **492**, 127.
5. M. Bocchio, A. P. Jones and J. D. Slavin, *Astron. Astrophys.*, 2014, **570**, A32.
6. C. Cecchi-Pestellini, A. Cacciola, M. A. Iatì, R. Saija, F. Borghese, P. Denti, A. Giusto and D. A. Williams, *Mon. Not. R. Astron. Soc.*, 2010, **408**, 535.
7. M. A. Iatì, R. Saija, F. Borghese, P. Denti, C. Cecchi-Pestellini and D. A. Williams, *Mon. Not. R. Astron. Soc.*, 2008, **384**, 591.
8. C. Cecchi-Pestellini, S. Casu, G. Mulas and A. Zonca, *Astrophys. J.*, 2014, **785**, 41.
9. C. Davoisne, H. Leroux, *et al.*, *Astron. Astrophys.*, 2008, **482**, 541.
10. A. Li, *Nature*, 2009, **459**, 173.
11. P. Ábrahám, A. Juhász, *et al.*, *Nature*, 2009, **459**, 224.
12. D. Vincović, *Nature*, 2009, **459**, 227.

CHAPTER 7

Information from Solar System Dust

7.1 Interstellar and Local Dust in the Solar System

Gas and dust in the interstellar medium may enter the heliosphere, the region around the Sun that is filled with the solar wind plasma. This flux of interstellar gas coming towards the Sun, called interstellar wind, has the direction of its velocity vector almost parallel to the ecliptic plane (the apparent path of the Sun on the celestial sphere, as viewed from the Earth in orbit round the Sun). This flux offers a unique opportunity for interstellar matter to be studied *in situ* from spacecraft. The flux of interstellar matter provides the only dust component in the Solar System that was not previously incorporated in larger objects, such as asteroids and comets. However, the dust conveyed by the interstellar wind into the heliosphere is only a small fraction of the circumsolar dust.

After the Sun and its planets and their moons were formed, a large amount of solid material remained, ranging from ice and rock balls to tiny dust grains. These remaining solids formed in the protoplanetary disk from gravitationally collapsing interstellar material which was mixed and heated and, finally, condensed and agglomerated into planetesimals that accreted into planets. (We shall discuss planet formation in more detail in Chapter 10.) Some of these remaining solids fell into the Sun, while others were cleared from the Solar System by gravitational encounters with Jupiter. Some were destroyed or removed by stellar radiation pressure, grain–grain collisions, and sublimation of grain material.

Nevertheless, some dust remained, orbiting in the Kuiper Belt, just outside Neptune's orbit, and in the much more distant Oort Cloud that extends

The Chemistry of Cosmic Dust
By David A. Williams and Cesare Cecchi-Pestellini
© David A. Williams and Cesare Cecchi-Pestellini 2016
Published by the Royal Society of Chemistry, www.rsc.org

towards the neighbouring stellar system. The Oort Cloud has never been directly observed, but is believed to have at least 10^{12} icy objects located between 3000 and 100 000 AU in a spherical distribution around the Sun. There is also a region roughly overlapping the Kuiper Belt, the scattered disk, composed of objects having highly eccentric orbits with large orbital inclinations. Close encounters with giant planets cause smaller objects to be flung to the outer Solar System. On the other hand, these bodies may also be launched towards the Sun, where they become comets. Comets are characterized by a fuzzy appearance; they have km-sized nuclei, and may be accompanied by tails that indicate some physical activity due to volatile loss under solar heating. The scattered disk objects are thought to be the main repository of the short-period comets, such as comet Halley, with orbital periods shorter than 200 years. Short-period comets display a tendency for their aphelia (their furthest distance from the Sun) to coincide with the orbital radius of a gaseous planet, thus forming a "family". The Jupiter family of comets is the largest such family. By contrast, the long-period comets generally have orbits that are randomly oriented and not necessarily close to the ecliptic. They are thought to originate in the Oort Cloud.

Much closer to the Sun, in a wide strip between the orbits of Mars and Jupiter (the Main Belt), there are millions of bodies including the asteroids, which are planetesimals that failed to grow to planets (probably because of strong perturbations from nearby Jupiter). They appear star-like when observed with a telescope, and have a very limited activity. On some of those (*e.g.* 24 Themis; see Campins *et al.* 2010[1]) water ice has been found on the surface. These objects may produce transient, comet-like comae (*i.e.* nebulous envelopes) and tails. In addition, other orbital families exist with significant populations, as gravitational perturbations caused by planets and collisional effects drive a continuous migration that brings Main Belt asteroids closer to the Sun. These bodies thus cross the orbits of the inner planets, including several thousand large ones that intersect the Earth's orbit. If one of these intruders hit Earth it could devastate life on our planet (as noted by Alvarez *et al.* 1980[2]). The largest known asteroid is about one-quarter the size of the Moon.

The location of the asteroids between the terrestrial and gas giant planets provides a clue about the different modes of planet formation in these two regions. The distribution of the chemical composition of asteroid types as a function of semi-major axis provides support for the presence in the early stages of the solar nebula of a heliocentric temperature gradient. Such a gradient would have resulted in a "snow line"[3] somewhere interior to Jupiter's orbit, beyond which water ice would have condensed out of the nebula. Thus, beyond the snow line, the increased surface density of solids—together with the presence itself of an ice coating—resulted in larger planetesimals that grew faster than in the inner nebula, providing cores big enough to accrete gaseous envelopes, and become the gas giant planets. The location of the asteroids between the gas giants and terrestrial planets suggests that they represent the boundary between two distinct reservoirs of planet-forming material: relatively anhydrous bodies from within the snow line and more water-rich bodies from outside the snow line.

Within the planetary region, asteroids and comets generate the majority of dust and solid objects, in a wide range of masses and sizes. Only a tiny part comes from surfaces of planets and satellites. Such space debris are generally classified according to their masses as meteoroids – mass m larger than 10^{-5} g, zodiacal dust – $10^{-12} < m < 10^{-5}$ g, β-meteoroids – $10^{-15} < m < 10^{-12}$ g, and nano-dust – $m < 10^{-15}$ g. Dust grains less than a few hundred micrometres in size are generally termed interplanetary dust particles (IDPs). The upper size of meteoroids is placed at about 10 m, while larger bodies are classified as asteroids. A meteoroid that reaches the surface of the Earth without being completely vaporized is called a meteorite. Micrometeorites are micrometeoroids which have survived entry through the Earth's atmosphere. They differ from meteorites in being smaller, much more numerous, and different in composition. Whereas most meteorites probably originate from asteroids, the different makeup of micrometeorites suggests that they originate from comets. Since destruction or clearing mechanisms, such as expulsion by radiation pressure, inward Poynting–Robertson radiation drag, solar wind pressure, sublimation, mutual collision, and the dynamical effects of planets limit the lifetime of dust debris severely, these solids must be continuously replenished.

The distribution of parent bodies in the Solar System is related both to an outer cold disk associated with bodies beyond the Neptune orbit (*trans-Neptunian objects*), and a warmer disk located within the Jupiter orbit. The freshly produced solid particles have comparatively small relative velocities with respect to their parent bodies and move in similar prograde near-ecliptic orbits. Since these particles are subjected to a number of perturbing interactions, they form a region with overall cylindrical symmetry about an axis through the centre of the Sun, perpendicular to the ecliptic: the debris disk. Debris disks have been identified around a significant fraction of main-sequence stars by mid or far infrared excesses in the spectral energy distribution of the stars. They are orders of magnitude brighter than the solar debris disk.

IDPs detected between Jupiter and Saturn arise mainly from the cometary components. Cometary grains are thought be a combination of silicates, organics and some minor constituents. These grains are embedded in the ices of the nucleus and released as the ice sublimes when warmed by the Sun. Dust grains form part of the coma around the comet and, later, the dust tail. The very first information on cometary dust came from instruments onboard the Halley missions (Giotto, VeGa 1 and 2) in 1986. The particles were found to be weakly constructed aggregates of Mg–Fe glasses, iron–nickel sulfides and metal phases of variable compositions co-mixed with carbonaceous materials. The cometary dust samples collected from comet 81P/Wild 2 during the STARDUST mission in 2006 showed that although comet ice condensed in cold regions beyond Neptune, the rocks, probably the bulk of any comet's mass, were probably formed in hot inner regions of the solar nebula. This suggestion followed the discovery of roundish particles similar in chemical and textural details—but not in shape—to chondrules,[4] the dominant material in many primitive meteorites. Chondrules are rounded droplets of rocks that melted and then quickly cooled as they orbited the Sun. In addition, white irregular particles known as calcium aluminium inclusions (CAIs) were also detected.[5] CAIs are much rarer than chondrules and are distinguished by their unusual chemical and isotopic composition. They are considered to be among the first solids to condense in the Solar System, approximately two million years before chondrules, and are composed of exotic refractory minerals, such as spinel, $MgAl_2O_4$, and very iron-poor olivine. Such discoveries provide strong support for the idea that matter originally formed in the inner Solar System was somehow transported to the cold, outer comet-forming regions. Wild 2 dust grains include coarse silicate grains rich in magnesium and iron, of sizes around 10 μm. Some grains are dominated by a single mineral species of density around 3–4 g cm^{-3}, while others are porous, low density aggregates from a few nanometres to 100 μm, with densities as low as 1 g cm^{-3}, composed of silicates and sulfides.

The zodiacal meteoroid cloud is the inner part of the Solar System debris disk. It is a thick circumsolar disk of small debris particles produced by asteroid collisions and comets, extending from the Sun's corona to the orbit of Jupiter. By reflecting sunlight, the dust particles produce the zodiacal light – a faint, roughly triangular, whitish glow seen in the night sky. The zodiacal light appears to extend up from the vicinity of the Sun along the ecliptic. One of the longest-standing controversies debated in the interplanetary dust community concerns the relative contributions to the interplanetary dust cloud from asteroid collisions and cometary activity. It is believed that at the present time comets produce most of the dust at 1 AU and asteroidal collisions contribute little to the zodiacal cloud, although it is likely that this proportion may have changed with time. The main contributors appear to be dust particles produced by Jupiter family comets. The small solids are scattered by Jupiter before they are able to orbitally decouple from the planet, and drift down to 1 AU, becoming zodiacal dust.[6]

Smaller β-meteoroids experience a radiation pressure force comparable to or exceeding gravity and the particles are channelled outward in hyperbolic orbits. Nanodust particles become charged due to the charging currents of the solar wind and photoionization as a result of solar radiation. Because of their large ratio of surface charge to mass, such particles are deflected from their initial paths under the influence of the Lorentz force, with sufficiently large deflections preventing such dust grains from reaching the inner Solar System.

In the outer Solar System, the flux of interstellar dust impacts the cylindrically symmetric debris cloud perpendicularly to the rotation axis. Interstellar dust is most likely the major dust component beyond the asteroid belt. Close to the Sun, the radiation pressure force deflects small (<0.1 μm) interstellar grains that are thus mostly excluded from reaching the planetary system by electromagnetic interactions in the heliopause (the boundary of the heliosphere) and in the inner heliosphere. Larger grains, for which radiation pressure is relatively small, are attracted by the solar gravity and the flux forms a focusing cone in the interstellar downwind direction. This cone is similar to that observed for interstellar neutral gas.

7.2 "True" Interstellar Dust Particles in the Solar System

As the preceding Section has shown, the Solar System evidently contains a variety of kinds of dust. This variety includes "true" interstellar dust as well as the dust formed during and after the creation of the Solar System. In this book, we are concerned with interstellar dust. Our purpose in this Section is to describe the nature of this interstellar—*i.e.* presolar—dust found in the Solar System, as an indication of the nature of true interstellar dust on which interstellar surface chemistry may occur.

Dust grains in the local interstellar medium entering the heliosphere can be studied *in situ* by spacecraft.[7] These grains are characteristic of the current interstellar conditions in the vicinity of the Sun. Meteorites and micrometeorites contain another interstellar dust component; this component condensed in ancient stellar outflows (through processes described in Chapter 6), and became part of the presolar nebula from which the Solar System formed some 4.6 billion years ago. These presolar dust grains survived destruction in the early Solar System because they were trapped in asteroids and comets, eventually reaching the surface of the Earth, where they can be studied in the laboratory. They are recognized as presolar by the striking differences from their body carriers in their isotopic ratios. Other pristine dust components that are measured in the Solar System are cometary dust, and nanodust found in anhydrous IDPs (considered to have cometary origin) consisting of glass embedded with metal and sulfides (GEMS[8]). While cometary dust is a remnant of the protoplanetary disk out of which the Solar System formed, GEMS grains have been proposed to be isotopically and chemically homogenized interstellar amorphous silicate dust. Finally, closing the circle, the STARDUST mission, launched in 1999, gathered circumsolar dust *in situ* and returned it to Earth in 2006 for laboratory analysis. Among the collected particles are seven which have features consistent with an origin in the contemporary interstellar dust stream.[9]

7.2.1 Measurements *In Situ*

The first direct measurements of interstellar dust in the heliosphere were carried out by the Ulysses mission. So far, Ulysses has been the only space mission that left the ecliptic plane and passed over the poles of the Sun. The spacecraft was launched in 1990 and was successfully operated until mid-2009. The orbital plane of the spacecraft was almost perpendicular to the ecliptic, and thus almost perpendicular to the flow direction of the interstellar wind. Ulysses rotated about an axis pointing towards the Earth, so the dust detector during each spacecraft rotation pointed into the direction of the interstellar wind. Ulysses had on board a highly sensitive impact ionization dust detector which measured impacts of micrometre and submicrometre dust grains. Interstellar dust particles pass on hyperbolic orbits through the Solar System, shifted by about 180° with respect to solid debris in bound prograde orbits. They can therefore be identified by their impact direction and their impact speed, *i.e.* the motion of the Sun through the local interstellar cloud (~26 km s^{-1}). The measurement of the masses of interstellar dust grains by the dust detectors on board the Ulysses allowed the determination of the grain mass distribution of the local interstellar dust component. Later measurements with the Galileo dust detector in the ecliptic plane confirmed the Ulysses results.[10] Interstellar grains observed with the spacecraft detectors range from 10^{-15} g to more than 10^{-10} g.

If we compare the mass distribution of these interstellar impactors detected *in situ* with the dust mass distribution derived from astronomical observations we find only marginal overlapping between *in situ* and remote sensing measurements. Smaller grains are missing and the larger ones have approximately equal mass per unit logarithmic interval out to the largest sizes that could be detected (~1 µm). The lack of small grains is not surprising because they are tightly coupled to the magnetic field of the heliosphere that substantially deflects incoming particles with masses lower than 10^{-13} g. The tail of large grains is, however, totally unexpected and completely at odds with the conclusions drawn from studies of interstellar extinction described in Chapter 3. The overall dust mass detected by the spacecraft would exceed the available elemental budget in the interstellar medium; and the interstellar extinction curve resulting from such a collection of grains would be totally unlike that which is actually observed[11] (see Chapter 3). It is, however, unclear to what extent the measured dust size distribution is representative of the undisturbed size distribution in the surrounding interstellar cloud. The charge-to-mass ratios for micrometre sized grains imply Larmor radii larger than the size of the heliosphere. Medium-sized grains have instead gyroradii on the order of a few AU. Thus, large grains flow into the inner Solar System and are concentrated near the Sun by its gravity, while trajectories of intermediate size grains depend strongly on the solar wind magnetic field polarity, because these grains are charged. The main sources of charge for interstellar grains in the heliosphere are electron impacts and sticking, and photoionization by the far ultraviolet radiation field. Depending on the sign of the charge, dust grains of suitable size may be either focused near the Sun or de-focused resulting in low densities in the inner heliosphere. In both cases, model results are inconsistent with elemental depletion data for the local interstellar medium. A different explanation may involve the nature itself of dust grains in the local interstellar medium: are they representative of the galactic dust? We shall discuss this point in the last Section.

7.2.2 Presolar Grains

The first presolar grains to be identified, diamond and silicon carbide (SiC), were physically and chemically isolated from primitive meteorites by Ed Anders and co-workers at the University of Chicago.[12–14] These identifications were followed by those of graphite, some refractory oxides, mainly Al_2O_3, $MgAl_2O_4$, and $CaAl_{12}O_{19}$, silicon nitride, and finally silicates in the following years. In IDPs, presolar grains occur in even higher abundances than in meteorites. With the exception of diamonds (only 2–4 nm in size), all these presolar minerals have sizes larger than 0.1 μm and sometimes even larger than 1 μm (Figure 7.1).[15] These sizes allow the determination of the isotopic compositions of single grains, which is the only reliable criterion by which they can be identified. As a rule, their isotopic content strongly deviates from the Solar System composition, with signatures pointing to nucleosynthetic processes occurring in the winds of evolved stars or in the ejecta of stellar explosions, giving rise to the more evocative name "stardust" for presolar grains (see Chapter 5).

7.2.2.1 *Diamonds*

Although diamonds were the first to be identified and, so far, are the most abundant of the presolar grains in meteorites, doubts are cast as to what fraction of them is truly presolar. Such uncertainty is due primarily to their sizes; they are so small that they cannot be studied individually. Typical nanodiamonds have 1000–2000 carbon atoms and about 20 atoms of nitrogen. Consequently, trace elements on which the characterization depends, such as xenon, occur with very low abundances – only about one atom per million diamond grains. Moreover, the isotopic compositions of the structural element itself is, on average, characteristic of the Earth and the Solar System ($[^{12}C:^{13}C]$ ~89), while the $^{14}N:^{15}N$ ratio differs from the terrestrial one, but it is similar to recently determined solar nitrogen isotope ratio (~441[16]) and to that of Jupiter's atmosphere (~448[17]). Evidence that at least a portion of the meteoritic nanodiamonds could be presolar arose from the isotopically anomalous xenon gas that is enriched in both the heavy (neutron-rich) and light (proton-rich) isotopes relative to those of intermediate mass. The unusual isotopic composition of xenon is reminiscent of the p- (giving rise to light isotopes) and r-processes (forming heavy isotopes) of nucleosynthesis occurring during supernova explosions. Bulk diamond residues show also r-process tellurium. Diamond is a typical high-pressure phase that may form *via* shock transformation from graphitic material,[18] and through condensation by a CVD process.[19] The solar nitrogen and carbon isotopic compositions suggest that most diamonds were formed in the Solar System and that the rare xenon-bearing nanodiamonds are from supernovae.

7.2.2.2 *The Reduced Component: Silicon Carbide, Graphite, and Silicon Nitride*

Data on Si-, C-, O-, and N-isotopic ratios may be displayed in diagrams in which each grain occupies a unique position. The space in the diagrams is mostly empty, but filled regions enable identification of distinct populations of grains and isotopic trends. All SiC grains in primitive meteorites are of presolar origin, and they are the best characterized because the SiC crystal structure accepts a number of minor and trace elements. Based on the C-, N-,

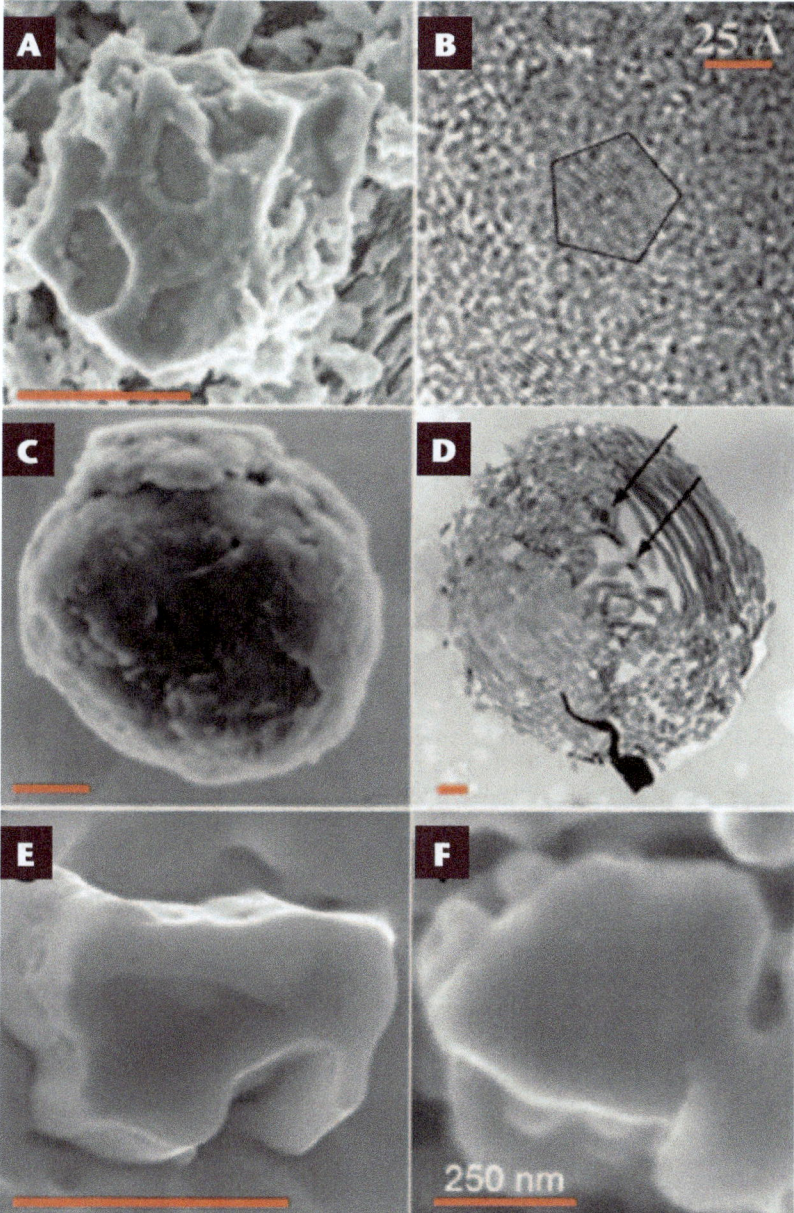

Figure 7.1 Images of presolar grain phases. Scale bars are 1 μm unless otherwise noted. (A) silicon carbide; (B) high resolution TEM image of presolar nanodiamonds (one of which is outlined) dispersed on amorphous carbon film; (C) graphite; (D) titanium carbide grains (arrows) within a graphite slice; (E) corundum; (F) silicate. The corundum and silicate grain surfaces are smooth from sputtering in the Nano SIMS ion probe. Reproduced with permission from Nguyen & Messenger (2011)[15] [photos courtesy of T. Bernatowicz (A), T. Daulton (B), S. Amari (C), K. Croat (D), L. Nittler (E), and A. Nguyen (F)].

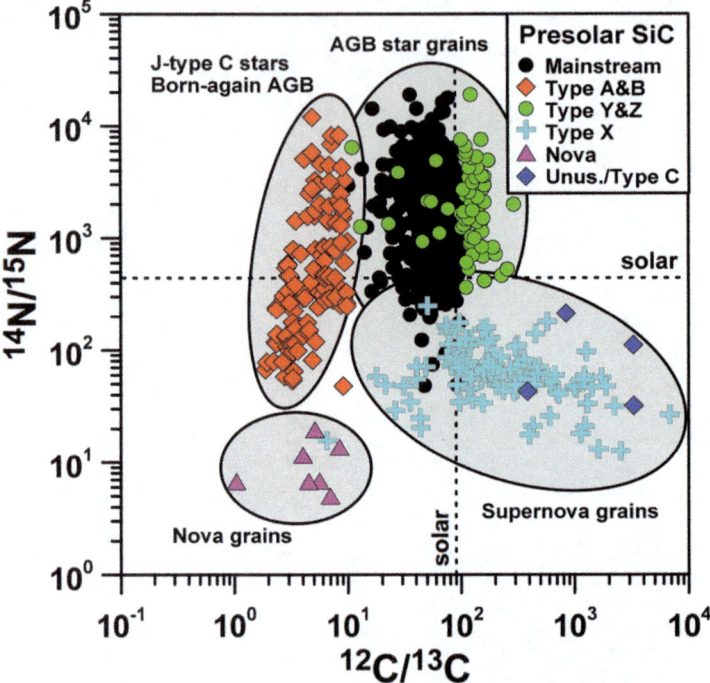

Figure 7.2   A plot of carbon and nitrogen isotopic ratios for presolar grains of SiC, graphite, and Si_3N_4. The ratios cover four orders of magnitude, but the grains plot into distinct groups, depending on their origin. Type X grains are from core collapse supernovae, and Types Y and Z are from low mass AGB stars. Some of the X grains exhibit an unusual Mo isotopic pattern. The origin of Type AB grains is unclear, Type C grains are from supernovae, and Type N grains are from novae. Reproduced with permission from Hoppe (2011).[20]

and Si-isotopic compositions, presolar SiC may be divided in several groups, with the mainstream (MS) grains being the largest (see Figure 7.2).[20] Grains with similar isotopic ratios are further selected by the isotopic distributions of minor and trace elements. Isotopic diagrams are then compared with evolutionary tracks of specific stellar sources. Among individual SiC grains, $^{12}C/^{13}C$ and $^{14}N/^{15}N$ span about four orders of magnitude while they vary a few percent in natural terrestrial materials.

MS grains carry radiogenic ^{26}Mg from the decay of radioactive ^{26}Al, with $^{26}Al:^{27}Al$ ratios typically in the range 10^{-4} to 10^{-3}, and show elements of intermediate and high mass exhibiting the signature of s-processes, suggesting an origin in low mass (1.5–3 solar masses) evolved stars of approximately solar metallicity. In these stars helium keeps building up until temperatures and pressures are high enough to initiate the triple-α reaction, in which helium is burned to carbon. The latter burning phase occurs episodically, causing a thermal pulse that dredges new ^{12}C and s-process heavy elements into the envelope. MS grains constitute about 90% of all SiC grain population. The grains identified in Figure 7.2 as Y and Z grains (about 8% of presolar SiC grains), residing mainly to the right of the MS region, are believed to come from low- and intermediate mass evolved stars with one third to half of the solar metallicity. Evolved stars of low metallicity are expected to dredge up more ^{12}C and more material that experienced neutron capture than do stars of solar metallicity. The envelopes of such stars are therefore expected to have higher $^{12}C:^{13}C$, $^{30}Si:^{28}Si$, $^{49}Ti:^{48}Ti$, and $^{50}Ti:^{48}Ti$ ratios than MS grains.

The type X SiC grains (see Figure 7.2 for the definition of X, Y, Z and other types) are characterized and defined by isotopic deficiencies in the heavy isotopes of both C and Si, by excess ^{15}N, lower than solar $^{29}Si:^{28}Si$ and $^{30}Si:^{28}Si$ ratios, and very high $^{26}Al:^{27}Al$ ratios of up to nearly one. They appear to be a small fraction, about 1.5% of the SiC grains. Type X grains have been subdivided into Types X0, X1, and X2 based on silicon isotopic compositions and several other isotopic systems, such as ^{49}V. The large ^{44}Ca excesses observed in a subset of X grains—coming from the decay of ^{44}Ti—suggest that this type of grain comes from core-collapse Type II supernovae. The isotope ^{44}Ti, which is produced only in supernovae, decays with a half-life of 60 years. Before becoming a supernova, a massive star has an onion-like structure consisting of concentric layers of decreasing temperature from the stellar core to the surface, containing the nuclear burning products. The element ^{44}Ti is produced in the inner region of the star that is composed mostly by ^{28}Si and ^{32}S. In the exterior layers ^{18}O and ^{26}Al are produced by the He burning of ^{14}N and H burning of Mg, respectively. Unlike the s-process elements Zr, Mo, and Ru, trace elements such as V, Cr, Fe, and Ni are primarily produced *via* explosive Si and O burning in Type Ia and II supernovae, and not in low mass stars that inherited these elements at birth. Supernova grains thus have large excesses of ^{28}Si and ^{18}O and large inferred $^{26}Al:^{27}Al$ ratios, determined from excesses in ^{26}Mg, the decay product of the short-lived radioisotope ^{26}Al. Even a strong enrichment in ^{16}O, the third most abundant isotope ejected by supernovae after hydrogen and helium, is a predominant signature of supernova grains.

Very rare grains with very low $^{12}C:^{13}C$ and $^{14}N:^{15}N$ ratios are SiC grains from novae. They also have enhanced ^{30}Si and high $^{26}Al:^{27}Al$ ratios. However, not all grains with low $^{12}C:^{13}C$ and $^{14}N:^{15}N$ are from novae since also some supernova grains carry this signature.

The origin of the type AB grains is still not well understood, and several alternatives have been suggested. These grains—approximately 4% of the collected SiC grains by number—have $^{12}C:^{13}C < 10$ and a large range of N-isotopic ratios with $^{14}N:^{15}N$ from 50 to more than 10 000. While the AB grains have Si-isotopic compositions similar to those of MS grains, they tend to have higher $^{26}Al:^{27}Al$ ratios of typically 10^{-3} to 10^{-2}. These grains may have or may not have been enriched in s-process isotopes. In the first case the proposed stellar sources are "born-again" evolved stars, while the latter might have originated in J-type C stars. About 10% of AB grains may have a supernova origin because of the lower than solar $^{33}S:^{32}S$ ratios.[21]

Silicon carbide grains of type C are even rarer (about 0.1% of all SiC grains) than SiC X grains. They have a large excess in ^{29}Si and ^{30}Si, some contain extinct ^{44}Ti, similar to SiC X grains. Their sulfur isotopic ratios show large ^{32}S excesses, with ^{32}S:^{33}S and ^{32}S:^{34}S ratios up to 20 times the solar value. Such grains probably originate in supernovae.

Like SiC grains, graphite grains are large enough to be individually analysed. Graphite grains are the most complex of presolar grains, as they contain populations of presolar grains trapped within presolar grains, such as refractory carbides and metal nuggets. They consist either of concentric layers of well-graphitized carbon (onions), and the most disordered "cauliflower" graphites, composed by aggregates of submicron grains, whose density is typically lower. Refractory inclusions, formed by direct condensation from the gas, probably served as a nucleation centre for graphite grains. The composition of refractory metal nuggets reflects the temperature of its last equilibration with the gas, giving precious information about the condensation temperature of the host graphite grain. The ^{12}C:^{13}C ratios of presolar graphite grains span a similar range to that observed in SiC, from ~2 to 7000, but with a different distribution. Most SiC grains have ^{12}C:^{13}C ratios lower than solar, while graphite grains typically have higher than solar ^{12}C:^{13}C ratios. Nitrogen isotope ratios tend towards terrestrial values (Figure 7.3), probably due to contamination with terrestrial nitrogen during the extraction from meteoritic material. Presolar graphite grains bear a similarity with SiC X grains in their C-, N-, and Si-isotopic signatures in addition to large excesses of ^{18}O, pointing to a supernova origin for these grains. Another key to revealing stellar sources of graphite grains is the degree of s-process element enrichment (Zr, Mo, and Ru relative to Ti) within carbides. Current estimates predict that 60% of graphite grains, in particular the low density grains, come from core-collapse supernovae.[22] Grains with lower ^{12}C:^{13}C ratios are probably formed in low mass AGB stars (~30%), while the remainder in other types of carbon-rich stars.[23]

Among the reduced components of presolar grains, Si_3N_4 can also be found as a separate phase. The rare silicon nitride grains show isotopic signatures similar to SiC X and supernova graphite grains, and probably derive from supernovae.

7.2.2.3   Oxides and Silicates

The main components of meteorites and IDPs are oxide and silicate grains that formed in the Solar System. Thus, the identification of O-rich presolar grains necessitates the isotopic analysis of a large number of grains. Presolar oxide grains, principally aluminium oxide and hibonite ($CaAl_{12}O_{19}$), have been known since the mid-1990s, but with sophisticated techniques that can identify submicron blocks, the list has expanded to include corundum, spinel ($MgAl_2O_4$), chromite (($Fe,Mg)Cr_2O_4$), and TiO_2. Corundum, the thermodynamically stable phase of Al_2O_3, is predicted to be the most abundant refractory dust species condensed in envelopes around oxygen-rich evolved stars. An important finding in the last decade was the discovery of presolar

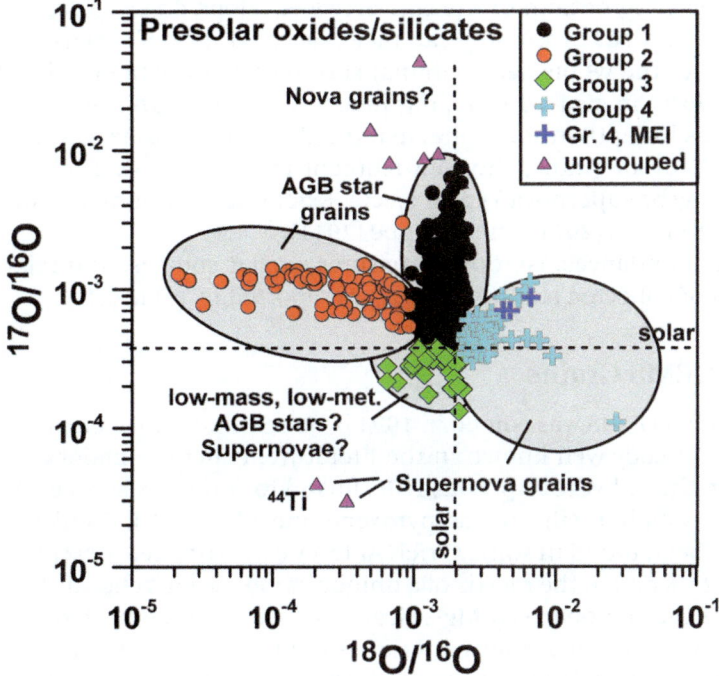

Figure 7.3 A plot of oxygen isotopic ratios for different groups of presolar oxide and silicate grains. The data fall into distinct groups, and the origins of the groups is indicated. Reproduced with permission from Hoppe (2011).[20]

silicates in IDPs[24] and primitive meteorites.[25] These grains remained unrecognized for a long time until technical advances allowed oxygen isotopic mapping with a spatial resolution of ~100 nm. Silicates turned out to be the second most abundant presolar mineral after diamonds.

Presolar silicates and oxides share the same oxygen isotopic range. Similarly to the reduced component, presolar oxygen-rich dust is divided into distinct groups[20,22] (Figure 7.3). Most abundant are the Group 1 grains containing 70% of all grains. These grains present significant enrichments in ^{17}O, and $^{18}O:^{16}O$ ratios near solar or slightly lower than solar ratios. The same isotopic trends are observed spectroscopically in red giant stars. The most likely stellar sources are evolved stars of 1–2 solar masses and approximately solar metallicity. Since the predicted maximum $^{17}O:^{16}O$ ratio that can be reached in low and intermediate mass evolved and other stars of solar metallicity is about 0.005, grains having $^{17}O:^{16}O$ ratios of much greater than this limit are probably of nova origin. Group 2 contains about 15% of all grains. The main signatures are strong depletions in ^{18}O and moderate enrichment in ^{17}O. The origin of Group 2 grains is unclear. The most popular explanation of their composition is some kind of extra-mixing mechanism in evolved stars, called "cool bottom processing", in which matter is transferred from the non-burning bottom of the convective envelope down to regions where some nuclear CNO processing can take place.[26] Grains of Group 3 (these are about 5% of the whole set of grains) show slight enrichments in ^{16}O. They probably form in evolved stars, but in environments with initial oxygen isotopic ratios lower than solar, and, marginally, in supernovae. Group 4 grains, about 10% of all grains, show enrichment in ^{17}O and ^{18}O. Their formation sources may be supernova Type II ejecta. Extensive reviews on presolar grains are given in Davis (2011)[22] and Hoppe (2011).[20]

Types, abundances, isotopic signatures, stellar sources, and relative contribution of collected stardust grains are reported in Table 7.1.

7.2.3   GEMS Grains

The acronym GEMS was coined in 1994 by John Bradley for a class of particles that were already well known in the literature, being the major amorphous silicate in the anhydrous porous IDPs linked to cometary sources. IDPs also contain crystalline silicates as pyroxene and olivine. GEMS-like materials have also been found in some C-rich Antarctic micrometeorites, (*e.g.* Dobrica *et al.* 2012[27]), and in the matrix of a unique carbonaceous chondrites.[28] These grains are submicron glassy Mg-silicate particles mantled with or embedded in amorphous carbonaceous material, containing in their interiors nanometre Fe sulfide crystals and kamacite (FeNi metal) inclusions, some or all of which may be super-paramagnetic. GEMS may contain in its interior at least a few percent carbon.[29] It is not clear if such internal carbon and the abundant carbon that occurs outside GEMS particles have a common origin.

The list of determined properties of GEMS grains is almost exactly what one would write down as desirable properties for interstellar dust, as estimated by remote sensing measurements. Their size range is needed to cause interstellar extinction at visual wavelengths. They show infrared spectra with bands at 10 and 18 μm, as observed—in absorption and emission—along many interstellar lines of sight.[30] Some GEMS exhibit a broad, featureless silicon–oxygen stretch band similar to those observed in interstellar molecular clouds and young stellar objects. Moreover, the super-paramagnetic inclusions are very close to what had been predicted by Mathis (1986)[31] as necessary to produce the partial alignment of large dust grains by the interstellar magnetic field. Finally, Bradley *et al.* (2005)[32] derived UV spectra of presolar grains embedded within IDPs, found an absorption band centred at 5.7 eV, that matches the prominent 217.5 nm astronomical feature, although with a broader bandwidth. The species implicated as possible carriers are carbonyl-containing (C=O) groups in carbonaceous materials and hydroxylated amorphous silicates, *i.e.* GEMS materials. In GEMS particles the strength of the absorption feature correlates with the hydroxyl content (OH$^-$). Laboratory spectra of hydroxylated amorphous Mg_2SiO_4 have been shown to exhibit an absorption feature at 217.5 nm that matches both the central wavelength and the bandwidth of the interstellar bump.[33] This absorption feature may be due to an electronic transition of OH$^-$ ions in sites of low coordination.

Table 7.1 STARDUST INVENTORY. Data are from Davis (2011)[22] and Hoppe (2011)[20].[a]

Mineral	Type	Size (μm)	Signature	Stellar source	Rel. contrib. (%)
Diamond		0.002	Solar $^{12}C:^{13}C$, $^{14}N:^{15}N$; Xe-HL	SN II; solar system?	
SiC		0.3–50			
	Main-stream		Low $^{12}C:^{13}C$; high $^{14}N:^{15}N$; s-elements	AGB (1.5–3 $M_\odot$)	90
	AB		Very low $^{12}C:^{13}C$; high $^{14}N:^{15}N$	J-stars; born-again AGB	<5
	C		High $^{12}C:^{13}C$; very high $\delta^{29,30}Si$; extinct ^{26}Al, ^{44}Ti	SN II	0.1
	X0		Low $^{14}N:^{15}N$; negative $\delta^{29,30}Si$; high $^{29}Si:^{30}Si$; extinct ^{26}Al, ^{44}Ti, ^{49}V	SN II	0.2
	X1		Low $^{14}N:^{15}N$; negative $\delta^{29,30}Si$; midrange $^{29}Si:^{30}Si$; extinct ^{26}Al, ^{44}Ti, ^{49}V	SN II	1
	X2		Low $^{14}N:^{15}N$; negative $\delta^{29,30}Si$; low $^{29}Si:^{30}Si$	SN II	0.3
	Y		High $^{12}C:^{13}C$; high $^{14}N:^{15}N$	~1/2 solar metallicity AGB	Few
	Z		Low $^{12}C:^{13}C$; high $^{14}N:^{15}N$; (mostly) negative $\delta^{29}Si$; high $\delta^{30}Si$	~1/4 solar metallicity AGB	Few
	Nova		Low $^{12}C:^{13}C$; high $\delta^{30}Si$; Ne-E(L)	Novae	0.1
Graphite		1–20			
			Low $^{14}N:^{15}N$, high $^{18}O:^{16}O$; extinct ^{26}Al, ^{41}Ca, ^{44}Ti, ^{49}V	SN II	60
			s-Elements	AGB (1.5–3 $M_\odot$)	30
			Low $^{12}C:^{13}C$	J-stars; born-again AGB	<10
			Low $^{12}C:^{13}C$; high $\delta^{30}Si$; Ne-E(L)	Novae	<10
Si_3N_4		≤1	Low $^{14}N:^{15}N$; high ^{28}Si; low $\delta^{29,30}Si$; extinct ^{26}Al	SN II	
Oxides		0.1–2	High ^{17}O, depleted or solar ^{18}O	AGB (1–2.2 $M_\odot$)	

(continued)

Table 7.1 (*continued*)

Mineral	Type	Size (μm)	Signature	Stellar source	Rel. contrib. (%)
Silicates		≤1			
	1		High $^{17}O:^{16}O$; low or normal $^{18}O:^{16}O$;	AGB (1–2.2 $M_\odot$)	70
	2		High $^{17}O:^{16}O$; very low $^{18}O:^{16}O$	AGB (<1.8 $M_\odot$); CBP	15
	3		Low $^{17}O:^{16}O$, $^{18}O:^{16}O$	AGB (low mass and metallicity); SN II	5
	4		Low $^{17}O:^{16}O$, $^{18}O:^{16}O$; extinct ^{44}Ti	SN II	10
	N		Very high $^{17}O:^{16}O$; low $^{18}O:^{16}O$	Novae	<1

[a]Xe-HL: heavy and light isotopes of Xe; δ values: deviations from normal values in per mill; Ne-E(L): neon highly enriched in ^{22}Ne coming from the decay of ^{22}Na; CBP: cool-bottom processing.

Chemical and isotopic analyses on GEMS grains, as well as laboratory experiments that synthesize amorphous silicates, are being conducted to establish how and where GEMS grains formed, with the crucial point being to determine whether they have a presolar or a solar nebula origin. Unfortunately, the submicron size of GEMS grains is near the limit of isotopic measurements, which unavoidably fuels a robust debate. In the presolar origin scenario envisaged by Bradley and collaborators GEMS grains began as free-flying, crystalline mineral grains that were exposed to ionizing radiation in the interstellar medium. During irradiation grains were chemically and isotopically homogenized by prolonged sputtering and redeposition, before being incorporated into the cloud of dust and gas from which our Solar System formed. However, such an interstellar origin is disputed by Keller and Messenger, who argue that GEMS grains display enormous chemical variability which is inconsistent with the idea of chemical homogenization. GEMS grains show order of magnitude variations in Mg, Fe, Ca, and S abundances. Their average element abundances differ significantly and systematically from solar abundances and from those inferred for interstellar silicates based on element depletion patterns, containing too little Mg, Fe, and Ca, and too much S. Moreover, GEMS grains have complementary compositions to the crystalline components in IDPs suggesting that they formed from the same reservoir in the inner Solar System before being transported out to the comet-forming region.[34] While a few GEMS grains have highly anomalous O isotopic compositions consistent with a condensation origin in evolved O-rich stellar envelopes and supernovae being thus surviving presolar grains, the bulk of these materials have solar isotopic compositions indicating the solar nebula as their designated provenance. However, normal, Solar System-like oxygen isotopic ratios do not preclude an origin in the interstellar medium because the range of isotopic composition currently measured in the interstellar medium overlap Solar System values. Keller and Messenger suggest that the observed heterogeneities point to non-equilibrium conditions in the early Solar System, and may reflect the changing composition of the gas from which they condensed. Primordial interstellar amorphous silicates may have been destroyed by nebular and parent body processes or they may have eluded detection because of their small sizes. In young stellar objects, silicates show a progressive change in their composition from being predominantly amorphous into crystalline grains, reflecting transformation and transport processes from the hot inner part of the disk to cooler regions. The large range in silicate abundances found in anhydrous IDPs may be the chemical fossil of such evolution.

GEMS grains have chemical and isotopic properties that reflect both presolar and Solar System origins, so that their nature remains highly controversial. Such particles still hide valuable information about ancient stardust, processes in the solar nebula, and the very beginnings of our Solar System.

7.2.4   The STARDUST Mission

The STARDUST probe was launched in the early 1999 to collect the first samples of a comet. The spacecraft flew by the asteroid 5535 Annefrank in November 2002, arrived close to Comet Wild 2 in January 2004, flying through the comet's coma where it collected particles from that object. Two years later STARDUST ejected its sample-carrying capsule to Earth. Subsequently, the probe had a rendezvous with the comet Tempel 1, already object of investigation during the NASA Deep Impact mission in 2005.

The primary objective of the mission was to capture cometary samples and possibly interstellar dust. The main challenges to accomplishing this successfully was how to slow down the particles from their high relative velocity with minimal heating or other effects that would cause physical and chemical alterations. The adopted dust collector media consist of blocks of 1 and 3 cm thick underdense, microporous silica aerogel mounted in modular aluminium cells. The aerogel has a porous, sponge-like structure in which 99.8% of the volume is empty space, thus greatly reducing the stress of impact on the particles.

Thousands of solid particles were collected during the mission. To catch interstellar dust grains, the collector was exposed to the interstellar stream for 195 days in two periods in 2000 and 2002. The comet particles returned by the STARDUST mission contained some stardust grains, but the majority of the solids were Solar System materials that appear to have formed over a very broad range of solar distances and perhaps intervals of time. The comet contained a significant amount of solid material in structures much larger than interstellar grains, even a million times more massive than typical stardust grains. Moreover, they contained abundant crystalline minerals and in most cases it was clear that they did not form by the mild heating of interstellar dust in the vicinity of a star that they may have been orbiting. Such samples appeared to have been heated during their formation to severe temperatures, temperatures high enough to melt or vaporize them. As a consequence comets—at least comet Wild 2—are a peculiar mix of both fire and ice. This implies that while comets contain ices that formed at the edge of the Solar System, the rocky materials that constitute the cores of comets actually formed close to the Sun, and were then distributed to all regions of the Solar System.

Seven particle tracks, among thousands recorded on the STARDUST dust collectors, have been identified as impacts of probable interstellar origin.[9] Such research has been supported by a massively distributed, volunteer-based search of optical micrographs of the aerogel collectors, using a virtual microscope programme called the Stardust@home programme. Citizen scientists—who call themselves "Dusters"—scanned more than a million images to look for dust tracks, eventually finding two. The number of interstellar dust particles that the STARDUST mission collected is unexpectedly small, based on observations of interstellar dust carried out by the earlier Galileo and Ulysses probes. The seven identified particles were much more diverse in terms of chemical composition and structure than expected. The smaller particles (four out of seven) differed greatly from the larger ones and appear to have varying histories. Two of the larger particles (named Orion and Hylabrook) have been described as having a low-density fluffy structure, similar to a snowflake, made up of an agglomeration of other particles. Fluffiness may explain the discrepancy with Ulysses measurements, as these particles experience more pressure from sunlight than compact grains. The third one is poorly characterized, and it is possible that little of the original projectile was retained in the track. The three biggest particles weigh about three pg, and have sizes larger than 1 μm, in agreement with the Ulysses findings. The other four are roughly a thousand times smaller. In total, the amount of interstellar dust STARDUST captured was less than a millionth the amount of cometary material it collected.

Three of the smallest particles showed elemental compositions within the range characteristic of GEMS materials, and two of them had Solar System oxygen isotopic ratios. Thus, it is unlikely that they are presolar. They contain sulfur compounds. This is significant, as sulfur has generally not been supposed to be locked in interstellar dust grains. The last small particle had lower Si and higher O abundances than GEMS materials.

The Orion and Hylabrook particles were distinct from GEMS grains in size, composition, and degree of crystallinity. Orion has a forsteritic olivine core, spinel-like amorphous oxides, and iron-bearing phases with minor elements calcium, chromium, manganese, and nickel. Hylabrook is a magnesium-, iron-, and silicon-rich particle with a partially amorphized, forsteritic olivine core. The presence of crystalline silicates—olivine—suggests that these particles came from disks around other stars and were subsequently processed in the interstellar medium. Unfortunately, all the chemical phases of which these particles are composed are common to IDPs. Due to this ambiguity, the strongest constraints on their interstellar origin came from the geometry of the STARDUST collection.

If these seven particles are truly interstellar they differ from any model of interstellar dust inferred by remote sensing measurements.

7.3 The Transition from Interstellar Dust to Planetesimals

What we see today in the form of IDPs, zodiacal dust, and meteoroids originated in larger bodies that formed at the dawn of our Solar System. These parent bodies suffered significant variations during the aggregation stage of the original interstellar building blocks, and further physical and chemical changes must have occurred within the parent bodies, and following fragmentation events. The formation of planetesimals and planets was accompanied in many cases by high temperatures and violent conditions, and most interstellar dust particles were destroyed.

The story of the Solar System begins in the relic of the stellar accretion disk which, for a solar-type star, is 99% gaseous hydrogen and helium by mass. Terrestrial planets, and probably the cores of giant planets, exploit the remaining 1% of mass, *i.e.* the dust grains, whose collisional accretion can eventually create planets like Earth. When a grain collides with another particle, its kinetic energy increases due to binding energy and because it is falling into a potential well. If some energy is dissipated, the total energy decreases. After the collision the energy is further reduced by the binding energy as the grain leaves the potential well. The two particles stick together, then forming a bigger particle (eventually a planetesimal), if the final energy is negative. Low velocity encounters, high binding energy, and efficient dissipative mechanisms facilitate particle growth. Binding energies are generally electrostatic in nature, typically van der Waals and Coulomb forces. Ice coatings also help sticking. Gravitational interaction becomes important when particles are very massive. The particle growth gets increasingly difficult with increasing sizes because binding energies decline while relative velocities increase. As planetesimals grow bigger, they tend to move faster in the disk, with their peak velocity—100 metres per second—for a size of about one metre. Once such objects are formed, they quickly drift towards the star and evaporate. Moreover, when they reach such high velocities any collision will become destructive. Both effects act to stop the growth, a problem which is usually called the "metre-size barrier" in the literature. However, if in some way they manage to cross such a barrier, the velocities go down again and the growth restarts with the planetesimal sweeping-up smaller fragments. We discuss the topic of planet formation elsewhere (see Chapter 10). The relevant points in this context are the transformations occurring to the body during and after the accretion process.

The evolution of small planetary bodies such as asteroids begins with the accretion from the dusty and unconsolidated material of the protoplanetary disk, *i.e.* the agglomeration of fine dust grains like those found in IDPs into aggregates of increasing size. Such a coagulation process is not subject to any pressure and forms loose structures, with very high porosity (~90% voids). These materials seem to be preserved in some comets.[35] Collisions of aggregates during the growth process of planetesimals provides initial compaction, decreasing the porosity of a factor of two or more. The first planetesimals to accrete out of the solar nebula were undifferentiated and presumably composed of silicates (pyroxenes and olivine) and iron–nickel–sulfur metals. These bodies were likely to have accreted cold, *i.e.* in thermal equilibrium with the nebular environment.[36] It is nevertheless possible that the small cm-size aggregates arising from the accumulation of ice-coated interstellar grains that formed the building blocks of these planetesimals experienced significant thermal alteration from the release of chemical energy.[37] Internal chemical energy is released when such dust aggregates are warmed enough to fuel rapid exothermic chemical reactions, leading to a thermal runaway. Such reactions may involve free radicals (expected to be abundant in ices), whose mobility driven by the warm-up may liberate sufficient energy to explode[38] (see Chapter 9). Considerable chemistry occurred both in the gas phase and on ice and mineral grain surfaces during the accretion phase, which lasted a few millions of years. Many of the products of these reactions were thus incorporated into growing asteroids and planetesimals. Heating by radioactive decay combined with the self-gravity led to the substantial compaction at sub-solidus temperatures and in case of icy bodies to crystallization of amorphous ice and phase transitions of volatile species. The decay of short-lived radioactive isotopes such as ^{26}Al and perhaps ^{60}Fe (although its abundance was probably too low[39]) was the most significant source of heat in the early Solar System. Since the half-lives of these isotopes are 0.74 and 1.5 Myr respectively, planetesimals must have accreted within a few half-lives of the parent isotope or, equivalently, within a few million years after the formation of CAIs. The formation of CAIs is an effective "time zero" for the history of the Solar System and is often assumed to be the time at which ^{26}Al and other short-lived radioisotopes were introduced. These very small objects, formed through some unknown flash heating processes in the solar nebula, were rapidly cooled and predate accumulation of planetesimals. Under such radiogenic heating the iron and silicate phases start to melt upon reaching the respective solidus temperature (1200 and 1400 K, respectively). In some cases, this heating may produce the differentiation of the planetesimals into an iron core and a silicate mantle. In case of icy bodies, melting of the ice started before iron and silicates giving rise to a third component that was possibly separated. Depending on the volatile content, iron could be carried to the surface by the increased buoyancy of partial melts, and ejected from the body instead of sinking gravitationally to the centre of the body. The same could also occur for silicates. Whether a planetesimal remains undifferentiated or melts and differentiates depends mainly on its chemical composition, the accretion time, the available amount of short-lived radioactive elements, and the size of the body. The typical cooling history shows that the planetesimals are first heated by radioactive decay and then cool down over an extended period. If the cooling time scale—roughly proportional to the characteristic time scale for thermal conduction—is lower than the heating time scale it would be unlikely that the object can sustain melting temperatures, and the body does not differentiate. Numerical models suggest that bodies as small as 15 km in diameter could reach the silicate solidus temperature and maintain this temperature for hundreds of thousands of years. The earlier the onset of accretion and the larger the body, the higher the interior temperatures and the more likely differentiation. The accretion time is known to be proportional to the heliocentric distance. This implies that in the inner Solar System planetesimals could incorporate higher abundances of live radionuclides than bodies formed at greater heliocentric distances. The latter planetesimals presumably accreted too slowly to reach large temperatures. The compositional gradient across the asteroid belt could be related to heliocentric zoning by ^{26}Al heating.[40]

Planetesimals that reached very high temperatures would have completely melted and could have differentiated into an "onion layer" structure with an Fe–Ni–S core, an olivine-rich mantle and a plagioclase/pyroxene-rich basaltic crust. The term plagioclase emphasizes a solid solution series more than a particular mineral with a specific chemical composition. The series ranges from $NaAlSi_3O_8$ to $CaAl_2Si_2O_8$. Plagioclase is a major constituent mineral in the Earth's crust, and one of the major components in lunar rocks. The final solidification of the silicate crust and mantle would follow a crystallization sequence with olivine condensing first, followed by low-calcium pyroxene, then high-calcium pyroxene and finally plagioclase. Fully differentiated bodies were the most altered planetesimals, other bodies experienced less extreme thermal histories.

The wide range in composition and structure of meteorites implies that these objects derived from more than 100 parent bodies, most of which were asteroids. The majority of meteorites are undifferentiated, showing the coexistence of phases formed at low temperatures and nebular phases formed at high temperatures, such as CAIs and chondrules. Only a minority of meteorites (and asteroids) are thought to have derived from planetesimals whose interior temperatures were high enough to facilitate the process of metal-silicate differentiation. Collisional disruption of a fully differentiated body would produce a suite of fragments representing its various compositional layers, although re-accretion may have followed disruption. Some of these meteorites are composed of iron, others mainly of silicates, and all of them are thought to originate from Vesta, the second major resident in the asteroid belt. The most massive body is Ceres, the very first minor planet to be discovered (January 1, 1801 by Giuseppe Piazzi). Vesta and Ceres orbit in a similar region of the Solar System, between 2.36 and 2.77 AU, and are of similar size, about 500 and 1000 km in diameter, respectively. Yet they are very different in nature. While Vesta is a rocky differentiated body with an iron-rich core and a silicate mantle topped by a basaltic crust, Ceres contains large quantities of ice, and might even harbour a subsurface ocean of liquid water. The profound differences in geology between these two protoplanets that formed and evolved so close to each other may result from slightly different formation times, when radioactive material was more abundant in the Solar System.

The two major classes of undifferentiated meteorites are ordinary chondrites—the most abundant class of meteorites falling on Earth—and carbonaceous chondrites. Carbonaceous chondrites show large amounts of water and carbon and contain a fine-grained matrix rich in volatile elements and fewer chondrules. Such a matrix contains hydrous minerals—primarily Fe and Mg-rich serpentines and tochilinite—resulting from the ancient presence of liquid water, which originated from a mild thermal heating of the parent body initiating fluid flow. The thermal and structural evolution of a planetesimal can vary substantially depending on the concentration of water. The presence of water ice moderates the temperature increase during heating because substantial amounts of heat may be absorbed during melting. In addition, the warm melt water redistributes heat within the body by hydrothermal convection and by porous flow.

The carbonaceous chondrites are considered to be the most primitive objects. The carbon in carbonaceous chondrites is present in several forms, such as silicon carbide, graphite, and nanodiamonds condensed in stellar outflows and supernovae explosions (see Section 3.2). Carbonate minerals were formed in the early Solar System, as aqueous alteration took place on asteroids. Organic matter is quantitatively the most important carbon-bearing component, and appears in two phases: firstly, solvent-soluble or "free" organic material (about 30%), such as amino acids, carboxylic acids, sulfonic and phosphonic acids, and even sugar-like molecules and nucleobases; and secondly, an organic macromolecular material mostly made up of aromatic hydrocarbons, which is insoluble in common organic solvents. The organic matter is present within the fine-grained inorganic meteorite matrix, with the association being highly specific to a particular inorganic phase – clay minerals.[41] Clays have the ability to adsorb organic molecules and catalyse mutual reactions. Such minerals may have trapped and concentrated organic matter in the early Solar System. Moreover, organic–inorganic interactions may have played a role in the assembly of increasingly complex organic entities 4.6 billion years ago.

Much of our current understanding of meteoritic organic matter has come from investigations of the Murchison meteorite that fell in Australia in 1969, providing us with a tremendous amount of meteoritic material (~100 kg). The amino acid suites of meteorites are generally very large, and include α-, β-, γ-, and δ-amino acids. The complete structural diversity, a predominance of branched chain isomers, and the exponential decline in amount with increasing carbon number in free organic matter, indicate an abiotic reaction mechanism with random chain synthesis. Moreover, such organic compounds have strong deuterium enrichments, and a sharp decrease of the ^{13}C content with increasing carbon number indicating a ladder synthesis of higher molecular weight compounds from lighter homologues. All these characteristics suggest that significant amounts of meteoritic organic matter, or its precursor materials, are synthesized in an interstellar environment. However, although the feedstock for meteoritic organic matter may predate the Solar System, the final molecular architecture of meteoritic organic matter must be affected by the processes occurring in the meteorite parent body. As these objects grew in size, they experienced secondary, post-accretion alteration events. This post-accretionary heating provided opportunities for chemical reactions to take place, both directly from thermal metamorphism (thermal energy) and from aqueous alteration resulting from the melting of water ice. Detailed information on the link between planetesimals and meteoritic material is found in Gail *et al.* (2014).[42]

7.4 Implications for the Composition of Interstellar Dust

Table 7.1 shows that a wide variety of materials exists in presolar dust. As we have seen in this Chapter, these presolar grains are found as tiny particles in primitive meteorites, IDPs, and in cometary dust. The patterns of isotopic abundances (sometimes of embedded and very rare inert gases) that are found in these presolar grains clearly identify their origins in the cool envelopes of evolved low mass stars and in supernovae ejecta. Therefore, these particles have survived through many episodes of possible destruction, including their ejection from the stellar envelopes in which they were formed, their passage through the interstellar medium where they are subjected to intense radiation fields and dynamical shocks, their incorporation into the molecular cloud that formed the Solar System, and through all the varied and violent processes involved in the formation of the Sun and its planets. Table 7.1 lists the stellar origins of many types of dust grain. These grains evidently existed prior to the formation of the Sun, hence their name: presolar grains, or—alternatively—stardust.

However, the list of types of presolar grains shown in Table 7.1 bears only a slight resemblance to the composition of interstellar dust as determined by remote observations. Modelling of observational data of interstellar extinction, polarization, continuum and line emission and absorption *etc.*, as described in Chapter 3, shows that there is convergence among several approaches on the chemical and physical nature of interstellar dust. All the observational evidence supports the view that interstellar dust in diffuse regions of space is mainly silicates without long-range order, amorphous carbon including both sp^2 (possibly in part as graphite) and sp^3 (probably as hydrogenated carbon rather than diamond) bonding, and some free-flying PAHs. This conclusion is consistent with the available elemental abundances in the interstellar medium. Of course, these components are essentially the main ones inferred, and the true composition of interstellar dust may be much richer with the addition of many other minor components.

From the perspective of surface chemistry on interstellar dust grains that we shall discuss in Chapters 8 and 9, it is important to know the composition of the dust material, and so the different conclusions for Solar System presolar dust and for remotely examined interstellar dust may seem at first sight to be a cause for concern. In fact, the apparent discrepancy between the compositions measured for presolar dust and those inferred from models of interstellar dust simply reflects the history of presolar dust. For example, Table 3.1 shows that diamond has a relatively high abundance in presolar dust, as might be expected for this extremely robust material. It has survived rather well through interstellar shocks and irradiation, and through the violent events associated with the formation of the Solar System. Silicates also appear to survive fairly well in their passage from interstellar to circumsolar existence. Graphite, SiC, and Si_3N_4 are of relatively low abundance in presolar dust. In any case, almost all of the silicon available in the interstellar medium is locked in silicates, so that the amount of silicon that can be present in other species (such as SiC and Si_3N_4) in interstellar dust can only be small.

The formation of the Solar System is, therefore, a filter that severely modifies the composition of interstellar dust; those materials passing through the filter are the most robust to destruction and erosion. Just as near-stellar dust is modified in the interstellar medium, and becomes interstellar dust, so interstellar dust is modified when it is incorporated in the gas that formed the Solar System. In general terms, the bulk of interstellar dust has a composition that is best determined through remote observation. Undoubtedly, interstellar dust is more complex than is inferred from the modelling methods described in Chapter 3. However, those conclusions must be broadly correct. More importantly, from the point of view of surface chemistry it is the materials of silicates, carbons, and PAHs that provide most of the materials on which chemistry may occur. This conclusion underlies the work described in Chapters 8 and 9.

Further Reading

1. H. Campins, K. Hargrove, N. Pinilla-Alonso, E. S. Howell, M. S. Kelley, J. Licandro, T. Mothé-Diniz, Y. Fernández and J. Ziffer, *Nature*, 2010, **464**, 1320.
2. L. W. Alvarez, W. Alvarez, F. Asaro and H. V. Michel, *Science*, 1980, **208**, 1095.
3. C. Hayashi, *Prog. Theor. Phys. Suppl.*, 1981, **70**, 35.
4. R. C. Ogliore, G. R. Huss, *et al.*, *Astrophys. J.*, 2012, **745**, L19.
5. M. Zolensky, M. E. Zega, *et al.*, *Science*, 2006, **314**, 1735.
6. D. Nesvorny, P. Jenniskens, H. F. Levison, W. F. Bottke, D. Vokrouhlicky and M. Gounelle, *Astrophys. J.*, 2010, **713**, 816.
7. E. Grun, H. A. Zook, *et al.*, *Nature*, 1993, **362**, 428.
8. J. P. Bradley, *Science*, 1994, **265**, 925.
9. A. J. Westphal, R. M. Stroud, *et al.*, *Science*, 2014, **345**, 786.
10. M. Landgraff, W. J. Baggaley, E. Gruin, H. Kruger and G. Linkerr, *J. Geophys. Res.*, 2000, **105**, 10343.
11. B. T. Draine, *Space Sci. Rev.*, 2009, **43**, 333.
12. T. Bernatowicz, G. Fraundorf, *et al.*, *Nature*, 1987, **330**, 728.
13. R. S. Lewis, M. Tang, J. F. Wacker, E. Anders and E. Steel, *Nature*, 1987, **326**, 160.
14. E. Zinner, M. Tang and E. Anders, *Nature*, 1987, **330**, 730.
15. A. N. Nguyen and S. Messenger, *Elements*, 2011, 7, 17.

16. B. Marty, M. Chaussidon, R. C. Wiens, A. J. G. Jurewicz and D. S. Burnett, *Science*, 2011, **332**, 1533.
17. M. M. Abbas, P. D. Craven, *et al.*, *Astrophys. J.*, 2004, **602**, 1063.
18. A. G. G. M. Tielens, C. G. Seab, D. J. Hollenbach and C. F. McKee, *Astrophys. J.*, 1987, **319**, L109.
19. R. S. Lewis, E. Anders and B. T. Draine, *Nature*, 1989, **339**, 117.
20. P. Hoppe, *Proc. Sci.: Nucl. Cosmos XI*, 2011, 21.
21. W. Fujiya, P. Hoppe, E. Zinner, M. Pignatari and F. Herwig, *Astrophys. J.*, 2013, **776**, L29.
22. A. M. Davis, *Proc. Natl. Acad. Sci. U. S. A.*, 2011, **108**, 19142.
23. D. D. Clayton and L. R. Nittler, *Annu. Rev. Astron. Astrophys.*, 2004, **42**, 39.
24. S. Messenger, F. J. Stadermann, C. Floss, L. R. Nittler and S. Mukhopadhyay, *Science*, 2003, **300**, 105.
25. A. N. Nguyen and E. Zinner, *Science*, 2004, **303**, 1496.
26. K. M. Nollett, M. Busso and G. J. Wasserburg, *Astrophys. J.*, 2003, **582**, 1036.
27. E. Dobrica, C. Engrand, H. Leroux, J.-N. Rouzaud and J. Duprat, *Geochim. Cosmochim. Acta*, 2012, **76**, 68.
28. H. Leroux, P. Cuvillier, B. Zanda and R. H. Hewins, *Lunar Planet. Sci.*, 2013, **44**, 1528.
29. D. E. Brownlee, D. J. Joswiak, J. P. Bradley, J. C. Gezo and H. G. M. Hill, *Lunar Planet. Sci.*, 2000, **31**, 1921.
30. J. P. Bradley, L. P. Keller, *et al.*, *Science*, 1999, **285**, 1716.
31. J. S. Mathis, *Astrophys. J.*, 1986, **308**, 281.
32. J. P. Bradley, Z. R. Dai, *et al.*, *Science*, 2005, **307**, 244.
33. T. M. Steel and W. W. Duley, *Astrophys. J.*, 1988, **315**, 337.
34. L. P. Keller and S. Messenger, *Geochim. Cosmochim. Acta*, 2011, **75**, 5336.
35. J. Blum, R. Schräpler, B. J. R. Davidson and J. M. Trigo-Rodriguez, *Astrophys. J.*, 2006, **652**, 1768.
36. R. E. Grimm and H. Y. McSween, *Icarus*, 1989, **82**, 244.
37. D. D. Clayton, *Astrophys. J.*, 1980, **239**, L37.
38. J. M. Greenberg, *Astrophys. Space Sci.*, 1976, **39**, 9.
39. M. Telus, G. R. Huss, *et al.*, *Meteorit. Planet. Sci.*, 2012, **47**, 2013.
40. R. E. Grimm and H. Y. McSween, *Science*, 1993, **259**, 653.
41. V. K. Pearson, M. A. Sephton, *et al.*, *Meteorit. Planet. Sci.*, 2002, **37**, 1829.
42. H.-P. Gail, M. Trieloff, D. Breuer and T. Spohn, in *Protostars and Planets VI*, ed. H. Beuther, R. Klessen, C. Dullemond and Th. Henning, 2014, p. 571.

Section III
Chemically Active Interstellar Dust

CHAPTER 8

Catalysis on the Surfaces of Bare Dust Grains

8.1 The Need for Surface Chemistry in the Interstellar Medium

As we noted in Chapter 1, many of the detected interstellar molecular species can be formed through networks of gas-phase reactions. These networks include ion–molecule reactions, dissociative and radiative recombinations, and neutral exchanges. The gas phase reaction networks are often very efficient; many of the processes that are included occur on almost every collision. These networks can account not only for the existence of many interstellar species but in many cases also for their relative abundances. Many of the reactions in currently used reaction networks (which typically may include several thousand reactions) have molecular hydrogen as a partner, and many more involve proton donations from H_3^+ which is derived directly from H_2. From a chemical point of view, therefore, molecular hydrogen is a very important component of gas phase interstellar chemistry. From an astronomical point of view, molecular hydrogen is observed to be very abundant in interstellar clouds of the Milky Way galaxy. So far, so good; but how is the molecular hydrogen formed?

Gas phase routes to form molecular hydrogen under the conditions in the interstellar clouds of the Milky Way were identified in the mid-20th century. The basic radiative association of two H atoms under interstellar conditions is very strongly forbidden by quantum mechanical selection rules, and this basic process can safely be discounted. However, three-body associations such as

The Chemistry of Cosmic Dust
By David A. Williams and Cesare Cecchi-Pestellini
© David A. Williams and Cesare Cecchi-Pestellini 2016
Published by the Royal Society of Chemistry, www.rsc.org

$$H + H + H \rightarrow H + H_2$$

or

$$H + H + H_2 \rightarrow 2H_2$$

are allowed and may be very efficient if the number density is sufficiently large. However, interstellar number densities are known to be many orders of magnitude less than this critical value, so this process, too, may be ignored for the general interstellar medium although it may be important in some special circumstances such as the envelopes of cool stars or the ejecta from novae and supernovae (see also Section 9.4).

Exchange reactions of H with precursor species H^- and H_2^+ to form H_2 *via* reactions

$$H + H^- \rightarrow H_2 + e$$

$$H + H_2^+ \rightarrow H_2 + H^+$$

are efficient, but those precursors can form only very slowly by radiative attachment

$$H + e \rightarrow H^- + h\nu$$

and radiative association

$$H + H^+ \rightarrow H_2^+ + h\nu$$

respectively, and are also subject to rapid photodetachment and photodissociation by the interstellar radiation field.

Molecular hydrogen in unshielded regions of the interstellar medium is rapidly destroyed by the ultraviolet radiation from bright stars, and detailed studies show that the overall rate of H_2 production through those two chemical intermediaries fails significantly to account for the H_2 abundances observed in typical interstellar clouds in the Milky Way. However, these routes to H_2 may have applications in rather special situations, for example, in the early Universe, and in rare dust-free interstellar environments in the present Universe.

Exchange reactions between hydride radicals such as CH with H can produce H_2. CH molecules are known to exist in the interstellar medium with abundances relative to molecular hydrogen of ~10^{-8}, but since CH requires H_2 for its formation, this is a "chicken and egg" problem! Which comes first? In any case, this type of reaction, *i.e.* CH + H $\rightarrow$ C + H_2, cannot account for significant amounts of H_2 formation since CH is of low abundance. Investigations showed that all the mechanisms proposed to form molecular hydrogen were too inefficient compared to the rapid photodestruction of H_2 by the interstellar radiation field to account for the high interstellar H_2 abundances that were observed.

By default, therefore (rather than any desire to introduce into interstellar chemistry a few highly uncertain reactions on the surfaces of poorly-characterised dust), the idea that molecular hydrogen could be formed in surface reactions between H atoms on grain surfaces began to be seriously considered. However, it was only towards the end of the 20th century that techniques in both the laboratory and in theory had advanced sufficiently to make serious progress in understanding such reactions in the interstellar context. These topics are the subjects of Sections 8.2–8.4, below. A speculative alternative route is considered in Section 8.5. Although some problems remain, those Sections indicate that the formation of interstellar molecular hydrogen through surface reactions on dust grains now appears to be a plausible mechanism and one that is reasonably well understood from both theoretical and experimental points of view. So one may ask the question: if H_2 can be formed this way, can other species also arise through surface reactions on bare grains? We address this topic in Section 8.6.

As we have described above, purely gas-phase schemes fail, in general, to provide the observed abundances of interstellar molecular hydrogen. But another serious problem with purely gas-phase networks is that they seem to fail to provide mechanisms that would be expected to create reasonable abundances of some other molecular species detected in interstellar clouds (even such simple species as, for example, water, or carbon dioxide). The most dramatic example of this failure of gas-phase networks is the variety of relatively large molecular species (*i.e.* up to about ten atoms) found in star-forming regions, the very dense and dusty regions associated closely with newly-forming stars and in the centre of the Milky Way. These detected species include methanol (CH_3OH), ethanol (C_2H_5OH), dimethyl ether (CH_3OCH_3), methyl formate ($HCOOCH_3$), formic acid ($HCOOH$), acetic acid (CH_3COOH), propynal (HC_2CHO), propenal (CH_2CHCHO), propionaldehyde (CH_3CH_2CHO), glycolaldehyde (CH_2OHCHO), ethylene glycol ($HOCH_2CH_2OH$), ethylene oxide (c-C_2H_4O), acetaldehyde (CH_3CHO), and ketene (H_2CCO). While viable gas phase routes to produce these species can usually be devised, it appears that these potential routes are unable to supply the substantial abundances of these species under the known physical conditions in a star-forming region. Attention has therefore focused on grain-related chemistries that might be able to generate large quantities of these relatively complex species. Therefore, the question arises: can dust grains act to promote the chemical complexity observed in star-forming regions from the relatively simple interstellar species that are available? Both laboratory-based and theoretical approaches to understanding the origin of these detected interstellar species are making progress. These ideas are currently the subject of much attention, and this developing area is described in Chapter 9.

8.2 Formation of Interstellar H_2 in Surface Reactions: Possible Mechanisms

The idea is basically very simple: a hydrogen atom collides with an interstellar grain and is retained on the surface until a second hydrogen atom arrives and locates the first; a reaction occurs between the two hydrogen atoms, forming a hydrogen molecule that is ejected from the surface with some kinetic and internal energy; the ejection also deposits some energy into the surface. Thus, the grains provide a means of overcoming the very low number density of H atoms in space by bringing two H atoms together on the surface, and then permitting them to react by absorbing some of the energy released in the reaction, so stabilising the newly-formed molecule. If this is supposed to be the mechanism that provides interstellar molecular hydrogen, then—without going into any detail concerning the chemical process—astronomers can compute the necessary efficiency of the process. Astronomical observations provide basic information on the number densities of atomic hydrogen and molecular hydrogen, the gas kinetic temperature, and the number density and size range of dust grains in a particular astronomical region. Assuming that the H_2 is destroyed by the local ultraviolet radiation field and that a chemical steady-state applies, a simple calculation based on this mechanism allows the inference to be made that the H_2 formation mechanism must be very efficient; that is, most of the H atoms arriving at the surfaces of dust grains eventually leave the surface as part of an H_2 molecule.

This is an important result because it severely constrains the details of the mechanism; these details, on closer examination, become rather complex. Firstly, the result implies that there must be a high probability that one H atom will stick to the surface of a dust grain, and be retained on the surface for a time that is at least long enough, on average, for a second H atom to arrive on the same grain. This constrains the grain temperature, assuming that thermal desorption determines the H atom residence time. Secondly, two surface H atoms must find one another before one of them is desorbed, and the atoms must react with high efficiency. If one H atom is strongly bound to the surface (with a bond with an energy on the order of a few eV), then it will not be mobile, so for a reaction to occur a second H atom must be able to search for and locate the first and form a bond with it. If the second H atom is weakly physisorbed (with bonding to the surface in the order of ~meV), then it may search the surface with high mobility in the Langmuir–Hinshelwood mechanism (see Figure 8.1), in which motion over the surface arises through quantum mechanical tunnelling through small barriers. This situation may apply in the limit of low H atom densities and low coverage. On the other hand, the second H atom may arrive at the surface sufficiently near to the first that a bond may form directly between the two atoms, forming an H_2 molecule. This is known as the Eley–Rideal mechanism (see Figure 8.1). It could be the case when the surface coverage is relatively high. In either case, the product H_2 molecule may be desorbed promptly, or at a later stage. A prompt desorption may leave the molecule with some kinetic and rovibrational energy, and will deposit some energy into the grain surface. These energies originate in the bond energy of the newly-formed H_2 (up to about 4.5 eV, depending on the rovibrational state). If the newly-formed H_2 is retained on the surface, then it will thermalise at the grain temperature (typically ~10 K) and be trapped until a later transient heating event causes it to desorb. A schematic diagram showing these processes is shown in Figure 8.1.

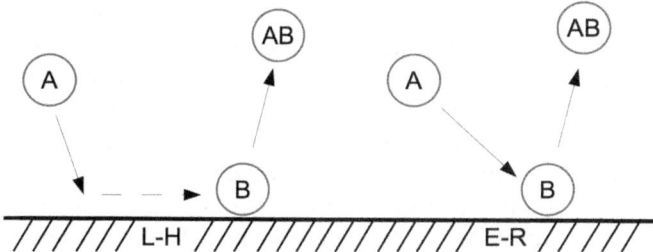

Figure 8.1 A diagram indicating the Langmuir–Hinshelwood and Eley–Rideal mechanisms in surface chemistry.

The Langmuir–Hinshelwood and Eley–Rideal mechanisms are, of course, idealised, and the actual process of H_2 formation may be a combination of both. Indeed, in most low density interstellar regions, a gas of H atoms is likely to have a much higher kinetic temperature (~100 K) than that characterising the dust grain (~10 K), and thermal accommodation of H atoms at the surface may occur through repeated interactions with the surface in which the kinetic energy of the H atoms is successively reduced. If so, then these "hot" H atoms will travel over the surface during the process of thermal accommodation, and may encounter another H atom while doing so. This mechanism, commonly called the "hot atom mechanism" (sometimes the "Harris–Kasemo" mechanism[1]) has some characteristics of both the Langmuir–Hinshelwood and Eley–Rideal processes. However, from a theoretical point of view it is convenient to consider these two processes as distinct.

Theoretical studies of these mechanisms have been attempted for many years, but it is only in the last two decades or so that computational techniques became available enabling detailed and accurate quantum mechanical studies to be made. The success of some of these approaches is described in Section 8.3. In addition, spectacular advances in laboratory techniques have permitted very successful experimental studies of the H_2 formation mechanism to be carried out over the last fifteen years. Results from several groups are reported in Section 8.4. An excellent and very comprehensive review of theoretical (H_2) and experimental (HD or D_2) studies of molecular hydrogen formation by surface reactions has been published by Gianfranco Vidali (2013).[2]

8.3 Theoretical Considerations

8.3.1 The H Atom–Surface Interaction

The physical and chemical nature of the interstellar grain surface is crucial in determining the interaction between an incident H atom and the surface. In theoretical treatments of the interaction between the H atom and the grain, a form of carbon is often selected as the model grain surface in preference to other candidates, such as magnesium/iron silicates or amorphous ice (see Chapter 5). Carbon is chosen partly because the modelled carbon surfaces can be accurately represented, and partly because reactions on carbon have a very wide application beyond the demands of astrochemistry such that a huge literature exists on this topic. Since all grain models (see Chapter 3) include carbon with at least some sp^2 (graphitic) bonding, many of the detailed theoretical studies that have been made of the H_2 formation process have taken the grain surface to be represented by this form of carbon. One can do this, for example, by adopting as the modelled surface one (or more) layers of graphene, or a piece of a perfect graphite lattice (repeated periodically), or a molecule such as coronene (the seven-ringed compact hydrocarbon $C_{24}H_{12}$) or circumcoronene (the nineteen-ringed compact hydrocarbon $C_{54}H_{18}$). Such structures are, of course, highly idealised representations of the actual surfaces of interstellar dust grains whose surfaces are likely to be irregular and defected. The results of such studies may give useful information, although this may possibly be more qualitative than quantitative. Nevertheless, these investigations introduce the general issues affecting the interaction of H atoms with a surface and subsequent molecular hydrogen formation. We shall describe some of the work involving the interaction of an H atom with carbonaceous surfaces, and then mention similar but more limited work on silicate surfaces.

The interaction of an H atom and a graphitic surface has been accurately computed by various authors using a variety of methods. The results of these studies are in general agreement about the nature of this interaction. Firstly, there is a weak, site-independent physisorption bond which holds the H atom at about 0.3 nm above the graphite surface. The bond strength (expressed in temperature units) is approximately 450 K, but depends on the adsorption site. In Figure 8.2 (ref. 3) possible adsorption sites for H on coronene are shown; these are directly above one of the carbon atoms of the central hexagon (labelled T for top), above the midpoint of a side of the central hexagon (labelled B for bridge), or above the centre point of the central hexagon (labelled H for hollow). The adsorption energies for physisorption for an H atom in the T, B, and H position on coronene are found by Ferullo *et al.* (2010)[3] from computational studies using van der Waals corrected density functional theory to be 32.1, 32.7, and 38.1 meV, respectively, while the experimental value measured on a single graphite layer is 39.2 ± 0.5 meV.

An H atom may also be chemisorbed above a carbon atom of the lattice. Many very detailed studies have shown that to establish the chemisorption bond requires the distortion of the graphitic plane so that the carbon atom is raised out of the plane by a few percent of one nm. Establishing this distortion creates a barrier to bonding of about 0.2 eV, and increases the sp^3 character of the carbon.[4] When established, the chemisorption bond energy is about 0.67 eV. The existence of a barrier of ~0.2 eV in the formation of the chemisorption bond calls into question the existence of chemisorbed atomic hydrogen on graphite in the cool interstellar medium where gas kinetic temperatures are in the range 10–100 K (since 0.2 eV is equivalent to a temperature of approximately 2200 K). However, recent experimental work, followed

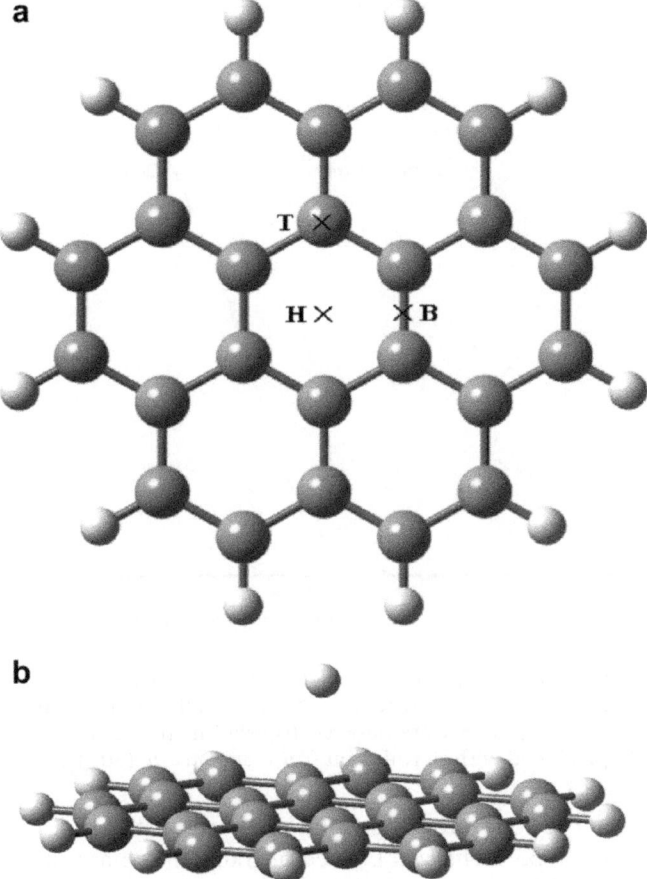

Figure 8.2 The coronene molecule as a representation of the graphite surface. (a) The crosses indicate the adsorption sites (T: top; B: bridge; H: hollow). (b) H atom physisorbed at a hollow site. Reproduced with permission from Ferullo *et al.* 2010.[3]

up by detailed quantum modelling[5] suggests that there is a striking reduction in adsorption barriers for H atoms near to pre-chemisorbed H atoms. As in the case of physisorption, the precise location of the pair of H atoms affects the result. In the most favourable case, the so-called *para* site, the interaction is barrierless (see Figure 8.3) on both rigid and relaxed graphene surfaces, and also when a second graphene layer lies beneath the surface layer. Other sites than the *para* site may also be barrierless. Thus, chemisorption on perfect carbon surfaces may occur at particular sites. (Here, the term *para*, and—below—*meta* and *ortho* refer to two H atoms on a single hexagon that are increasingly distant from each other.) Goumans (2011)[6] suggested on the basis of theoretical work that edge sites on PAHs may in fact be more favourable to chemisorption than interior sites of the PAH plane.

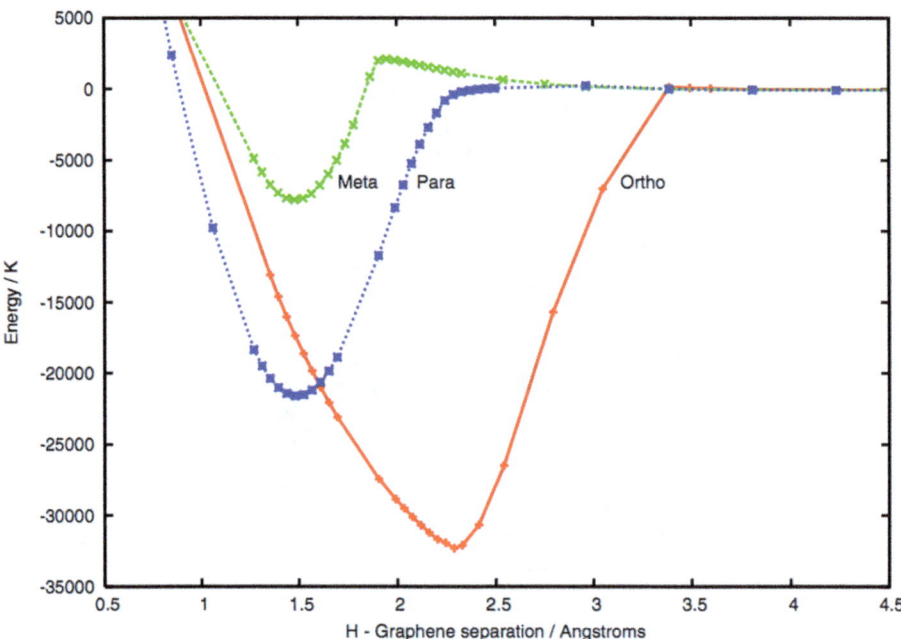

Figure 8.3 Potential energy curves for the adsorption of a single H atom on graphene in the vicinity of a chemisorbed H atom. The carbon atoms in the graphene were allowed to relax in this double chemisorption. Reproduced with permission from Irving *et al.* (2011).[5]

In the interstellar medium, it is unlikely that the material of which dust grains are made is a perfect crystal. It seems likely that materials are defected and contain impurities. Recent work has considered some of these possibilities and found that adsorption properties can be greatly modified. When the possibility of substitutional defects in the carbon lattice is investigated, it is found that—depending on the dopant—chemisorption can be barrierless (Irving & Meijer, private communication). These authors have also studied the effect of swapping a single carbon atom from a graphene layer for a single H, O, Fe, N, Si, and Mg (these being abundant interstellar species). For a single gas-phase H atom adsorbing onto a defect of this type, a range of behaviours is found. At one extreme, H and Mg defects open up edge sites in the graphene layer that are strongly binding to incident H atoms, so that subsequent reaction with another H atom to form H_2 may have a barrier. On the other hand, O atoms as defects are only physisorbed, and therefore these sites may be short-lived. Defects created by substitutions of carbons by Si and N atoms create barriers to incident gas phase H atoms, while Fe may be barrierless. However, if the incident gas phase H atom approaches a carbon atom *neighbouring* the defect, then barrierless adsorption may occur in many cases. This is the case for *para* sites (as in the simple non-doped case) for all the substitutions considered, and although *meta* adsorption has a barrier, it is much smaller than for the non-substituted case.

Interstitial atoms may also be present in interstellar graphitic carbon. The role of interstitials in a double layer of graphene has been explored. It appears that chemisorption of a gas phase H atom is in many cases barrierless (Irving & Meijer, private communication). Adsorption on to *ortho*, *meta* and *para* sites has also been investigated, and it appears in many cases to be barrierless. We conclude that chemisorption of H atoms on *realistic* carbonaceous surfaces in the interstellar medium is predicted by these computations to be barrierless in many cases.

8.3.2   The H + H Reaction

The H_2 formation process begins with the sticking of an H atom to the surface. Sticking probabilities are determined by computing a transfer of energy from the incident H atom into the phonon modes of the surface. Theoretical methods that have been used include molecular dynamics and classical trajectories. Evidently, the adopted model of the grain surface determines the phonon spectrum and consequently the computed sticking probability is highly model-dependent. Sticking may be efficient when the kinetic energy of the incident H atom is comparable to peaks in the phonon spectrum. Therefore, sticking probabilities may vary significantly at low collision energies.

Figure 8.4 shows the computed sticking probability for a physisorbed H atom with collision energy in the range 0–25 meV on graphite for which the surface temperature is either 10 K or 300 K.[7] The sticking probability varies considerably with collision energy and is larger at higher surface temperature, but is always in the range of a few percent. Calculations of sticking probability for an H atom on other model materials at low collision energy have also been computed to be a few percent. As discussed in Section 8.2, such values of the sticking probability appear to be too low for efficient interstellar H_2 formation. Using more realistic models of grain materials would be expected to affect the computation of the sticking probability in the physisorption case quite significantly. However, sticking in the chemisorption case is efficient.

The next step in H_2 formation in the Langmuir–Hinshelwood picture requires that at least one of the H atoms be mobile. The mobility of physisorbed H atoms on carbonaceous surfaces through quantum tunnelling has been found in various computational studies to be high, so that at the typical temperature of interstellar dust (~10 K) the time scale for a physisorbed H atom to scan the surface of a grain is very short compared the H atom arrival time on that grain. These computations suggest that the Langmuir–Hinshelwood model of H_2 formation from physisorbed H atoms should be viable.

In the Eley–Rideal model, it is assumed that an H atom from the gas is incident directly on (or sufficiently near to) an adsorbed H atom, but does not itself adsorb on to the surface. There is general agreement among many calculations using various methods that the interaction of a gas phase H atom with a chemisorbed H atom is strongly attractive so that reaction occurs between the two H atoms. It is of considerable interest to know how the energy budget is allocated in this H_2 forming reaction. Since the H–H bond

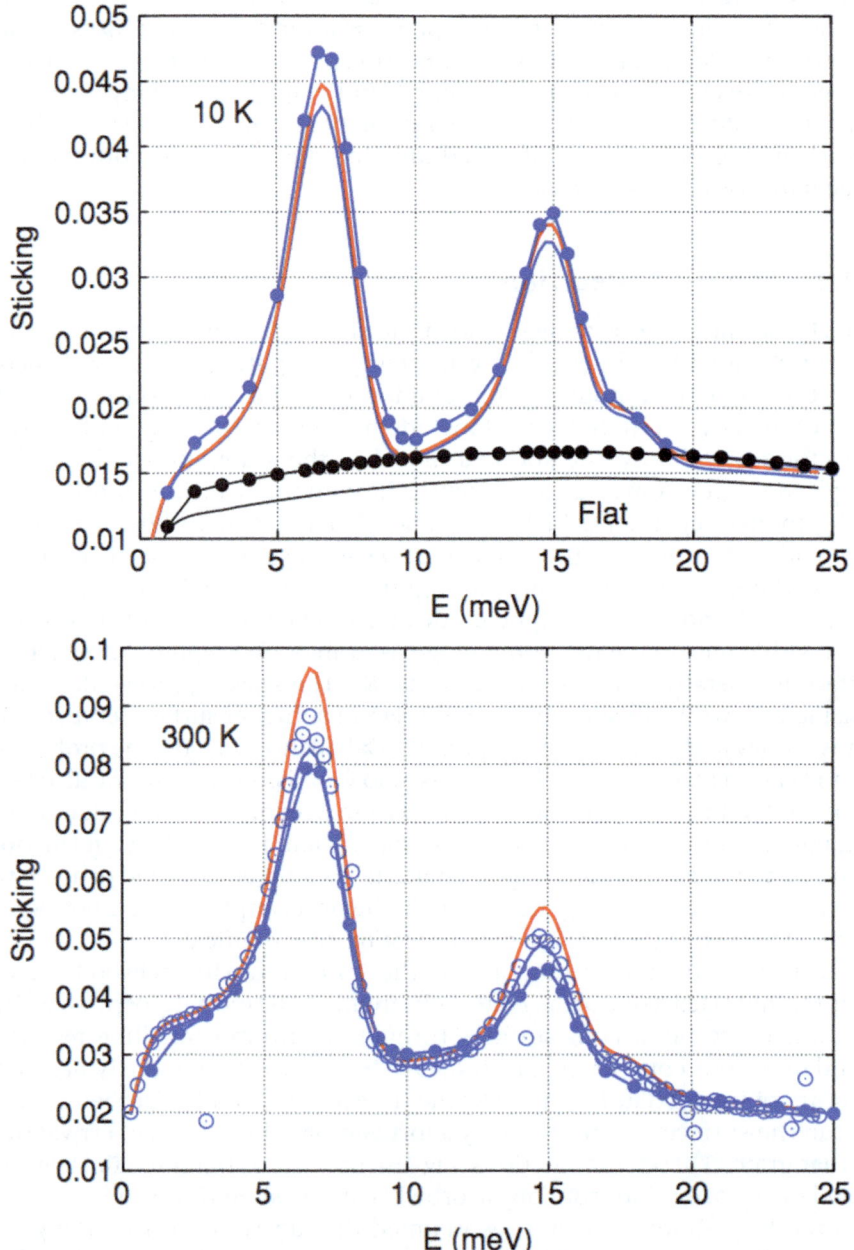

Figure 8.4 Sticking probabilities of an H atom on graphite at 10 K and 300 K. Results from several different theoretical methods are compared. Reproduced with permission from Lepetit *et al.* (2011).[7]

that is formed makes about 4.5 eV available and only 0.67 eV needs to be found to release the chemisorbed H atom, up to about 3.8 eV is available for the product molecule and the surface. This energy may appear as rovibrational energy and as kinetic energy in the H_2 molecule, and as energy transferred to the lattice. From the astronomical perspective, these are potentially important energy sources for the gas and for the grain.

8.3.3 The Energy Budget

Many theoretical studies have been made of the energy budget of the H_2 formation on the surface of carbonaceous dust grains. These include both the Langmuir–Hinshelwood mechanism involving physisorbed H atoms and the Eley–Rideal mechanism involving one adsorbed H atom, usually chemisorbed but possibly physisorbed. The studies use either graphene or graphite as model surfaces, and may allow for lattice distortion.

For physisorbed H atoms reacting to form H_2 on graphene or graphite, the reaction probability is found in all studies to be large. All studies also predict that these reactions produce H_2 with a high amount of vibrational excitation. Figure 8.5 (ref. 8) suggests that formation of H_2 into vibrational level $v'' = 8$ may dominate for gas kinetic temperatures above about 20 K. The nature of the perfect surface probably does not play a significant role in this process, so that a 2D model (as used in the calculation leading to Figure 8.5) may be adequate. The main role of the surface in these reactions is to transfer energy from the hindered vibration–rotation of the nascent H_2 into translational energy. This may lead to a prompt direct desorption of the molecule, in an excited rovibrational state. Alternatively, the nascent molecule may be weakly trapped in a state from which it will be later released. Distortions and other imperfections of the lattice may have a significant role in determining the vibration–rotation excitation.

Similar studies have been made of the Eley–Rideal process in which a gas phase H atom impinges directly on a chemisorbed H atom bonded with a graphene or graphite surface. In this reaction, the role of the surface is greater than in the Langmuir–Hinshelwood case. A recent study shows that the incident gas phase H atom causes an additional puckering of the lattice, and that this distortion is only released after the reaction forming H_2 occurs. The probability of this reaction is very high. This study[9] suggests that most of the energy released ends up as rovibrational, about a quarter is given to the surface, and rather less than a quarter is translational. The details of the surface and of the interactions with it are clearly of greater consequence in the Eley–Rideal case. If the gas phase H atom approaches one of two *para*-chemisorbed H atoms on graphite, rather than a single chemisorbed atom, then the reaction still occurs with very high probability, but the vibrational excitation—while still high—may be slightly reduced.[10] The consequences of including the effects of zero point energy of the chemisorbed H atom appear to be a slight reduction in the predicted vibrational energy of the nascent H_2 (towards $v'' < 5$) and a broader range in the rotational excitation ($J'' = 0–16$).[11]

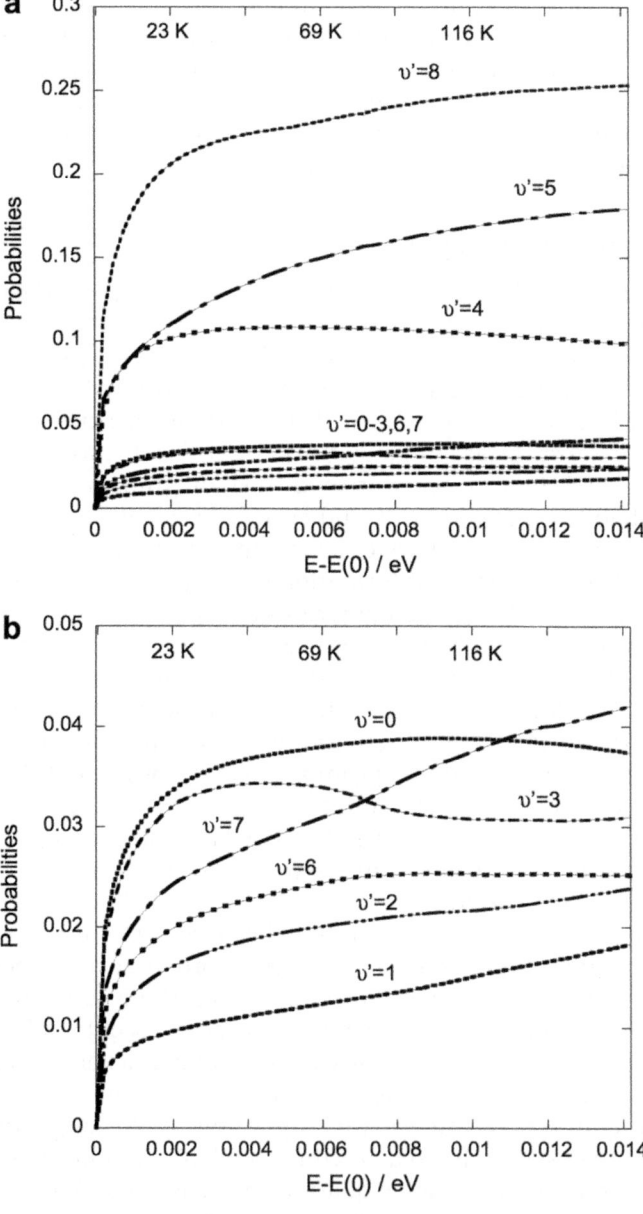

Figure 8.5   (a) Theoretical probabilities of H_2 formation from physisorbed H atoms on graphite into specific vibrational levels. (b) Expanded view of the lower probabilities. Reproduced with permission from Kerkeni & Clary (2007).[8]

While it could be misleading to lay too much stress on any individual calculation, nevertheless there is general agreement that H_2 formation reactions either by the Langmuir–Hinshelwood or the Eley–Rideal processes occur with very high probability, and that a newly-formed H_2 molecule will be in a highly excited internal state. As well as the excited rovibrational state, the molecule is expected to have appreciable translational energy, probably making it a significant source of energy for the ambient gas. Equally important is the general conclusion that a significant amount of energy is transferred to the surface. In the interstellar context, this heat source for the grain may be important either in controlling the bulk temperature or in causing local heating that may lead to the desorption of weakly adsorbed species other than hydrogen.

8.3.4 Theoretical Studies of H_2 Formation on Silicate Surfaces

A detailed study of the adsorption of H atoms and subsequent H_2 formation on a silicate surface has been reported by Goumans *et al.* (2009),[12] see Figure 8.6. The silicate chosen for these calculations is the magnesium-rich olivine Mg_2SiO_4 (forsterite, 010 surface). The physical adsorption energy of an H atom on this ionic surface (equivalent to 1240 K) is computed to be much stronger than in the case of physisorption on a graphitic surface. However, barriers to migration over the surface are found to be comparable to the physisorption energy so that H atom mobility by hopping or by quantum tunnelling may be high, even at surface temperatures as low as 10 K. This implies that H_2 formation from physisorbed H atoms on perfect silicate surfaces may be efficient.

However, it appears that there is no barrier to the chemisorption of an H atom on this surface because the energy released in physisorption is found to be equal to the barrier for chemisorption from the physisorbed state. In chemisorption, the H atom binds to the O atom site, protonating the O atom to form an OH^- site, and donating an electron to the neighbouring Mg^{2+} site. Therefore, H atoms may chemisorb directly on to the pristine olivine surface and may not exist in the physisorbed state. The chemisorption energy is high, ~12 200 K, in agreement with previous results (~1 eV).

When chemisorption occurs a surface electron is produced and this electron site attracts an incident H atom to the Mg^{2+} ion adjacent to the chemisorption site where it is strongly bound by 39 800 K. If this energy can be released as the two H atoms recombine, then H_2 formation should be efficient. Further, the strong interaction of both H atoms with the surface suggests that the coupling of the excess energy in the reaction should be easily dissipated into surface phonon modes, so that the rovibrational energy of the product molecule should be less than in the case of physisorption on silicates or of any reaction on graphitic surfaces.

It is possible that the energy of 39 800 K from the second chemisorbed atom may be lost to the surface and not available to promote the H_2 formation from the two chemisorbed H atoms. However, Goumans *et al.* find that a third H atom approaching the two chemisorbed H atoms can react with either of them to form H_2. This reaction is exothermic and barrierless, and the molecule should desorb on formation. These strong predictions for H_2 formation on silicates remain to be tested for silicates with defects, steps, kinks, and substitutions. However, Goumans and Bromley (2011)[13] have shown that the reactivity of nanosilicates is expected to be high.

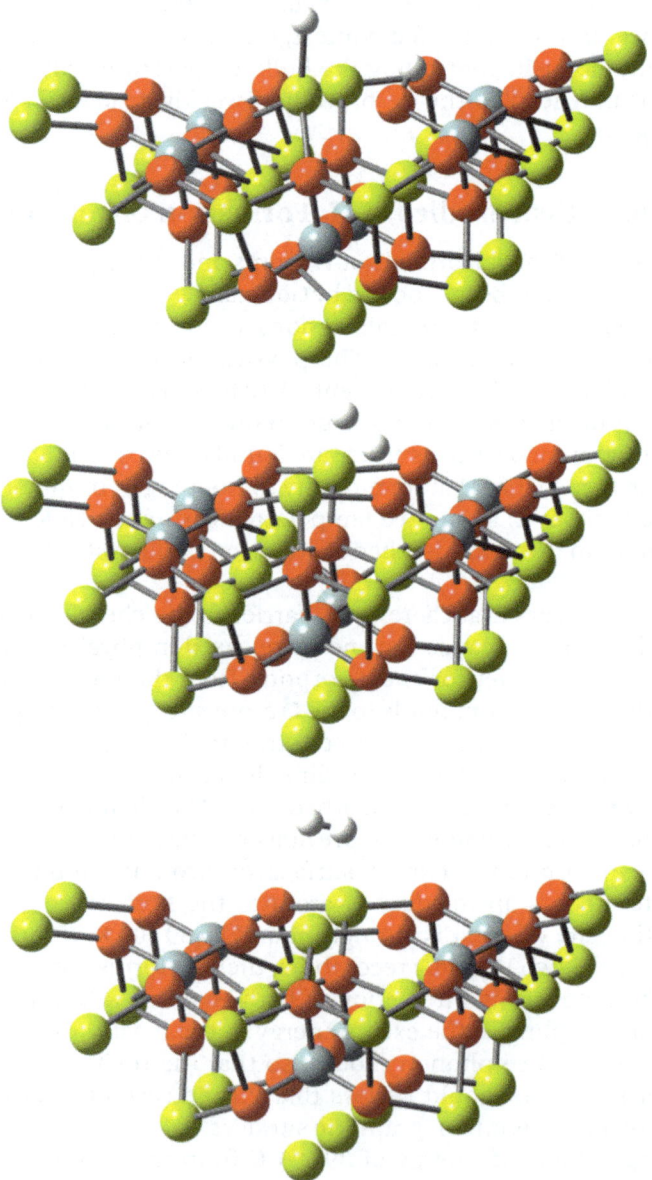

Figure 8.6   The positions of two chemisorbed H atoms (top) and physisorbed H_2 (bottom) on forsterite (010) surface; the transition state is also shown (middle). Gray: Si atoms; yellow: Mg atoms; red: O atoms; and white: H atoms. Reproduced with permission from Goumans *et al.* (2009).[12]

8.3.5   A Negative Ion Route

Interstellar grains in diffuse and dark clouds are expected to carry negative charges under most conditions. The charges will typically be equivalent to up to a hundred electrons. David Field (2000)[14] made the interesting suggestion that H atoms incident on the grain surfaces may be converted to H^- ions that are ejected into the gas phase, where they associatively detach in reactions with other H atoms to form H_2 molecules:

$$H + (grain)^- \rightarrow grain + H^-$$

$$H^- + H \rightarrow H_2 + e^-$$

If this mechanism is to compete with or replace the kinds of surface chemistries discussed so far in this chapter, then observational constraints would require that both of the above reactions be very efficient, occurring in most collisions. The second reaction is considered to be fairly well determined and efficient; however, the H^- has other routes that may be competitive. Certainly, photodetachment of H^- by infrared photons in the interstellar medium is likely to be relatively fast, and may constrain the H_2 formation rate. If photodetachment suppresses the H^- abundance significantly—as seems likely in unshielded regions of interstellar clouds—then the H^- route would be unable to compete effectively with the surface chemistry routes envisaged elsewhere in this chapter.

The main uncertainty is the binding energy of electrons, as compared to the electron affinity of H atoms (0.75 eV), *i.e.* the energy that could be released on H^- formation. For insulators such as silicates, the conduction band may be significantly (~1 eV) below the vacuum level. For conductors such as pure graphite, the situation would be even worse. The conduction band would be located at the Fermi level, and the binding energy would be given by the work function for graphite, or about 4.5 eV. However, interstellar grains may be heavily processed, and some evidence suggests that in such materials conduction bands may lie close to the vacuum level. In such a case, then the formation of H^- may be possible, and may even be sufficiently exothermic to enable the H^- to be detached from the surface.

It is of interest that H_2 formed *via* H^- reactions will have an emission spectrum that is easily distinguished from other formation or excitation mechanisms. Emission from H_2 formed *via* H^- will have a strong emission from high rotational levels in $v'' = 1, 2$, and 3.

8.3.6   Conclusions for Theoretical Studies of H_2 Formation on Surfaces

We can make some general conclusions based on theoretical studies of H atoms on graphite or graphene, as reported in the preceding paragraphs. This may not be too great a restriction for the interstellar application: some researchers (including the authors of this book) support the view that most interstellar dust grains—whatever the nature of their bulk material—are likely to be coated with carbon. Much of this carbon may be graphitic and, if so, most of the theoretical work described above would apply directly to the interstellar case. However, some of this surface carbon may be polymeric, with sp^3 rather than sp^2 bonding. There has been little study of the H–sp^3 interaction and we can make no clear conclusions for that case. Other than carbon, interstellar dust grains are believed to be composed mostly of amorphous silicates. Much less theoretical attention has been given to the interaction of hydrogen with bare silicates, but predictions of H$_2$ formation on perfect silicates appear to be very positive.

For the Langmuir–Hinshelwood mechanism, all studies agree that two physisorbed H atoms on a perfect graphite or graphene surface should be able to explore that surface, and in the case of interstellar grains with perfect surfaces should be able to meet and react to form H$_2$. However, the computed sticking probabilities for H atoms physisorbed on a graphitic surface are low. For those H atoms that do stick then the hydrogen molecule formed from two physisorbed H atoms should appear in a highly excited vibrational level, with significant kinetic energy, and a small energy transfer to the surface which may be sufficient to ensure that desorption occurs. The main difficulty for the interstellar situation is the sticking probability of an H atom to perfect graphite or graphene surfaces which appears to be too low to account for the abundance of H$_2$ observed in interstellar clouds. The problem would be exacerbated if the carbon sp^2 surface available in the interstellar medium is only a minor fraction of the total. It seems likely, however, that the sticking probability would be enhanced if the surface was imperfect, with lattice defects and atomic substitutions.

Most theoretical studies of H$_2$ have focused on the Eley–Rideal process in which a gas phase H atom interacts with a chemisorbed H atom on the surface. Apparent problems arising from a significant barrier in the formation of the chemisorption site have been resolved by more detailed studies in which the effects of adsorption at neighbouring sites, and of the consequences of imperfections in the lattice caused by substitutions and other defects. It appears that there are many realistic pathways to establishing the chemisorbed H atom. All studies agree that the reaction of the incident gas phase H atom with the chemisorbed atom is efficient and that the hydrogen molecule should be formed in a significantly excited internal state, with substantial kinetic energy, and with a significant amount of energy being transferred into the surface.

We conclude that the process of H$_2$ formation on interstellar dust grains whose surfaces can be described as sp^2 carbon is reasonably well understood from a theoretical point of view, and that the more realistic the models the greater clarity of the effectiveness of the process. The most effective process appears to be Eley–Rideal. The Langmuir–Hinshelwood process is hampered by the low sticking probability of H atoms on perfect graphitic surfaces, although surface defects may resolve this difficulty.

The theoretical prediction for silicates—at least, for pristine forsterite, 010—is that H$_2$ formation should proceed efficiently, probably by the

Eley–Rideal mechanism in which the incident H gas phase atom is guided to the appropriate site by electrostatic effects. Finally, there have been several studies of H_2 formation on amorphous ice surfaces. Such surfaces are important only in the interiors of dense clouds where dust grains are observed to accumulate mantles of dirty ice; in these regions, the hydrogen is almost entirely molecular. Theoretical studies apparently show that formation may occur by either Langmuir–Hinshelwood or Eley–Rideal mechanisms, with high efficiency in each case, leaving the molecule in highly excited rovibrational states. However, these studies have been carried out using classical molecular dynamics calculations rather than much more accurate quantum mechanical methods, and it is unlikely that predictions of these classical methods for hydrogen atoms and molecules on ices can have comparable accuracy to those quantum mechanical studies of H_2 formation on carbons and silicates. It is clear from the study of Le Bourlot *et al.* (2012)[15] that surfaces that include a wide range of binding sites will allow H_2 formation to occur over a broad temperature range (as discussed long ago by Gould & Salpeter 1963).[16]

8.4 Formation of Interstellar H_2: Laboratory Work

8.4.1 Methods

It is difficult, if not impossible, to replicate accurately interstellar surface processes in the laboratory. The first problems concern the surface: the nature of the interstellar dust that provides the surfaces is as yet not accurately characterized (see Chapters 3–6). Spectroscopic information certainly confirms that silicates, carbons and dirty ices are present in the dust, but that still leaves the experimenter with many choices to make. Are the interstellar surfaces pristine? If not, to what extent are they defected? Have substitutions in the lattice taken place? Are they crystalline or amorphous? Are they porous? Is the surface electrically charged? Further, interstellar grains are believed to range in size from nanometres to micrometres, while experiments are necessarily conducted on bulk materials. The finite size of interstellar surfaces may affect the chemical processes occurring on them. We shall return to this problem in Section 8.7.

What about the interstellar gas? We know from observations that in typical low density interstellar clouds the number density of hydrogen atoms may be on the order of ten or one hundred atoms per cm^3 and that the gas kinetic temperatures are about or less than 100 K. Such clouds therefore have pressures that are orders of magnitude lower than the lowest pressures that can be attained in ultra-high vacuum (UHV) laboratory experiments which may operate at pressures of 10^{-11}–10^{-10} Torr. There are, of course, denser regions of the interstellar medium that have pressures rather closer to those attainable in the laboratory. But the huge disparity in pressure between laboratory and much of interstellar space means that experimental results may not always be directly applied to the interstellar case without further interpretation.

Ensuring realistic temperatures presents less of a problem. The gas temperatures may be maintained near 100 K by refrigerators around the atomic beam. For the sample surfaces, it is comfortably within present techniques to hold the sample temperature to very low values near or below 10 K, comparable to the temperature of interstellar dust, by mounting the sample on a cold finger that is cooled using closed-cycle liquid helium refrigerators. In the experiments, it is important to keep the reacting surface clean. This may be done by thermally decoupling the sample from the cold finger and heating the sample to a high temperature.

Finally, there's a huge disparity in the available timescales between laboratory and interstellar space. In the laboratory, typical experiments may take a long time to prepare but the actual experiment may take only minutes or hours, while in space the "experiments" may run for many millennia. Experiments in the laboratory are usually made for irradiations of short enough duration so that a low surface coverage on the sample is maintained, as expected for the interstellar case.

The conventional arrangement is to have two atomic beamlines that impinge on the surface. Although the reaction to be studied is the formation of H_2 from two H atoms, it is customary to have one beam of H atoms and one beam of D atoms and to study the formation of HD rather than H_2. The H and D beams are conventionally generated by radio frequency (RF) dissociation of the parent species H_2 and D_2. However, the dissociation is never complete, and so the beams certainly contain H_2 and D_2. However, any HD must arise in the reaction $H + D \rightarrow HD$, which is analogous to the $2H \rightarrow H_2$ reaction. It is assumed that the description of HD formation emerging from experiments applies with only minor modification to the formation of H_2. Hence, detector systems used must be able to distinguish HD from H_2 and D_2.

Commonly-used detector systems include the following: (i) a quadrupole mass spectrometer capable of measuring densities of both H and D beams and densities of the released gaseous product HD; (ii) temperature programmed desorption (TPD) experiments which release and measure the amounts of any retained HD in a thermally controlled manner; (iii) reflection absorption infrared spectrometry (RAIRS), usually at grazing incidence over the sample to monitor changes in the species adsorbed on the sample surface; (iv) resonance-enhanced multi-photon ionisation (REMPI) which is used to interrogate the state of the HD molecules as they emerge from the surface and to determine their rovibrational energy distributions; and (v) time-of-flight experiments to measure the kinetic energy of HD molecules departing from the sample surface. Different authors have used a combination of detectors that suits best the precise purpose of their investigations.

A large number of experiments of this type have been performed since the first realistic investigations of interstellar molecular hydrogen formation were successfully made by Valerio Pironello and Gianfranco Vidali, and their collaborators in 1997. Those experiments were performed at Syracuse University, New York. They have been followed by further experiments in the same laboratory on an extensive programme of research studying many aspects of the problem, and providing prolific results. In Section 8.4.2 we shall describe some work carried out by this group, and we shall follow that by descriptions of several other investigations by a few other leaders in this developing research area, to demonstrate the range of the studies that have been undertaken and to indicate the current state of research in this area. In Section 8.4.3 we attempt to summarise our present understanding of interstellar molecular hydrogen formation based on laboratory experiments, and mention areas requiring further attention.

8.4.2 Selected Results from Specific Laboratory Studies

8.4.2.1 Some Research at Syracuse University, New York

Experiments conducted by Valerio Pirronello and Gianfranco Vidali and colleagues at Syracuse University, New York since 1997 demonstrated what can be achieved with modern laboratory techniques and stimulated much further work both at Syracuse and by researchers in other laboratories. The Syracuse experiments were the first to accumulate data that could be reliably applied to the formation of molecular hydrogen *via* surface reactions on interstellar dust grains. The apparatus used in the Syracuse experiments is indicated schematically in Figure 8.7.[17]

The hydrogen and deuterium beam lines are triply differentially pumped, and the atoms are created by the RF dissociation of the H_2 and D_2 molecules,

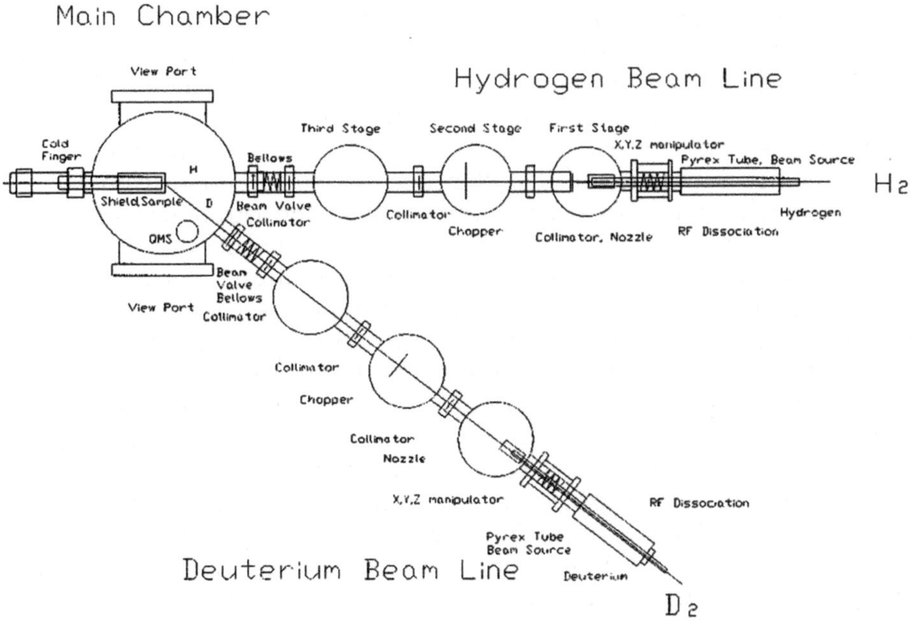

Figure 8.7 Schematic diagram of the Syracuse apparatus. Reproduced with permission from Pirronello *et al.* (1997a).[17]

with a dissociation fraction in the range 70–85%. The beams pass through a 1 mm capillary cooled by liquid nitrogen, and the flux is $\sim 10^{12}$ atoms cm^{-2} s^{-1} (about five orders of magnitude greater than in a typical interstellar diffuse cloud). A mechanical chopper can reduce the beam flux by one order of magnitude. The sample surface is attached to a helium-cooled cold finger, and the sample surface temperature is measured by calibrated thermocouples. The lowest sample temperature reached in the experiments is around 5 K. The incident and reflected beams are detected by a differentially pumped quadrupole mass spectrometer. After each experiment, the sample temperature can be raised quickly (at a rate $\sim$1 K s^{-1}) for TPD measurements.

In the original (1997) experiments, the sample was of polycrystalline olivine, a mixture of Fe_2SiO_4 and Mg_2SiO_4, cleaned then baked in a UHV at 150 °C. Before each experiment, the sample was baked again at $\sim$200 °C, cooled to the selected surface temperature ($\sim$5 K) and exposed for a specified time to the H and D beams, during which period the amount of HD desorbed during the irradiation is measured. After the beams were terminated, the TPD measurements were made to determine the amount of HD formed on the surface and retained until the TPD experiment released these molecules. Desorption rates measured in the 1997 experiments *via* TPD are shown in Figure 8.8a.[18] These TPD curves are conventionally interpreted using the Polanyi–Wigner equation.

By measuring the yield of HD during irradiation and also in the post-irradiation TPD phase and comparing to the incident flux one obtains the recombination efficiency, shown in Figure 8.8b. We see that in this experiment the prompt emission of HD molecules does occur but most of the HD is released in the TPD phase. The results are consistent with reactions occurring between weakly adsorbed atoms in the Langmuir–Hinshelwood mechanism; the Eley–Rideal mechanism does not appear to be important. The analysis shows that the reactions are second order at low temperature.

This first measurement of the efficiency was an important astrochemical achievement, but caused some difficulty from the astronomical point of view. The experiments suggested that the reaction efficiency (using the open squares data in Figure 8.8b) at the assumed temperature of interstellar grains (say, $\sim$15 K) was rather low, $\sim$0.05, whereas models suggested that it was required by astronomical observations to be about an order of magnitude larger. This apparent discrepancy promoted discussion about the possible importance of the physical nature of the surface on the reaction, and also the need to consider other sample materials. However, a later theoretical (rate equation) study of the TPD data on the olivine sample (and other materials) by Katz *et al.* (1999)[19] was able to infer a rather complete description of the hydrogen surface interaction, including activation energy barriers for atomic hydrogen diffusion and desorption and for desorption of surface molecular hydrogen, and the spontaneous desorption probability of newly-formed H_2. This work showed that the reaction efficiency on olivine was in fact high but only over a very limited range of surface temperature depending on the irradiation flux, the range for high efficiency being from about 6 to 10 K for

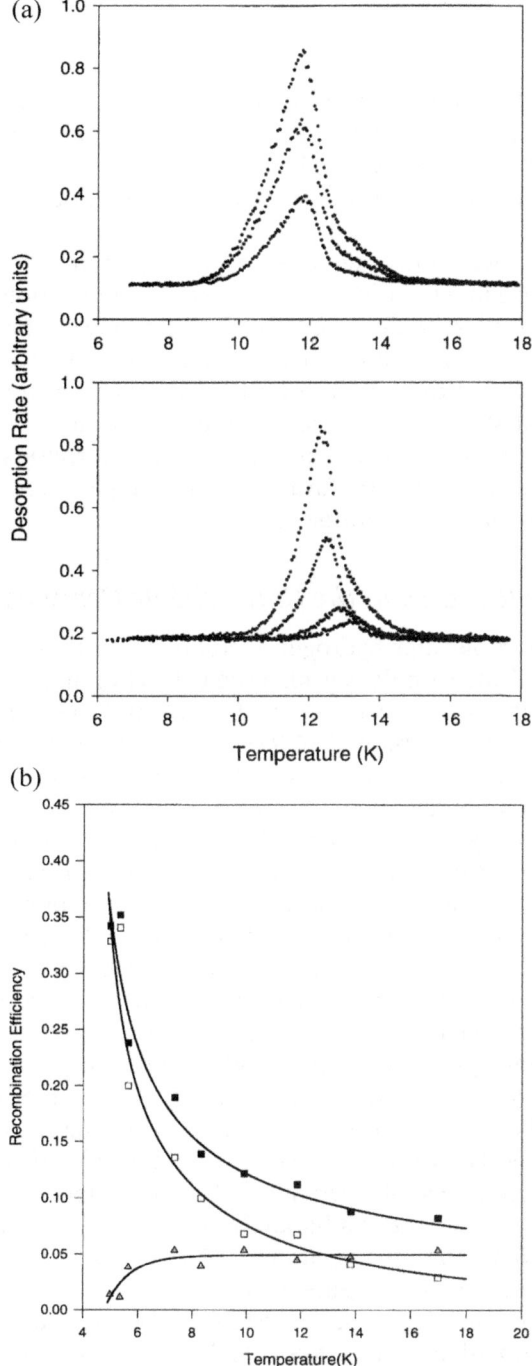

Figure 8.8 (a) Desorption rates of HD molecules during TPD from an olivine slab at 6 K: (bottom panel) after irradiation for (bottom to top) 0.07, 0.10, 0.25, and 0.55 minutes; (top panel) after irradiation for (bottom to top) 2.0, 5.5, and 8 minutes. (b) Recombination efficiency of hydrogen on an olivine slab as a function of temperature; triangles – during the irradiation time; open squares – from the TPD; filled squares – sum of both contributions. The irradiation time is about one minute. Reproduced with permission from Pirronello *et al.* (1997b).[18]

a flux corresponding to the diffuse cloud case. Thus, the laboratory experimenters suggested in effect that the problem should be thrown back to the astronomical modellers. However, it was expected that samples with different chemistry and different physical natures would modify the conclusions very significantly.

This expectation was confirmed when a closely similar experiment was performed in Syracuse in which the sample used was amorphous carbon.[20] The inferred energy barriers to diffusion and desorption were found to be larger than in the case of a crystalline silicate sample. The formation efficiency on carbon was also found to be high for a limited temperature range, but this range occurred at slightly higher temperatures than for silicates. Similarly, further experiments at Syracuse[21] showed that an amorphous silicate sample was an efficient catalyst for HD formation at a range of temperature somewhat higher than for the crystalline sample.

8.4.2.2   Some Research at the Université de Cergy Pontoise

The formation of molecular hydrogen (actually D_2 in these experiments) on an amorphous silicate sample was also the subject of an investigation led by Jean-Louis Lemaire of the Observatoire de Paris and the Université de Cergy Pontoise.[22] The apparatus (named FORMOLISM: FORmation of MOLecules in the InterStellar Medium) had some similarities to the Syracuse equipment, but some important differences. A single atomic beam was used, in which D_2 was dissociated (~80%) by RF excitation. After cooling and collimating, the beam impacted on the cold surface of the sample. In these experiments, the sample was an amorphous olivine film obtained by thermal evaporation of olivine and deposition on a gold substrate. Reactions occurring at the sample surface produced D_2 molecules which were regarded as the signal of reaction; non-dissociated D_2 in the beam was detected and allowed for in the analysis. Newly-formed D_2 molecules were detected in three ways: using a conventional quadrupole mass spectrometer (QMS), by REMPI coupled with a time-of-flight (TOF) mass spectrometer (REMPI-TOF), and by the conventional TPD method. REMPI-TOF was used to probe a particular rovibrational level of the D_2 molecule. In this experiment, the level chosen was $v'' = 4, J'' = 2$ of the ground electronic state of D_2. The REMPI signal was followed during both the irradiation phase and the TPD phase (during which the irradiation is switched off). A schematic diagram of the FORMOLISM apparatus is shown in Figure 8.9 and some results from this experiment are shown in Figure 8.10.

At time $t = 0$, the sample temperature fell to 12 K and the irradiation with D atoms began. Immediately, the QMS signal rose, showing that D_2 molecules were being formed and promptly released. The REMPI-TOF signal shows that some of these molecules are in the $v'' = 4, J'' = 2$ rovibrational state (later work by Gavilan *et al.* 2014 [23] also detected the $v'' = 0, J'' = 0$–5 states). At $t = 600$ seconds the irradiation ceased and the TPD phase began, with the temperature rising to 70 K by $t = 1200$ seconds. The REMPI-TOF signal disappears throughout the TPD phase but there is a sharp QMS signal.

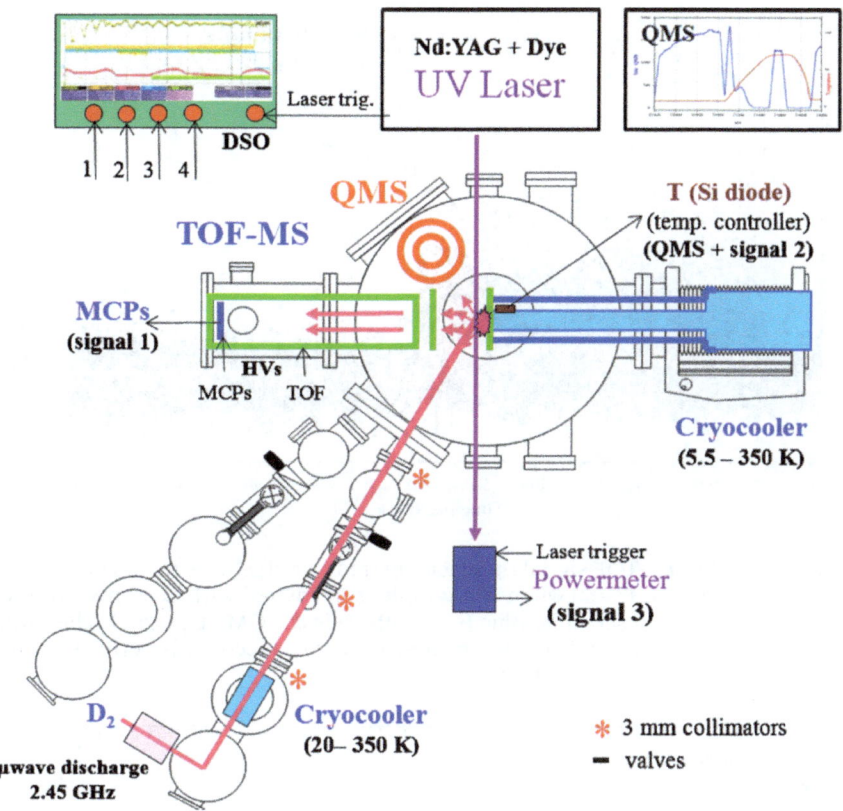

Figure 8.9 Schematic diagram of the "FORMOLISM" apparatus. Reproduced with permission from Lemaire *et al.* (2010).[22]

These results show that rovibrationally excited molecules are formed and released as soon as D atoms arrive at the sample surface. There is no REMPI-TOF signal during the TPD phase, so the D_2 molecules detected by the QMS in this phase are rovibrationally cool. The experiment was repeated for a range of sample temperatures from 12 K up to 70 K, at which temperature the REMPI signal disappears, so that it appears that molecule formation and prompt release continues up to these quite elevated temperatures. This is an important and rather unexpected result. The REMPI-TOF and QMS signals are shown as a function of sample temperature in Figure 8.11.

The REMPI and TPD signals both attain maxima for sample temperature around 10 K, but as the sample temperature increases the two signals behave in different ways, suggesting that there are two mechanisms involved. The QMS signal falls rapidly to low values at sample temperatures of about 15 K, while the REMPI signal falls more slowly until around 70 K. It appears that rovibrationally excited molecules can be produced over a much wider temperature range than molecules detected through QMS. Lemaire *et al.* note that at the elevated sample temperatures the residence times of D atoms on the

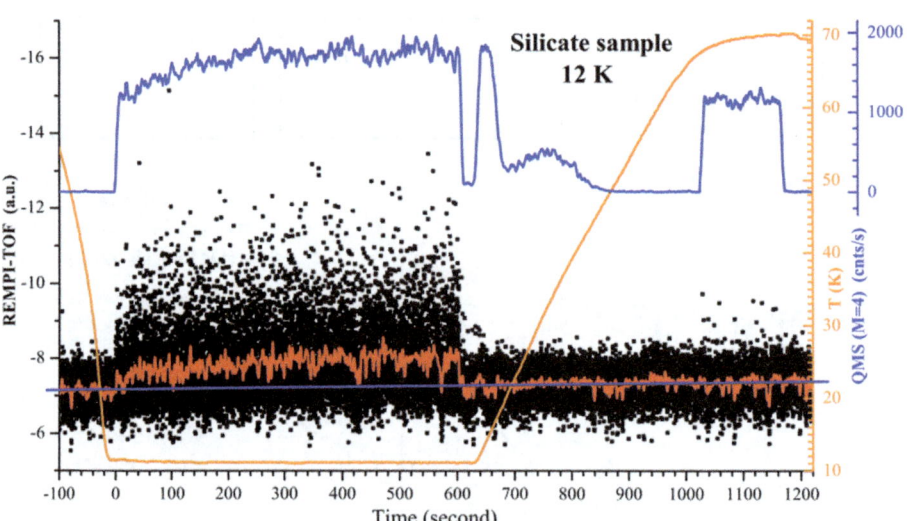

Figure 8.10   REMPI-TOF signal (in black, filtered in red) shows the formation of D_2 ($v'' = 4, J'' = 2$) when the sample is irradiated with D atoms from time 0 to 600 seconds. Blue (scale offset) is the QMS D_2 signal. The sample temperature is shown in orange. Reproduced with permission from Lemaire *et al.* (2010).[22]

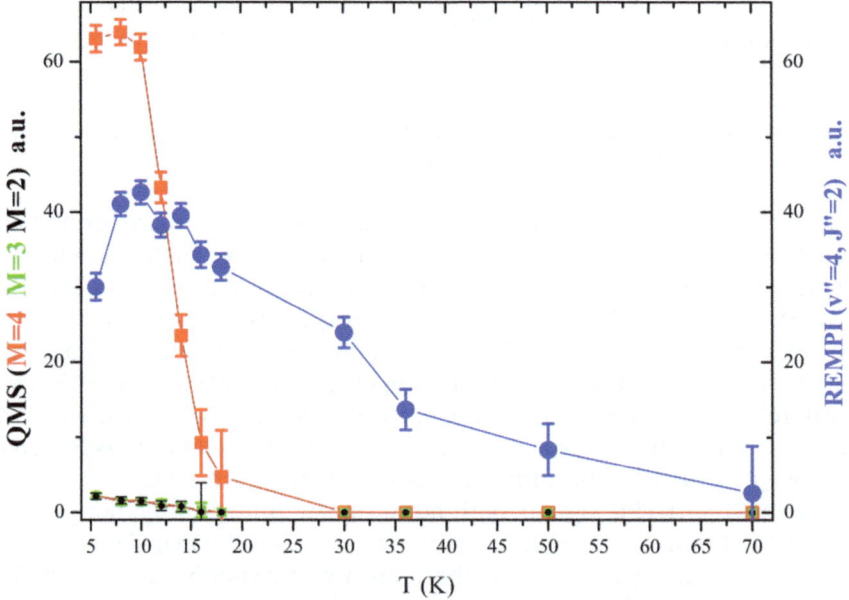

Figure 8.11   Integrated intensities of the REMPI (D_2) signal during irradiation and of the QMS (D_2) signal of the TPD after irradiation, both as a function of sample temperature. A small signal of HD is due to exchange reactions on the sample and walls of the apparatus. Reproduced with permission from Lemaire *et al.* (2010).[22]

surface is so short that very few molecules can be produced by the Langmuir–Hinshelwood mechanism, and they propose that the continuing production of D_2 at high surface temperatures occurs through the "hot atom" mechanism in which the fast motion across the surface allows atoms to locate each other and react, sharing some of the energy released with the lattice, while some energy is retained within the molecule and in its kinetic energy.

The identification of two distinct mechanisms for the formation of D_2 on amorphous silicates, involving Langmuir–Hinshelwood and "hot atom" mechanisms, was an important advance in laboratory astrochemistry. However, the amount of D_2 produced in the Langmuir–Hinshelwood mechanism is probably larger than that from "hot atom" reactions.

This experiment also demonstrates that internal excitation in the reaction product can be regarded as a useful signal of molecule formation. From an astronomical perspective, of course, it is much more than this: internal and kinetic energy of the product molecule is an important contributor to the heat balance of interstellar clouds, and to enhancing the gas-phase chemistry. It is, therefore, of interest to understand as fully as possible the energy budget of molecular hydrogen formation on interstellar dust grains. While Lemaire and his colleagues used a single rovibrationally excited state ($v'' = 4$, $J'' = 2$) as an indicator of surface reaction and prompt release from the amorphous silicate surface, there had been an earlier extensive programme of research to probe in detail the entire rovibrational population of the product molecules (see below).

The present lack of astronomical detection of emission from newly-formed interstellar H_2 remains a concern, especially as experiments for surfaces of silicates and carbons show that excitation is prominent. In further work at the Université de Cergy Pontoise, Congiu *et al.* (2009)[24] show that H_2 formed on water ice surfaces can be rapidly de-excited, and that, regardless of whether the ice is compact or porous the excitation is reduced when the coverage is high, as may be expected in dense clouds. Nevertheless, all experiments involving silicate and carbon surfaces—which are more appropriate for diffuse interstellar clouds—indicate that nascent H_2 should be excited.

A study in the same laboratory[25] of sticking coefficients of H_2 and D_2 on silicate surfaces at 10 K has been used to derive sticking coefficients for atomic H and D on silicates at gas temperatures up to 350 K. At 10 K, these coefficients are near unity. They decline with increasing temperature and are ~0.1 at ~350 K.

8.4.2.3 Some Research at University College London

A Centre for Cosmic Chemistry and Physics was established at University College London (UCL) in 1996 to promote theoretical and experimental studies relevant to astrochemistry. The experimental work of this programme related to molecular hydrogen formation (the Cosmic Dust Experiment) was—and continues to be—carried out in the laboratory of Stephen Price.[26,27] The motivation was to explore the formation pumping of rovibrational states of molecular hydrogen, with a view to characterizing fully the rovibrational state of nascent molecules and to detecting from the astronomical infrared emission spectrum of interstellar clouds an unambiguous signal of the molecular formation process in the interstellar medium. The sample surface used was graphitic (freshly cleaved highly oriented pyrolitic graphite; HOPG). The populations of the rovibrationally excited molecules were measured using REMPI-TOF. The experiments are capable of dealing with both H_2 and HD. A schematic diagram of the experimental arrangements to study HD formation on a cold HOPG surface is shown in Figure 8.12.[28]

The experiments are carried out in UHV with a base pressure 10^{-10} Torr in two stainless steel chambers. Incident H and D beams are generated by microwave dissociation of H_2 and D_2 in separate cells and piped in polytetrafluoroethylene (PTFE) tubes to an aluminium cooling channel, usually cooled to ~100 K. There is a low probability of hydrogen recombination on the surface of PTFE. The H and D beams are differentially pumped and piped through PTFE tubing to the target chamber. The HOPG sample is mounted on a closed-cycle cryostat that maintains the sample temperature to a specified value in the range 10–20 K during experiments, while the sample temperature can be raised to high values for sample cleaning. The beam fluxes are measured by REMPI spectroscopy and are on the order of 10^{12} cm^{-2} s^{-1}. Some recombination occurs during the passage of the atomic beams through the PTFE tubes to the chamber, and the overall dissociation efficiency is ~20%.

At the start of an experiment, the sample is heated to ~500 K to remove any adsorbates from its surface. When the sample has cooled to the chosen temperature and atom beams are running continuously, the laser is scanned

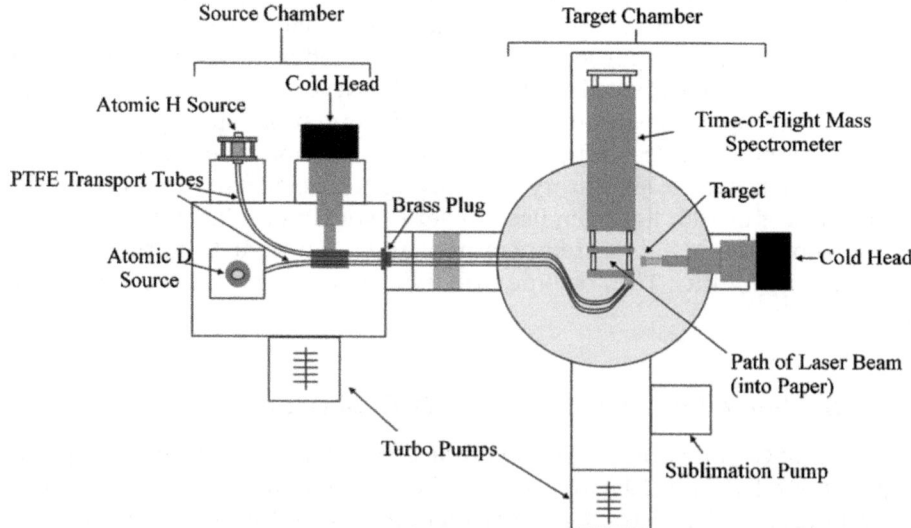

Figure 8.12   Schematic diagram of the UCL Cosmic Dust Experiment. Reproduced with permission from Islam *et al.* (2007).[28]

over the required rovibrational line while the counts are accumulated by the mass spectrometer of the HD$^+$ ions created by state-selective ionisation.

Molecules of HD newly-formed on HOPG surfaces at 15 K have been detected in states up to $v'' = 7$ and $J'' = 4$. Merging the results for the rovibrational excitation from two different experiments leads to the distributions shown in Figure 8.13.[29]

We see that the experiments find considerable vibrational excitation in the nascent HD, somewhat smaller than might be expected from calculation of molecular hydrogen formation by the Langmuir–Hinshelwood mechanism, but larger than that predicted for the Eley–Rideal mechanism, for a graphitic surface. The widths of the TOF mass spectra constrain the kinetic energy of the nascent molecules, and the data from the UCL experiments and other studies suggest an upper limit to the kinetic energy of 1 eV. Allowing for the internal excitation in the newly-formed HD molecules implies that about 40% of the available energy must flow into the surface. This amounts to a heating rate for an interstellar dust grain of more than 1 eV per hydrogen molecule formed. This may cause local heating of the surface of an interstellar dust grain, with consequent desorption of neighbouring adsorbed species.

The direct detection of infrared emission from H$_2$ molecules relaxing in a rovibrational cascade from levels populated in the formation process would bring valuable confirmation of all the experimental and theoretical work on interstellar molecular hydrogen formation. Therefore, the predictions of the

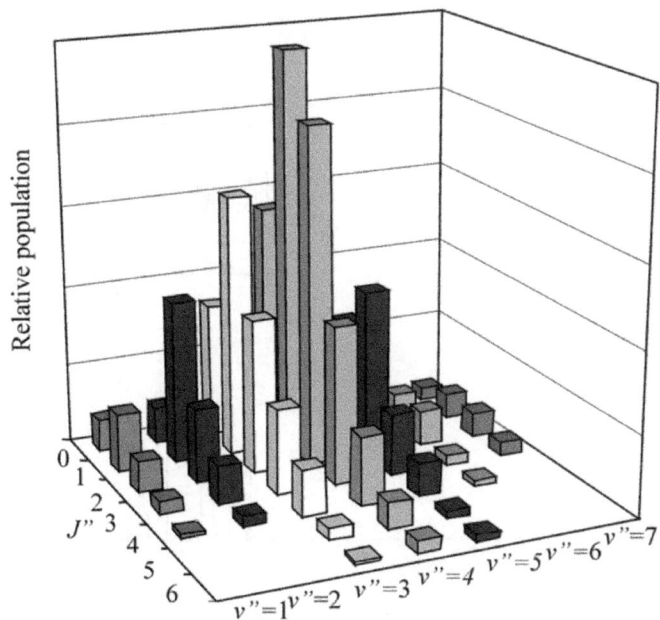

Figure 8.13   Derived relative rovibrational populations of HD formed on HOPG surface at 15 K in vibrational states $v'' = 1$–7. Reproduced with permission from Latimer *et al.* (2008).[29]

formation pumping populations obtained from these experiments have been used by Islam *et al.* (2010)[30] in computations of the infrared emission intensities to be expected from molecular hydrogen in interstellar clouds, as those molecules relax to the ground states. Those molecules are of course subject to excitations and de-excitations from collisions with electrons, pumping by ambient ultraviolet starlight, and other mechanisms. In Figure 8.14 we show a comparison of the predicted H_2 infrared cascade spectrum from a dense dark cold interstellar cloud subjected to the various pumping mechanisms, with the formation pumping turned on and off.

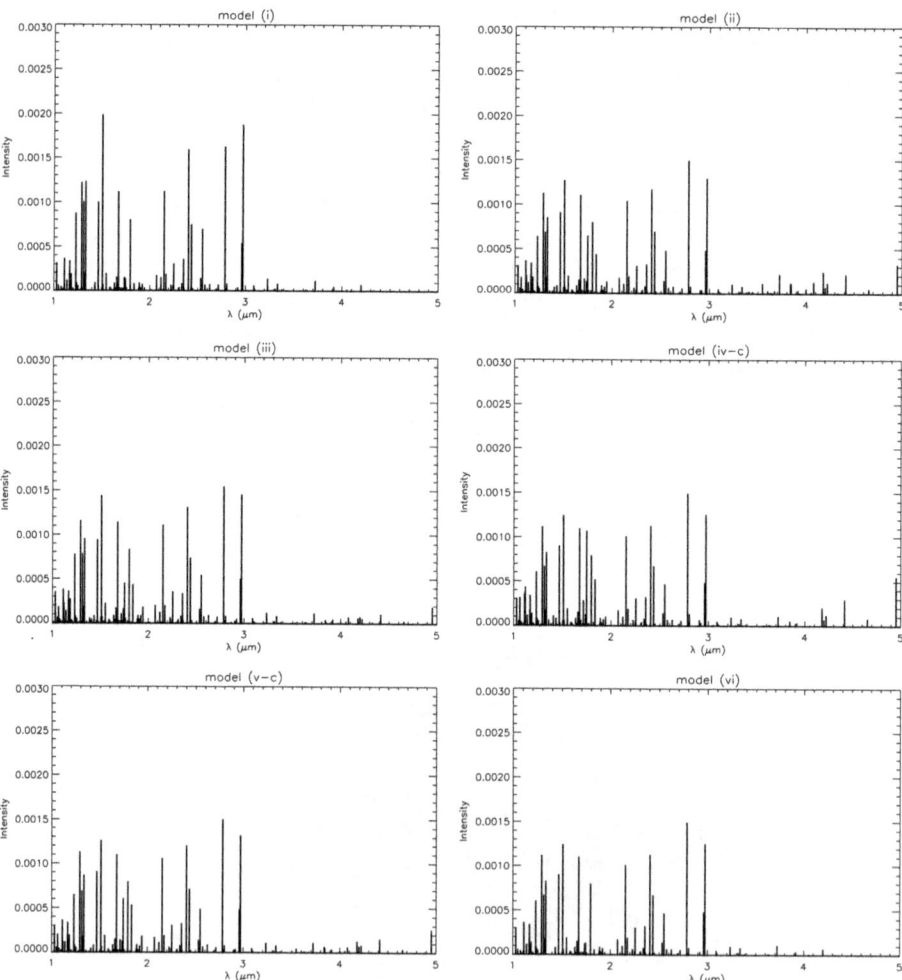

Figure 8.14   Computed IR emission (in erg s^{-1} cm^{-2} sr^{-1} μm^{-1}) from a dense, very dark interstellar cloud irradiated by standard cosmic ray and starlight radiation fields; model (i): formation pumping of H_2 rovibrational levels as determined by the Cosmic Dust Experiment; models (ii) (iii), (iv-c) and (v-c): theoretical estimates; model (vi): no formation pumping. Reproduced with permission from Islam *et al.* (2010).[30]

Evidently, the formation pumping makes some difference to the predicted spectrum. However, the intensities are weak and differences are slight. It is instructive to examine the variations of the predicted contributions from the various pumping mechanisms as a function of depth into the cloud. Islam *et al.* (2010)[30] have shown that the formation pumping is strongest at greatest depths in dense clouds. Islam *et al.* predict that the best opportunity to detect lines from formation pumping of H_2 is in very dense regions of interstellar gas dominated by X-ray emission. The high density (and high optical depth) would ensure that ultraviolet pumping is suppressed, but—since X-rays penetrate much further and destroy H_2 creating abundant atomic gas—the formation of H_2 should be rapid and subsequent formation pumping should be intense. This observation has not yet been made and therefore, the much desired, crucial direct detection of H_2 formation in space has yet to be achieved.

However, as the next experiment to be discussed will make clear, the nature of the sample surface may be the over-riding factor in the rovibrational excitation of the nascent molecule. It shows that if molecule formation occurs within the pores of a porous material, then collisional relaxation ensures that the molecule leaves the surface in thermal equilibrium with the surface. In that case, the internal excitation of the molecule will have been quenched. It seems that amorphous water ice deposited in the laboratory at low temperature ($T \sim 10$ K) is likely to be porous. Whether ice deposited in the interstellar medium is also porous remains to be seen.

8.4.2.4 Some Research in Denmark

We highlight here two research programmes led by Liv Hornekaer that have advanced our understanding of H_2 formation on interstellar dust grains. The first[31] uses the now familiar TPD approach to explore the importance of surface morphology of the sample; the second[32] uses a technique not previously discussed in this chapter, scanning tunnelling microscopy, to probe in detail the clustering of hydrogen atoms on a graphite surface. The results of this laboratory study seem to support the detailed predictions of fundamental quantum mechanical studies of the H–graphite interaction (see Section 8.3.1).

The surface morphology experiments study the formation of HD on the surface of amorphous ices, a type of surface not previously discussed in this chapter. Ices and molecule formation are the main topic of the following chapter, but it seems appropriate to consider all aspects of molecular hydrogen formation in this chapter.

It is worth noticing, however, that the most severe constraints on H_2 formation occur in low density interstellar clouds of low opacity, the diffuse clouds. In them, starlight in the wavelength range 100–110 nm readily penetrates the cloud and dissociates H_2 very rapidly. The lifetime of an unshielded H_2 molecule in the average interstellar radiation field is as short as a thousand years. Therefore, it is in such clouds that an efficient H_2 formation is required to maintain the observed H_2 abundances. This short lifetime is what requires almost all H atoms that arrive at a grain to leave as part of a molecule (Section 8.2, above). On the other hand, in the interiors of dark clouds where it is observed that dust grains become coated with ice mantles, those far-ultraviolet radiations are almost totally suppressed by dust extinction, and the rate of loss of H_2 (mainly through ionisation by cosmic rays, *i.e.* fast protons and electrons) is much smaller, unless there is a nearby powerful X-ray source. Most hydrogen molecules in a dense dark cloud survive as long as the cloud itself, possibly as long as tens of millions of years. So the existence of molecular hydrogen inside dense dark clouds does not provide a severe constraint on H_2 formation mechanisms. However, since amorphous ices can easily be created in the laboratory with varying morphologies, it is an excellent opportunity to be able to investigate experimentally the dependence of formation efficiency on varying morphology.

It is well known that non-crystalline water ice, commonly referred to in this context as amorphous solid water (ASW), takes different forms if the constituent water molecules are deposited on a cold surface at different temperatures. At low surface temperatures, ~10 K, a modestly porous ASW is created, while at higher temperatures, ~120 K, non-porous ASW is formed. Liv Hornekaer and colleagues (2003)[31] prepared ASW samples with different porosities in this way. The samples were then placed in an UHV chamber (less than ~10^{-10} Torr), cooled to low temperatures and exposed to separate atomic beams of H and D to allow HD to form. After dosage, temperature programmed desorption and laser-induced thermal desorption experiments were performed on each sample.

The results for the porous ASW and the non-porous ice are quite different. The TPD spectrum from the porous ASW shows a strong signal peaking at a surface temperature of 28 K. If either one or both of the H and D beams is turned off, then the large HD peak disappears, so it is associated with the recombination of H and D atoms on external and internal surfaces of the porous ice. However, the TPD signal from the similar experiment on non-porous ASW is extremely weak.

Detailed investigation suggests that prompt desorption dominates for the non-porous ASW surfaces, while product molecules tend to be retained on porous ASW surfaces. They find that H and D atoms are mobile on the ices even at temperatures as low as 8 K. The recombination on porous ASW surfaces takes place predominantly on the walls of the pores and the newly-formed HD is retained until desorption occurs, induced by the TPD. The translational and internal energy of HD in this case is initially high, but much of the energy is dissipated in collisions with the pore walls so that the molecule desorbs in thermal equilibrium with the surface, having lost nearly all of its translational and internal energy. On the other hand, the dominant fraction of nascent HD formed on non-porous ASW desorbs with a high internal and kinetic energy. The disposition of the energy available when a molecule is formed (~4.5 eV) between the surface and the kinetic and internal modes of the molecule when released from the surface therefore depends crucially on the morphology of the ice surface. Presumably, this important conclusion must apply to all possible sample surfaces. It must affect arguments about the detectability of infrared emission from the radiative relaxation of initially rovibrationally excited molecules. More recent experimental work by Gavilan *et al.* (2012)[33] and Hama *et al.* (2012)[34] supports the conclusion that the surfaces of ASW will be active catalysts.

The second Danish experiment (Hornekaer *et al.* 2006)[32] that we shall describe uses scanning tunnelling microscopy (STM) to explore how chemisorbed D atoms arrange themselves on graphitic surfaces. The experiments were carried out in an UHV chamber (base pressure $\sim 3 \times 10^{-10}$ Torr) in which is contained the STM system. The sample is freshly cleaved HOPG annealed at 1300 K *in situ* by electron bombardment. The sample is dosed with an atomic hydrogen beam at a temperature of ~2000 K. As discussed in Section 8.3.1, there is a barrier for H to chemisorb on graphite of around 0.2 eV; this is easily overcome by atoms from this beam. The graphitic surface is at 210 K.

In the experiments, the sample is dosed and the STM images taken. Then the sample is allowed to warm up to room temperature for 10 minutes and then cooled again and a new STM taken. Figure 8.15 shows these stages for low ((a) and (b)), and high ((c) and (d)) fluxes.

At low fluxes, the annealing reduces the number of adsorbed species, and the comparison of (a) and (b) is consistent with the theoretically-derived desorption barrier of 0.9 eV. At higher fluxes, see (c) and (d), the reduction by annealing is much less and closer inspection reveals that the structures are more complex than those in (a) and (b) which—as a close-up in Figure 8.16(a) reveals—have a three-fold symmetry, as expected for a D atom located over a C atom of the surface.

These more complex structures are shown in Figure 8.16(b), a close-up of structures similar to Figure 8.16(d), and are attributed to D atom clusters. Figure 8.16(b–d) show increasing coverages. In (d), the coverage of the graphitic surface is almost complete. The diffusion barrier for chemisorbed D atoms on graphite (1.14 eV) is larger than the desorption barrier (0.9 eV), so that when the sample is heated the D atoms will desorb rather than move. Therefore, the clustering observed in these experiments represents the true nature of the graphitic surface, and supports the theoretical picture described in Section 8.3.1, in which the adsorption of additional D atoms at a site near to the first have a reduced—or even vanishing—adsorption barriers.

In the interstellar medium there are regions such as photon dominated regions (PDRs) where the gas temperatures are high enough ($\sim 10^3$ K) for barriers to H atom chemisorption to be overcome. These experiments suggest that clusters may then form, leading to high coverages from which H_2 formation may occur by recombinative desorption or by the Eley–Rideal mechanism. Therefore, these experiments suggest a mechanism by which H_2 formation in PDRs may occur, as required by observations.

After Deposition (Low Dose)   After Annealing (Low Dose)

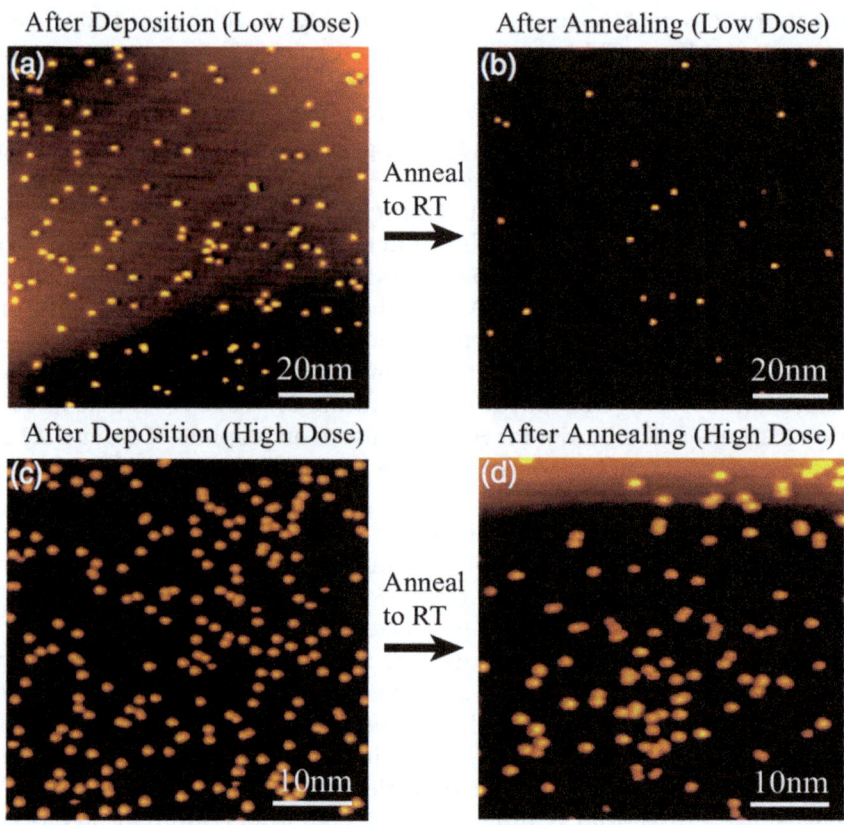

Figure 8.15   STM images of a graphite surface dosed with D atoms and subsequently annealed to room temperature (RT), (a) and (b) are for low
dosages; in (b), annealing removed much of the D coverage, (c) and (d)
are for high dosages; the effect of annealing is much reduced. Reproduced with permission from Hornekaer *et al.* (2006).[32]

8.5   A Speculative Approach to H₂ Formation: Grain Explosions

The suggestion that an interstellar dust grain may, under suitable conditions,
have a high coverage of H atoms, as discussed in the previous paragraph, is
an interesting idea. In the experiments of Hornekaer *et al.*, each hydrogen
atom is bound to the surface by an energy that is more than a physisorption bond energy but much less than a full chemical bond of several eV. In
principle, this coverage represents a large amount of "chemical energy"; it
amounts to about 4 eV per hydrogen molecule formed. If it could be released
abruptly, then it would amount to an explosion of the grain.

   Walt Duley and David Williams noted that this type of situation in the laboratory has often been described. For example, in some experiments carbyne
molecules formed in the laser ablation of graphite were condensed in an

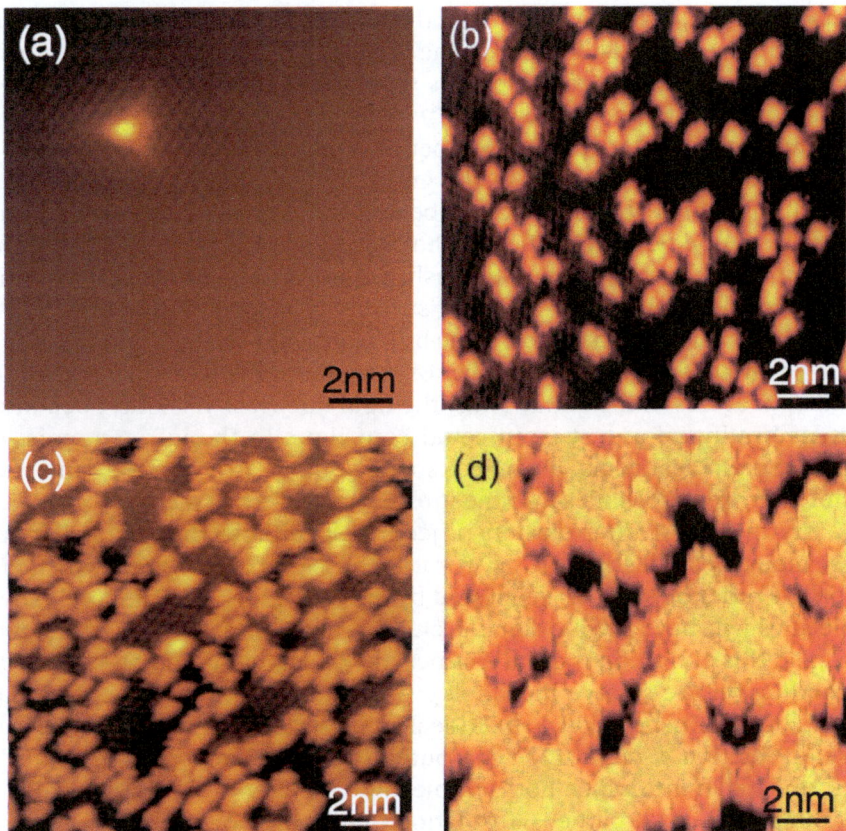

Figure 8.16 STM images of the graphite surface after increasing dosages of D atoms. (a) A close-up image of one of the bright protrusions in Figure 8.16(a), and shows a three-fold symmetry. (b)–(d) Show the effects of increasing dosages, indicating the presence of D atom clusters on the surface. Reproduced with permission from Hornekaer *et al.* (2006).[32]

inert gas on to a surface at a temperature of 6 K. When the temperature of the surface and carbyne layer was slowly raised to ~40 K there was an explosion with the emission of visible and near-infrared radiation. This abrupt energy release was attributed to the conversion of sp bonded carbon molecules such as alkynes (containing triple carbon bonds) and cumulenes (double-bonded carbon chains) to sp^2 bonded carbon as found in amorphous carbon. In these laboratory experiments, the runaway process was initiated by the warming of the material to a critical temperature at which the first reaction spontaneously occurred, the energy release from this reaction triggering all the rest of the reactions to occur very quickly and raising the temperature of the material to high values.

Could such a process occur on or in interstellar dust? While it is certainly possible that carbon chains are part of the carbonaceous material in interstellar dust, it seems likely from the preceding discussions that atomic hydrogen may also exist in and on the dust. Duley & Williams (2011)[35] estimated that a runaway process of this type, caused by the association of H atoms bound to grain, could raise the temperature of an entire canonical dust grain to about 10^3 K if the number of H atoms involved were about 5% of the total number of atoms in the grain. For canonical grains, the maximum number of H atoms that could be accommodated on the surface is on the order of one percent of the number of atoms in the grain; if the grains were porous or irregular then obviously more H atoms could be accommodated. Smaller grains have a bigger surface-to-volume ratio; if the grains were smaller than the canonical value by a factor of three in radius, then they could easily accommodate a number of H atoms equal to ten percent of the number of atoms in the grain. Thus, the figure of 5% is realistic. Such intermittent heating of a dust grain could contribute to the infrared emission from the grain as it cools. It would also generate intermittently an injection of hot hydrogen molecules into the interstellar gas.

If intermittent grain explosions inject hydrogen into the interstellar gas, these molecules will share the temperature of the exploding grain which may be on the order of 10^3 K (as suggested by the laboratory experiments). These molecules will have rovibrational excitation characteristic of that temperature, and this excitation is quite distinct from the excitation that may arise—depending on the surface morphology—in the continuous processes such as Langmuir–Hinshelwood, Eley–Rideal, and hot atom, previously described. Cecchi-Pestellini *et al.* (2012)[36] computed the emission spectrum from diffuse interstellar into which hot H_2 molecules are intermittently injected with an average rate of 1% of the H_2 formation rate. They show that such a process creates a distinct near-infrared spectrum, comparable with that observed. However, the spectrum generated in this way is closely similar to that generated by turbulence in the interstellar gas. Therefore, we do not have an unambiguous signature of the hot H_2 injection.

However, vibrational excitation in H_2 may open reaction paths that would otherwise be closed. Exchange reactions of C^+ ions with cold H_2 do not proceed rapidly, but become exothermic if the H_2 is rovibrationally hot. Thus, chemical routes to the formation of CH^+ become possible if hot H_2 is intermittently injected from exploding grains into interstellar gas. Cecchi-Pestellini *et al.* show that with a similar assumption about the hot H_2 injection rate, *i.e.* an average rate of 1% of the H_2 formation rate, the predicted abundances of these diatomic molecules are consistent with the abundances observed in diffuse interstellar clouds.

8.6 Formation of Other Species on Bare Dust Grains

Models in which both gas-phase and surface reactions contribute have been explored for many years. There are two approaches to the way in which surface chemistry is described. The first is to take a simplified view that if H atoms can form H_2 on the surface, then hydrogenation of other atoms that strike the surface may also be possible. Thus, oxygen atoms may form OH and H_2O which may be retained on the surface or ejected into the gas phase, and similarly, nitrogen may generate the species NH, NH_2, and NH_3. All these species have been detected in the interstellar medium, and all have competing gas-phase formation mechanisms. Models of chemistry in diffuse clouds usually rely on the fact that ices are not detected in these regions to infer that the saturated, hydrogenated species are ejected into the gas phase. This approach therefore ignores many potential contributions of surface reactions to interstellar chemistry, and—though plausible—is certainly over-simplistic (*e.g.* Cecchi-Pestellini *et al.* 2009).[37]

The second approach is to include a detailed description of how atoms collide and stick, move over the surface, and interact with other species. This second method is therefore a full generalization of the picture that has been developed for molecular hydrogen. In principle, this is a more desirable approach but in practice it means that the energy barriers for sticking, for mobility across the surface, and for reaction and desorption need to be known for many species. Models of this kind therefore require the input of a huge amount of data, most of which are unknown. The best known models of this kind are those of Eric Herbst and his collaborators (*e.g.* Garrod *et al.* 2008[38]). We discuss this approach in the following chapter.

Accurate studies in the laboratory and on the computer to provide this huge set of surface chemical data are rare. Cuppen *et al.* (2010)[39] have made a laboratory study of water formation in surface reactions, and propose that hydrogenation of O_2 may be more effective than hydrogenation of OH by H or H_2.

8.7 Discrete Grains

In theoretical modelling of surface reactions, it is often convenient to assume that the number of H (or other) atoms on the surface of a grain is a continuous quantity. Then it is straightforward to write down mathematical equations (usually differential equations) that describe how the number of atoms on the surface varies as new atoms arrive from the gas and stick to the surface, move over the surface, become bound at special sites, and finally react with other atoms and possibly leave the surface as part of the molecule. Differential equations of this kind, *i.e.* rate equations, have been used for many years (see for example Pickles & Williams 1977[40]) in exploring the consequences of different models of interstellar chemistry. Detailed models of this kind used in conjunction with laboratory data have been very effective in deducing values of energy barriers for atomic diffusion over the surface or for molecular desorption from the surface.[19,21,41,42]

However, if the coverage on a grain surface is low, then this approach may not be satisfactory. After all, if one H atom is on one grain and a second H atom is on another grain, then these two atoms will never react, but a rate equation approach that simply treated surface coverages as a continuous variable would predict that they could form a molecule. Obviously, there will be serious inaccuracies in the rate equation method when surface coverages on individual grains are low – as may well be the case in the low density interstellar medium.

There are two approaches that have been developed to deal with the problem. The first is the "master equation" method to describe the probability distributions of reactive species on grains[43,44] and the second is the Monte Carlo method.[45-47] The two methods give the identical results when using the same parameters.

For a small system of surface reactions the master equation approach can be coupled to the rate equations for gas-phase chemistry. For example, Green *et al.*[44] applied the master equation method to the formation of H_2, the oxygen system O, OH, H_2O and O_2, and the nitrogen system N, NH, NH_2, NH_3, and NO (all are known interstellar molecules; all also have formation routes in gas-phase reactions). However, large-scale calculations require approximations to be introduced to make the equations tractable, and the methods do not necessarily work for all regions of the parameter space.

The Monte Carlo method is conceptually appropriate but is difficult to couple to the system of rate equations describing gas-phase rate equations in gas–grain models of interstellar chemistry; this difficulty was overcome by Vasyunin *et al.* (2009)[45] but in practice the approach is computer-intensive. For smaller scale problems a related method, the continuous-time random walk (CTRW) method, or—more simply—the kinetic Monte Carlo method, can be used.[46]

The CTRW method has wide applicability. It can treat both Langmuir–Hinshelwood and Eley–Rideal mechanisms; it can accommodate grains that may be smooth, rough, or porous. Iqbal *et al.* (2012)[47] find that if strong chemisorption sites occur together with physisorption sites on the surface, then there may be a reasonable efficiency for H_2 formation up to very high temperatures (~700 K, far beyond anything likely to occur in molecular regions of interstellar space); see Figure 8.17. This is a conclusion that would undoubtedly have satisfied Gould & Salpeter (1963)[16] who—half a century ago—investigated theoretically the nature of the H-surface physical and chemical bonds and computed the efficiency of the H_2 formation reactions on grain surfaces; they did this even before the H_2 molecule had been identified in space!

8.8   Conclusions

There has been continuing interest in the formation of interstellar molecular hydrogen for more than half a century. The first suggestion that H_2 might be formed on the surfaces of dust grains was made by van de Hulst (1948).[48] Astronomers established its loss rate through photodestruction by starlight[49,50] and so—after the first detections of interstellar H_2 were made (in a rocket experiment by Carruthers 1970;[51] and from the Copernicus satellite observatory; Spitzer & Cochran 1973[52])—the necessary formation rate in diffuse interstellar clouds was inferred.[53] The range of values of the necessary formation rate obtained in this way confirmed that surface reactions on dust

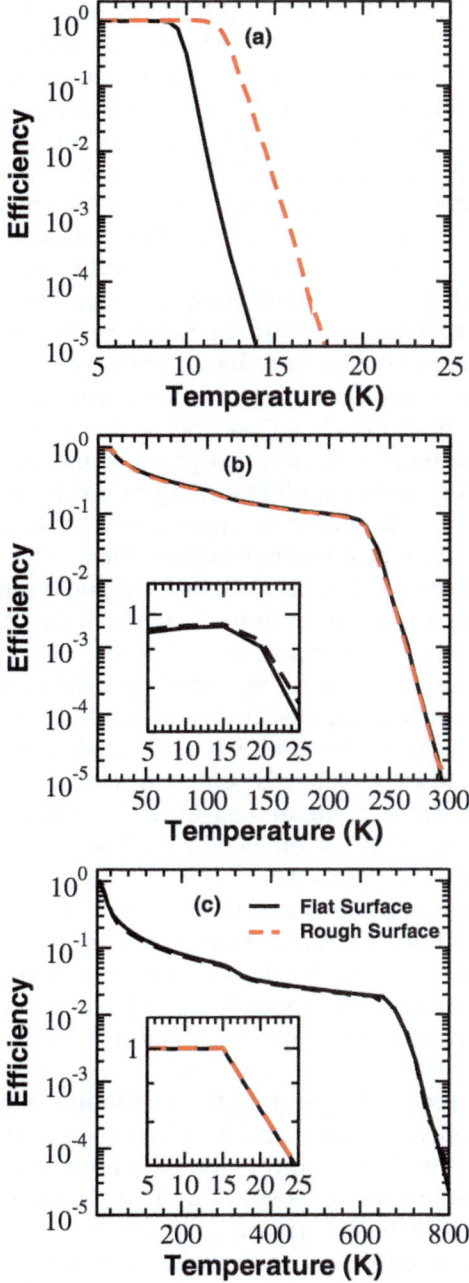

Figure 8.17 Efficiency of H_2 formation on surfaces. Panel (a) is for physisorption only, panel (b) is for chemisorption on olivine with a relatively low chemisorption energy, and panel (c) is for olivine with a higher chemisorption energy. In all panels, results for both flat and rough surfaces are given. Reproduced with permission from Iqbal *et al.* (2012).[47]

surfaces were likely to be the only viable H_2 formation mechanism in the interstellar space of the Milky Way. This formation mechanism, and the various fundamental processes that lie behind it, and the inferred interstellar formation rate have all been the targets of enormous efforts in astrochemical research. However, it was not until the last years of the 20th century that laboratory techniques and computational abilities had developed sufficiently to make possible realistic attacks both in the laboratory and *via* computational modelling of detailed quantum mechanical descriptions.

In less than two decades from the 1997 ground-breaking experiments of Pirronello and Vidali and their colleagues, there has been exceptional progress on this problem which is of fundamental interest to astronomers and surface scientists. Experiments and theory both confirm the viability of H_2 formation by physical and chemical adsorption on surfaces provided by plausible materials of interstellar dust grains. For graphitic surfaces, early indications that the barrier to chemisorption would inhibit H_2 formation have been removed by better understanding of the influence of increasing coverage and of the consideration of more realistic materials. The contrasting behaviour of amorphous and crystalline silicates has been made clear. As more and more research has been carried out and our understanding has improved, any possible difficulties in forming H_2 on dust have been lessened or even entirely removed. The importance of porosity in increasing the active surface has been established and the role of porosity in quenching rovibrational and kinetic energy of the nascent H_2 molecule has been demonstrated. It would be highly desirable to establish a direct observational signature of interstellar H_2 formation. As yet, this goal eludes astronomers.

Further experimental and theoretical work on dust grain analogues will certainly continue, not only for molecular hydrogen formation but also for the formation of other simple species. It is at present unclear, for example, whether energy released when O and H atoms combine on a surface to form OH cause the OH to desorb before the OH continues to form H_2O; and under what circumstances is the H_2O released from the surface and when is it retained? Is hydrogenation the most favourable surface reaction for atoms other than hydrogen, as it appears to be? What role does grain charge play in surface reactions?

More attention needs to be given to the astronomical scenarios in which these chemistries operate. For example, while the general nature of interstellar dust grains is well known, the details remain uncertain and may be affected by their environment. These details probably matter a great deal! Therefore, the balance of amorphous/crystalline and smooth/rough and porous/solid needs to be better determined from observations. Modellers should move away from the implicit assumptions that dust grains and their chemistry can be considered as if in steady-state. For example, dust grain temperatures are unlikely to be steady. Grains undergo temperature fluctuations, and the smaller the grains the larger the fluctuations.[53] Small grains undergo fluctuations in temperature so large that desorption will be greatly enhanced. Even larger grains are unlikely to be uniform but may be composed of poorly-connected pieces of material such as smaller grains. Therefore, temperature fluctuations may be occurring within grains and may affect desorption processes.[54,55]

Further Reading

1. J. Harris and B. Kasemo, *Surf. Sci. Lett.*, 1981, **105**, 281.
2. G. Vidali, *Chem. Rev.*, 2013, **113**, 8762.
3. R. M. Ferullo, N. F. Domancich and N. J. Castellani, *Chem. Phys. Lett.*, 2010, **500**, 283.
4. L. Jeloaica and V. Sidis, *Chem. Phys. Lett.*, 1999, **300**, 157.
5. B. J. Irving, A. J. H. M. Meijer and D. Morgan, *Phys. Scr.*, 2011, **84**, 028108.
6. T. P. M. Goumans, *Mon. Not. R. Astron. Soc.*, 2011, **415**, 3129.
7. B. Lepetit, D. Lemoine, Z. Medina and B. J. Jackson, *Chem. Phys.*, 2011, **134**, 114705.
8. B. Kerkeni and D. C. Clary, *Chem. Phys.*, 2007, **338**, 1.
9. D. Bachellerie, M. Sizun, F. Aguillon, D. Teillet-Billy and N. Rougeau, *Phys. Chem. Chem. Phys.*, 2009, **11**, 2715.
10. D. Bachellerie, M. Sizun, D. Teillet-Billy, N. Rougeau and V. Sidis, *Chem. Phys. Lett.*, 2007, **448**, 223.
11. M. Sizun, D. Bachellerie, F. Aguillon and V. Sidis, *Chem. Phys. Lett.*, 2010, **498**, 32.
12. T. P. M. Goumans, C. R. A. Catlow and W. A. Brown, *Mon. Not. R. Astron. Soc.*, 2009, **393**, 1403.
13. T. P. M. Goumans and S. T. Bromley, *Mon. Not. R. Astron. Soc.*, 2011, **414**, 1285.
14. D. Field, *Astron. Astrophys.*, 2000, **362**, 774.
15. J. Le Bourlot, F. Le Petit, C. Pinto, E. Roueff and F. Roy, *Astron. Astrophys.*, 2012, **541**, A76.
16. R. J. Gould and E. E. Salpeter, *Astrophys. J.*, 1963, **138**, 393.
17. V. Pirronello, C. Liu, L. Shen and G. Vidali, *Astrophys. J.*, 1997, **475**, L69.
18. V. Pirronello, O. Biham, C. Liu, C. Shen and G. Vidali, *Astrophys. J.*, 1997, **483**, L131.
19. N. Katz, I. Furman, O. Biham, V. Pirronello and G. Vidali, *Astrophys. J.*, 1999, **522**, 305.
20. V. Pirronello, C. Liu, J. Roser and G. Vidali, *Astron. Astrophys.*, 1999, **344**, 681.
21. H. Perets, A. Lederhendler, O. Biham, G. Vidali, L. Li, S. Swords, E. Congiu, J. Roser, G. Manico, J. R. Brucati and V. Pirronello, *Astrophys. J.*, 2007, **661**, L163.
22. J.-L. Lemaire, G. Vidali, S. Baouche, M. Chehrouri, H. Chaabouni and H. Mokrane, *Astrophys. J.*, 2010, **725**, L156.
23. L. Gavilan, J. L. Lemaire, G. Vidali, T. Sabri and C. Jaeger, *Astrophys. J.*, 2014, **781**, 79.
24. E. Congiu, E. Matar, L. E. Kristensen, F. Dulieu and J. L. Lemaire, *Mon. Not. R. Astron. Soc.*, 2009, **397**, L96.

25. H. Chaabouni, H. Bergeron, *et al.*, *Astron. Astrophys.*, 2012, **538**, A128.
26. J. S. A. Perry, J. M. Gingell, K. A. Newson, J. To, N. Watanabe and S. D. Price, *Meas. Sci. Technol.*, 2002, **13**, 1414.
27. S. C. Creighan, J. S. A. Perry and S. D. Price, *J. Chem. Phys.*, 2006, **124**, 114701.
28. F. Islam, E. R. Latimer and S. D. Price, *J. Chem. Phys.*, 2007, **127**, 064701.
29. E. R. Latimer, F. Islam and S. D. Price, *Chem. Phys. Lett.*, 2008, **455**, 174.
30. F. Islam, C. Cecchi-Pestellini, S. Viti and S. Casu, *Astrophys. J.*, 2010, **725**, 1111.
31. L. Hornekaer, A. Baurichter, V. V. Petrunin, D. Field and A. C. Luntz, *Science*, 2003, **302**, 1943.
32. L. Hornekaer, E. Rauls, W. Xu, Z. Sljivancanin, R. Otero, I. Stensgard, E. Laegsgard, B. Hammer and F. Besenbacher, *Phys. Rev. Lett.*, 2006, **97**, 186102.
33. L. Gavilan, J. L. Lemaire and G. Vidali, *Mon. Not. R. Astron. Soc.*, 2012, **424**, 2961.
34. T. Hama, K. Kuwahata, *et al.*, *Astrophys. J.*, 2012, **757**, 185.
35. W. W. Duley and D. A. Williams, *Astrophys. J., Lett.*, 2011, **737**, L44.
36. C. Cecchi-Pestellini, W. W. Duley and D. A. Williams, *Astrophys. J.*, 2012, **755**, 119.
37. C. Cecchi-Pestellini, D. A. Williams, S. Viti and S. Casu, *Astrophys. J.*, 2009, **706**, 1429.
38. R. T. Garrod, S. L. Widicus Weaver and E. Herbst, *Astrophys. J.*, 2008, **682**, 283.
39. H. M. Cuppen, S. Ioppolo, S. Romanzin and H. Linnartz, *Phys. Chem. Chem. Phys.*, 2010, **12**, 12077.
40. J. B. Pickles and D. A. Williams, *Astrophys. Space Sci.*, 1977, **52**, 443.
41. S. Cazaux and A. G. G. M. Tielens, *Astrophys. J.*, 2010, **715**, 698.
42. H. Perets, O. Biham, G. Manico, V. Pironello, J. Roser, S. Swords and G. Vidali, *Astrophys. J.*, 2005, **627**, 850.
43. O. Biham, I. Furman, V. Pironello and G. Vidali, *Astrophys. J.*, 2001, **553**, 595.
44. N. J. B. Green, T. Toniazzo, M. J. Pilling, D. P. Ruffle, N. Bell and T. W. Hartquist, *Astron. Astrophys.*, 2001, **375**, 1111.
45. A. I. Vasyunin, D. A. Semenov, D. S. Wiebe and T. Henning, *Astrophys. J.*, 2009, **691**, 1459.
46. H. M. Cuppen, E. F. van Dishoeck, E. Herbst and A. G. G. M. Tielens, *Astron. Astrophys.*, 2009, **508**, 275.
47. W. Iqbal, K. Acharyya and E. Herbst, *Astrophys. J.*, 2012, **751**, 58.
48. H. C. van de Hulst, *Harv. Obs. Monogr.*, 1948, **7**, 73.
49. T. P. Stecher and D. A. Williams, *Astrophys. J.*, 1967, **149**, L29.
50. J. H. Black and A. Dalgarno, *Astrophys. J.*, 1973, **184**, L101.
51. G. R. Carruthers, *Astrophys. J.*, 1970, **161**, L81.
52. L. Spitzer, Jr. and W. D. Cochran, *Astrophys. J.*, 1973, **186**, L23.
53. M. Jura, *Astrophys. J.*, 1975, **197**, 575.
54. W. W. Duley, *Nat. Phys. Sci.*, 1973, **244**, 57.
55. W. W. Duley and D. A. Williams, *Mon. Not. R. Astron. Soc.*, 1988, **231**, 969.

CHAPTER 9

Ice Formation on the Surfaces of Interstellar Dust Grains: Chemical Processing of the Ice

9.1 Ice Mantles on Dust Grains

Interstellar ice is a remarkable material: it enables a dramatic chemical transformation to take place in the interstellar medium. This transformation converts relatively simple chemical species (such as CO, H_2O, and NH_3) into some that are relatively complex (such as C_2H_5OH, CH_2CHCN, or $HCOOCH_3$). Much of this chapter will be concerned with how this transformation occurs. But first we need to be clear where the ice exists in the interstellar medium and where it is absent. An excellent review of interstellar water chemistry has recently (2013) been published by Ewine van Dishoeck, Eric Herbst, and David A Neufeld.[1]

Water ice is readily detected in interstellar space by its absorption in the pure O–H stretching mode, in a broad feature peaking near 3 μm. The first detections of water ice in the interstellar medium were made in 1973 by Gillett & Forrest.[2] The rotational structure of this feature that would appear in the spectra of *gaseous* H_2O molecules is, of course, suppressed in the solid ice, since molecules are not free to rotate. However, a number of other modes exist, from 4.5 to 63 μm. Detections of interstellar water ice at longer wavelengths, corresponding to various bending, combination and librational modes have also been made. In the interstellar medium, these broad line absorptions are normally detected in foreground material observed against a cool background star that has a prominent infrared continuum emission. Detections are also made in material observed towards an embedded protostar which

The Chemistry of Cosmic Dust
By David A. Williams and Cesare Cecchi-Pestellini
© David A. Williams and Cesare Cecchi-Pestellini 2016
Published by the Royal Society of Chemistry, www.rsc.org

heats dust in its close environment to hundreds of kelvin; this hot dust provides an infrared continuum against which line absorption by cooler dust in the same line of sight and further from the protostar is detected. Most of the detected interstellar ice seems to be in a high density amorphous form that does not occur naturally on Earth; the detected interstellar ice is believed to be mostly amorphous because a sharp laboratory spectral feature at 3.1 μm seen in the laboratory from crystalline ice is rare in the interstellar medium. Porous ice has a feature at 2.7 μm originating in a dangling OH bond; this feature has not been detected in interstellar space, so ices of high porosity (see the discussion in Section 8.4.2) may be rare (a conclusion supported by the experimental study of Accolla *et al.* 2011[3]). Interstellar ice is not pure water ice, but contains species such as CO, CO_2, and other less abundant molecular species (see Section 9.2, below), and these impurities tend to modify the central positions and line profiles of the pure water ice features.

Gaseous water has a rich spectrum that ranges from electronic transitions in the ultraviolet to pure rotational transitions in the millimetre and submillimetre region. However, absorption in the ultraviolet spectrum is not observed in the interstellar medium, because H_2O molecules are very readily photodissociated where ultraviolet radiation is intense. Interstellar detections of gaseous water are usually made in the rotational spectrum, in the millimetre and submillimetre wavebands. The H_2O molecule is an asymmetric rotor with spectra existing in *ortho* and *para* ladders.

The first detection of water vapour in the interstellar medium was famously made in 1969 by members of Charles Townes' group.[4] The emission was at 22 GHz (a wavelength of 1.38 cm) and was intense. It was attributed to maser emission in the 6_{16}–5_{23} rotational transition. Emissions from gaseous *ortho* water molecules in the 1_{10}–0_{01} transition (557 GHz) and *para* molecules in the 1_{11}–0_{00} (1113 GHz) and several other rotational lines have been studied in diffuse and translucent interstellar clouds using the HIFI and PACS instrument on the Herschel Space Observatory, and by the Spitzer Space Telescope in highly excited pure rotational transitions in the mid-infrared. The typical gas phase water abundance relative to hydrogen molecules in the same interstellar location is ~10^{-8}.

Non-detections of the 3 μm feature confirm that interstellar ice is absent from low density interstellar clouds through which starlight penetrates readily, *i.e.* those diffuse clouds through which the optical depth caused by dust extinction in the visual part of the spectrum is less than or about unity. Indeed, it was the absence of this 3 μm feature in much of the interstellar medium that made it clear that grains could not, in general, be composed of ice (see Chapter 1). However, the grains can—under suitable circumstances—provide substrates on which ice may be deposited. As Figure 9.1 shows, ice begins to be deposited on dust grains at an optical depth into dense clouds that is somewhat greater than the total depth of diffuse clouds. This threshold depth is typically about several magnitudes of extinction in the visual. The greater the visual extinction, the greater the optical depth in the water ice feature, and the two are linearly related above the threshold for ice deposition. Figure 9.1 shows observational data from Chiar *et al.* (2011)[5] for two different lines of

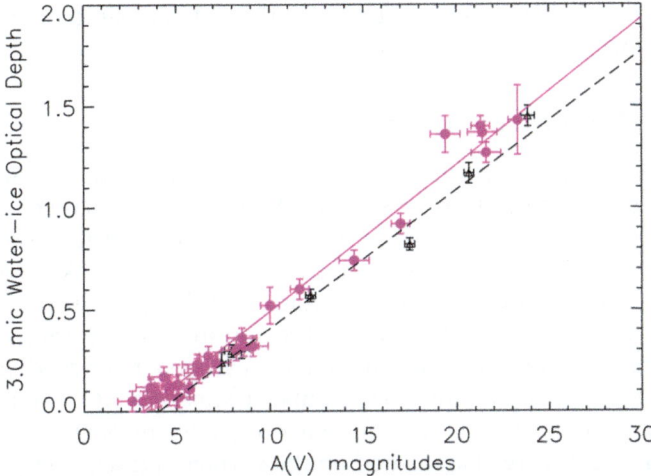

Figure 9.1 Optical depth of the 3.0 μm ice feature plotted against visual extinction, A_V, for field stars behind the dark interstellar clouds IC 5146 (black symbols) and Taurus (purple symbols). Reproduced with permission from Chiar *et al.* (2011).[5]

sight in the interstellar medium of the Milky Way galaxy. These data show how the optical depth in the 3 μm ice absorption feature varies as a function of extinction in the visual (defined to be at a wavelength of 550 nm); in the figure, both depths are expressed in magnitudes (see Chapter 2). In other words, for each line of sight, the amount of ice, as measured by the optical depth in the 3 μm feature, varies linearly with the amount of material along that line of sight, as measured by the optical depth in the visual.

Figure 9.1 shows that the line of sight towards the quiescent cloud IC 5146 possesses a threshold extinction for the onset of ice deposition on dust grains of 4.03 ± 0.05 magnitudes. However, there is foreground extinction (between the Sun and IC 5146) included in this amount. When that is allowed for, the actual threshold for deposition of ice in the IC 5146 cloud occurs at 3.2 magnitudes, the same value as for the threshold for ice deposition in the Taurus Dark Cloud (the data for Taurus are also shown in the Figure 9.1). However, this apparent agreement is fortuitous and the threshold extinction for the onset of ice may vary widely. In the Serpens and ρ Ophiuchus clouds the thresholds are about six and about 13 magnitudes, respectively.

We have mentioned that CO is an important constituent of the ice mantles and can be detected through absorption in C–O stretch modes at a wavelength of 4.7 μm. The correlation that exists for water ice optical depth with the visual optical depth exists in a similar fashion for CO in the ice. The threshold for the onset of CO ice is, however, somewhat larger than for water ice; in Taurus, CO ice onset threshold occurs at five visual magnitudes. The interpretation that is inferred from this information is that water ice deposition is mostly completed before the CO ice deposition occurs. The initial CO ice is closely associated with the H_2O molecules and the C–O stretching mode is affected by their dipole moments, while the CO deposited later shows the stretching mode of nearly pure CO ice.

9.1.1 The Water Ice Threshold

The gas phase abundance of water molecules (observed, see Section 9.1, above, or computed in suitable gas–grain chemical models) is so low that the ice mantles cannot be accumulated on grain surfaces within the lifetime of the clouds.[6] Therefore, the ice is inferred to form by the sticking of O atoms to the grain surface and their hydrogenation and retention there as water molecules. In fact, the 1973 detection of ice on interstellar grains was, after the 1970 detection of interstellar molecular hydrogen, the first definitive evidence for surface chemistry on dust grains (see Section 8.6). Chemical pathways to form water ice may also include hydrogenations of O_2 and O_3,[7] where these oxygen species are also formed on the surfaces (and are very rare in the gas phase). It is important, therefore, that free atomic or molecular oxygen is available to form water molecules. For the elemental abundances available in the Milky Way, the abundance of oxygen exceeds that of carbon by a factor of about two. In gas phase chemistry of interstellar molecular clouds, carbon takes up all the oxygen it can, so that carbon monoxide is the most abundant molecule after molecular hydrogen. The considerable amount of excess oxygen is available to be incorporated into other molecules, and some is incorporated into silicate dust. But under most conditions the silicate dust and these molecules take up rather small amounts of the remaining oxygen. Therefore, in most molecular clouds there is a considerable reservoir of atomic oxygen (about 10^{-4} relative to H atoms by number) that—under the right circumstances—may be converted to ice.

Why are interstellar grains free of ice in regions of space with visual extinction below the critical threshold, while grains in regions with extinction above the critical threshold are coated with ice? What causes this threshold? What are the processes that keep grains clean in some regions and allow ice mantles to grow in others? The absence of ices in low extinction regions of interstellar space and the inhibition of ice deposition in them are still not fully understood, but must depend on the balance between various desorption processes and the incidence of O atoms at the grain surfaces. Several ideas concerning the desorption of molecular species from grain surfaces have been discussed in the literature (see for example Roberts *et al.* 2007[8]). Thermal desorption seems a possible cause; it becomes sufficiently rapid in interstellar conditions when the ice temperature rises to about 100 K, but is totally negligible for typical ice temperatures of 10 K. Thus, other non-equilibrium processes may play a role in establishing the threshold for ice deposition. We mention some of these processes briefly here, but the ultimate cause of the threshold is currently unknown.

Photodesorption by ultraviolet radiation of water molecules from surfaces of grain analogues can be studied both in the laboratory and through classical molecular dynamics simulations. The process is found to be efficient, and to operate through an initial dissociation to H + OH, both species having excess energy. The most likely outcome is that the energetic H atom desorbs, leaving an OH trapped which will rapidly be converted to H_2O by reaction with adsorbed H. However, energetic OH desorption occurs in about one percent of these cases, effectively desorbing one H_2O molecule. Therefore, photodesorption by the ultraviolet component of the interstellar radiation field may tend to keep grains clear of ice in the outer parts of clouds in which starlight still readily penetrates. While photodesorption may have an effect at depths when the extinction is only a few magnitudes (as in the case of IC 5146 and Taurus) it is not clear that photodesorption can be effective when the extinction approaches, say, 13 magnitudes as in the ρ Oph cloud. In the ultraviolet (where photodesorption occurs) extinction is several times more effective than in the visual, so 13 visual magnitudes may be equivalent to around 30 magnitudes in the ultraviolet, so that ultraviolet is effectively totally suppressed. Evidently, photodesorption by starlight is not the whole story.

Cosmic rays (fast atomic nuclei with energies above about 1 MeV) deposit a small amount of heat when they pass through dust grains, causing some local transient thermal desorption on large grains and complete desorption on small grains. This process should operate throughout the cloud, regardless of depth, as cosmic rays—unlike ultraviolet photons—react only weakly with the cloud material and penetrate to very great extinctions. Direct impacts by cosmic rays can drive desorption from dust deep inside dense clouds, but are unlikely to determine the threshold extinction for ice deposition. Cosmic rays also induce a weak ultraviolet radiation field (with an intensity typically 10^{-4} of that of the mean interstellar radiation field) inside clouds; first, hydrogen is ionized, creating energetic electrons which relax collisionally and then excite electronic transitions in H_2. The relaxation from these excited states of H_2 creates the induced radiation field. This field can be an important contributor to desorption, but is unlikely to control the onset of ice deposition.

As we have seen in Chapter 8, surface reactions may deposit significant amounts of energy at the reaction site and so cause a local heating. This may also cause desorption of weakly bound species adjacent to that site. This process is weighted towards the edge of a cloud, because the frequency of surface reactions depends on the abundance of gas phase atoms, and atomic abundances decline with depth into the cloud.

A lone H_2O molecule on the surface of a bare grain experiences a different environment to that of an H_2O molecule that is part of a monolayer. While in the former case, the binding is simply with the surface, in the latter case, the binding of the H_2O to the grain is enhanced by H-bonding to adjacent water molecules. Thus, the cause of the onset of ice may simply be that while there are desorption mechanisms that tend to keep the surface clear of molecules in low extinction regions, these desorption mechanisms are simply swamped in denser regions. In these denser regions the flux of O atoms incident on the grain surfaces is so great that an H_2O monolayer tends to build up and H bonding among the constituent H_2O molecules dominates most desorption processes and ensures ice mantle growth on the initial monolayer.

Finally, as we have seen in Part B of this book, it is misleading to think of dust as unchanging from the moment of its formation until its final destruction. The physical and chemical nature of dust may change in response to the local physical conditions. It is possible that such physical and chemical changes may make dust grains more or less favourable to the deposition of ice mantles, depending on the local conditions.

9.1.2   The Ice: Depth Correlation

Figure 9.1 shows that the plots of the 3 µm optical depth (above the thresholds) with the optical depth in the visual are linear, and that both IC 5146 and Taurus show essentially the same slopes. The data imply that—above a critical threshold—the greater the amount of material in the cloud, the greater the amount of water ice in that cloud. The linearity and the commonality of slope appear to suggest that all available oxygen is rapidly converted into ice for all regions inside the threshold.

These conclusions remain untested by detailed models. One issue is the timescale for conversion of atomic oxygen to water ice. To build up a typical ice mantle on a dust grain in a cloud of total hydrogen density n_H takes about $(10^4/n_H \text{ cm}^{-3}) \times 1$ million years. Thus, clouds at the lower end of the density range in which ice mantles are found, with $n_H \sim 10^3 \text{ cm}^{-3}$, may not have sufficient time to form their ice mantles before the clouds disperse. Therefore, one might expect a wide variation in slopes in plots such as that shown in Figure 9.1, with lower density regions showing rather smaller slopes. Both plots in Figure 9.1 have very nearly the same slope.

The issue of density in the previous paragraph also raises the question: are the clouds of uniform density? High angular resolution studies of (gaseous) molecular lines in several quiescent molecular clouds appear to show that they are clumpy, and arguments guided by magnetohydrodynamics (MHD) considerations suggest that the clumps must be transient.[9] Densities within these transient clumps may be high, possibly $n_H \sim 10^5 \text{ cm}^{-3}$, so that the grains in each transient clump would attain their ice mantles relatively rapidly. In this picture, a molecular cloud consists of a large number of high density transient clumps embedded in a low density background gas. The grains in the clumps bear the maximum ice mantle. Therefore, lines of sight through such a cloud would intersect a number of clumps; if all clumps are of a similar size then the number of clumps intersected determines both the strength of the ice feature and the visual extinction. In other words, the linearity and uniformity of slope arise naturally from such a model.

The formation of water on the surfaces of interstellar grains is initiated by the arrival of O atoms at the surface. Routes to H_2O may involve O, O_2 and O_3.[10,11] However, it remains unclear how much of the product of a surface reaction is retained on the surface and how much is ejected from the surface by the excess energy in the reaction. In a laboratory study on dust grain analogues, He & Vidali (2014)[12] find no evidence of desorption of water on formation, while in their experiments Dulieu *et al.* (2013)[13] find that direct release to the gas phase may occur. However, the implication of the astronomical evidence is clear, as stated earlier: a high proportion of the oxygen atoms arriving at the surface of dust grains in dark clouds must contribute to ice formation, otherwise the ice mantles cannot be constructed within the lifetime of the clouds.

Many theoretical studies have been made of the formation of interstellar ice, some by the master equation approaches and some by the Monte Carlo methods (as discussed in Section 8.7). These studies may also be linked to gas-phase chemistry. Monte Carlo methods enable the nature of each layer of the deposited ice to be described. Figure 9.2 (ref. 14) shows the composition of each monolayer occupied by assorted molecules computed using

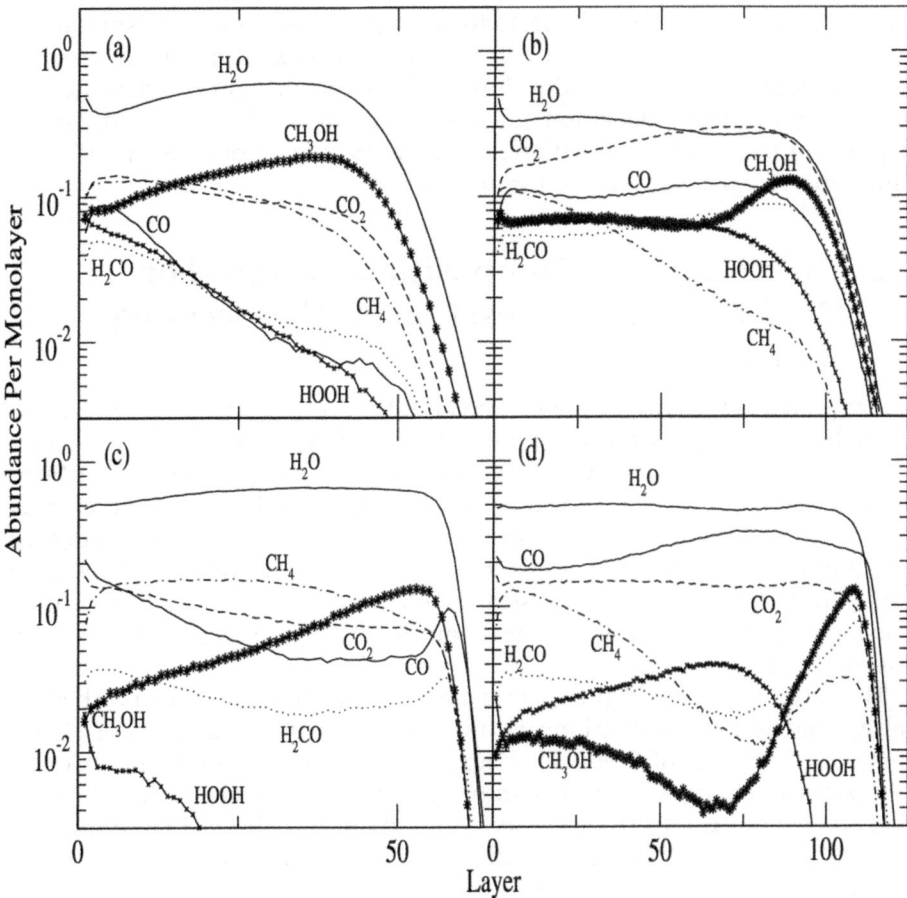

Figure 9.2   Fraction of each monolayer occupied by water and other molecular species in ice mantles on dust grains at an evolutionary time of 2×10^5 years; in (a) and (c) the H atom number density in the cloud is 2×10^4 cm^{-3} and in (b) and (d) it is 1×10^5 cm^{-3}. In (a) and (b) the temperature is 10 K while in (c) and (d) it is 15 K. Reproduced with permission from Chang & Herbst (2012).[14]

gas–grain models of interstellar chemistry, for cold gas clouds at two characteristic densities and two temperatures.

However, such models require the input of large amounts of surface science data, which are—unfortunately—poorly known.

The correlation between ice optical depth and visual extinction, and the similar slopes measured in different clouds, seem to imply that interstellar grains in interstellar clouds above the threshold tend to take up the maximum amount of ice possible. This supports the view that the freeze-out times are generally short compared to the cloud lifetimes (but see Boogert *et al.* 2013 [15] for observations of the Lupus clouds which may have lower water to hydrogen ratios). Some molecular line evidence suggests that molecular clouds are clumpy, where the clumps are transient and exist in a diffuse ice-free background.[16] If the timescale for clump formation and dissipation is too short, then ice production will not attain a maximum. Otherwise, ice measurements simply count the clumps in the line of sight. Neither the ice/visual depth linear correlation nor the discovery of the threshold for ice deposition has received much attention, yet both phenomena should contain important astrophysical information.

9.2 Interstellar Ice Composition Observed on Different Sight-Lines: Chemical Processing in Ices

A comprehensive study of the chemical nature of interstellar ice has been made by Karin Öberg and colleagues (2011)[17] using Spitzer Space Telescope archived data (at wavelengths 5–30 µm) supplemented by some data from ground-based telescopes. The data refer to a total of 63 lines of sight towards both low mass and high mass young stellar objects, encompassing a range of conditions from molecular clouds to disks near massive young stars, a huge range in density and other physical conditions. The data come from various papers, but have been homogeneously re-analysed. Some spectra were shown in Figure 2.3.

The various features in the spectra are identified by comparison with laboratory data obtained for various ice mixtures. The Öberg *et al.* analysis shows that there is a range of chemical composition of interstellar ice, and that this composition changes—if rather modestly—in a consistent way, indicating that the chemical nature of the ice responds to the physical environment. Table 9.1 gives the abundance medians relative to H_2O and upper and lower quartiles of the various spectral features analysed for lines of sight towards low mass and high mass stars. The features are associated both with pure material and with species in association with other materials, as indicated by laboratory data.

Towards low mass stars, most of the CO is pure, while towards high mass stars most of the CO is associated with H_2O. A small amount of CO is associated with CO_2, in each case. It is clear that although pure materials certainly exist in the ices, there is also considerable mixing. It is also useful—when considering the potential for chemistry that might occur in or from the ice—to know the mean relative abundances of various species, and these are shown in Table 9.2. In both tables, features at 2165 and 2175 cm^{-1} are attributed to OCN$^-$ and this species is listed as XCN.

There are a number of important conclusions that can be drawn from the Öberg *et al.* results. Some species seem to be well correlated with H_2O. These include CH_4 and NH_3, suggesting that these species are formed by the hydrogenation of C and N atoms, along with the hydrogenation of oxygen that forms the H_2O ice. The mixture CO_2/H_2O is also in this category, suggesting that CO is being oxygenated on the surface while C and N are hydrogenated, so that the resulting CO_2 may form a bond with the H_2O if suitably located. However, the more weakly bound pure species CO and CO_2 show wide variations relative to H_2O, suggesting that the influence of thermal and ultraviolet processing affects their abundances in the ice. Surprisingly complex consequences may occur following irradiation. Chen *et al.* (2014)[18] have studied the photodesorption of CO ice by vacuum ultraviolet emission and suggest that excited CO may react with another CO to form CO_2 (another appearance of the Boudouard reaction; see Chapter 5). Other species usually show a modest variation of a factor of 2 or 3 with respect to H_2O, but methanol can vary widely. Boogert *et al.* (2008)[19] show that CH_3OH/H_2O can reach values as high as 0.3 in some locations.

Table 9.1 Abundance medians and (lower, upper quartiles) of components of ices relative to water ice measured along 63 lines of sight towards both low mass stars and high mass stars (data extracted from Öberg *et al.* 2011[17]).

Ice feature	Low mass stars	High mass stars
H_2O	100	100
CO	38(20,61)	13(7,19)
CO_2	29(22,35)	13(12,22)
CH_3OH	7(5,12)	8(8,16)
NH_3	5(4,6)	16(10,17)
CH_4	5(4,7)	4(2,4)
XCN	0.6(0.2,0.8)	0.8(0.4,1.4)

Table 9.2 Typical relative abundances of ice components on lines of sight towards low mass, high mass, and background stars (data from Öberg *et al.* 2011[17]).

Component	Low mass stars	High mass stars	Background stars
H_2O	100	100	100
CO	29	13	31
CO_2	29	13	38
CH_3OH	3	4	4
NH_3	5	5	—
CH_4	5	2	—
XCN	0.3	0.6	—

These conclusions are, as we will see in the following section, consistent with laboratory data and with chemical gas–grain models. The molecule CO is the second most abundant molecule in molecular clouds (after H_2), and can be hydrogenated on surfaces to form two species identified in interstellar ice: formaldehyde and methanol. However, there are barriers (~2000 K) to the hydrogenation of CO and H_2CO that will normally prevent the complete conversion of all CO on the dust to CH_3OH (see Section 9.3.2, below). In some locations, elevated temperatures do, however, seem to drive this hydrogenation towards completion. The formation of CO_2 may be more complex; it may occur as indicated in the previous paragraph or through a reaction such as $CO + OH \rightarrow CO_2 + H$ which therefore requires the creation of OH radicals within the ice.

9.3   Cloud Evolution and Star Formation: Hot Core Chemistry

The chemistry of interstellar molecular clouds is determined by their physical condition. Low density clouds provide little obscuration to starlight, and—apart from H_2 and CO—the abundances of simple diatomic and triatomic molecules are really rather small and their chemistry is controlled by photo-processes as well as by cosmic ray ionization. (Grain surface reactions in these clouds are also important for the formation of H_2 and possibly other species, as we have seen in Chapter 8.) In translucent clouds where the gas number densities may be higher and extinction of starlight somewhat greater, starlight plays a reduced role and gas phase chemistry is largely controlled by the ionization created by cosmic rays. This chemistry is notable for the presence of unsaturated hydrocarbons. In still denser regions, the timescale for freeze-out of gas phase species on the surfaces of dust grains is so short that ice formation occurs, gas phase abundances are reduced, and a fairly simple solid state chemistry occurs, as described in the previous section, with ices largely dominated by H_2O, CO and CO_2. Can astronomers detect yet denser material? If so, what is the state of this gas?

The process of star formation involves the gravitational collapse and fragmentation of interstellar gas clouds. The collapse is opposed by thermal and magnetic pressures, by turbulence and by angular momentum. Chemistry during these dynamical processes produces molecules that are effective radiators even at low temperatures, so that gravitational energy released as heat during the collapse is radiated away and the clouds remain cool during the collapse. The chemistry producing the molecules also tends to suppress the fractional ionization so that magnetic pressures (which act only on the ions) are reduced. The remaining factors of turbulence and angular momentum fragment the gas, and denser clumps contract faster. The end-point of this evolutionary process is the formation of a star. Thus, the densest gas should be found in association with very young stars, protostars, still embedded in the gas from which they were formed. Initially, these stars are heated by the release of gravitational energy that occurs at an increasing rate as the star is forming. In this phase, the star is a powerful source of infrared radiation, and heats the gas in its immediate vicinity to temperatures of a few hundred Kelvin. Eventually, the compression of the gas by gravitational forces creates conditions in the interior of the protostar that enable thermonuclear processes to switch on, and the star rapidly heats up and becomes a powerful radiator in the visible and ultraviolet. During this phase, the surrounding material will be swept away by powerful winds originating in the star, and chemical complexity in this ejected material declines.

Thus, the densest gas should be found in association with the youngest stars, and should be warm. This is indeed what is observed. The detections are made in tiny regions, around one percent or so of the size of a typical molecular cloud, and are of rotational lines of gaseous molecules. The striking result of decades of observations and molecular line detections is that the molecular species detected are of a complexity not normally found in interstellar chemistry, and are present in substantial abundance. These tiny molecular regions are called "hot cores" when associated with massive and powerful protostars, and "warm cores" (or sometimes "hot corinos") when associated with less massive (and less powerful) protostars. The temperatures in these regions are typically a few hundred Kelvin, and the number densities of hydrogen molecules are typically ten or a hundred million per cm^3.

The range of species found in hot cores, but not present in molecular clouds, is remarkable. They are generally familiar saturated organic species such as alcohols, acids, aldehydes, esters and ethers. There are hydrocarbons, oxygen-containing and nitrogen-containing species in similar abundances, while sulfur-containing species are rare. In size they range up to around ten atoms. Probably the richest hot core in the Milky Way is found in the centre of the galaxy, within a region in the constellation of Sagittarius known as Sgr B2 that contains many of the observed molecular species. The emissions from hot core molecules peak within the hot core Sgr B2(N) a relatively tiny region comparable with the size of the Solar System but containing thousands of solar masses of material. Because of the concentration of large molecules in this region it is given the name Large Molecule Heimat, or Sgr B2(N-LMH). There are many hot cores in the Milky Way galaxy, possibly as many as $\sim10^4$.[20] In Figure 9.3 (ref. 21) we show the contour map of molecular line emissions associated with one particular hot core.

Hot and warm cores are not uniquely responsible for large molecules; they are also found in some other types of object, such as the extended outflowing envelopes around cool stars and planetary nebulae. A cold interstellar cloud—the Taurus Molecular Cloud-1 (TMC-1)—has been shown to be exceptionally rich in large carbon chain species. There is also a peculiar extended region in the centre of the Milky Way galaxy called the Central Molecular Zone (CMZ). It displays many molecular species that we would regard as typical of hot cores, but the gas number densities of the CMZ are much lower than in hot cores. While the chemistries of cool star envelopes and of TMC-1 have been thoroughly explored, the CMZ chemistry remains to be fully explained. In this chapter, we shall discuss only the chemistry of hot and warm cores and the ideas currently under discussion to account for their chemistries.

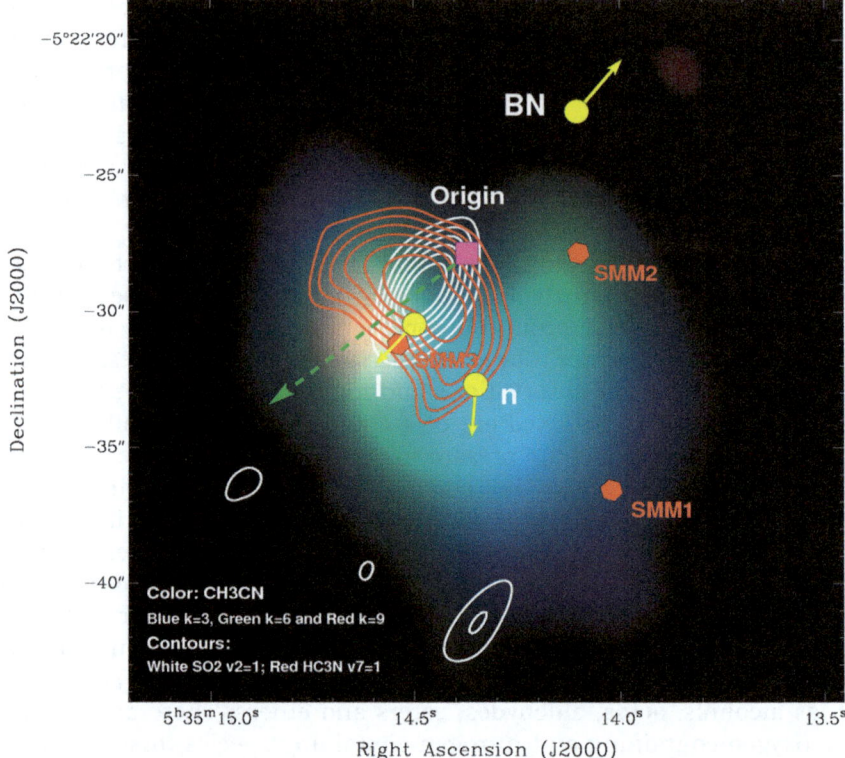

Figure 9.3   Millimetre and submillimetre molecular rotational line emissions from the dense core known as Orion KL. The coloured region shows emission in three lines of CH_3CN (red, green and blue). White and red contours show integrated emissions from HC_3N and SO_2. The red hexagons mark the positions of three submillimetre continuum sources. The pink square shows the origin of molecular outflows and of three ejected radio and IR sources (yellow spots); the green arrow indicates the position and orientation of filamentary structure revealed by the HC_3N emission. Reproduced with permission from Zapata *et al.* (2011).[21]

We list in Table 9.3 (ref. 22 and 23) detections of interstellar and circumstellar molecules of size equal to or larger than five atoms.

We have already emphasized in Chapter 8 that conventional gas-phase interstellar chemistries generally fail to account adequately for the abundances of many of the species listed in Table 9.3, and—in some cases—even their presence. How can they be formed? Since the freeze-out timescale varies inversely with gas density, that timescale becomes very short (~100 years) in the high density gas associated with a protostar. For conventional Milky Way parameters, desorption processes do not compete effectively with freeze-out and therefore much of the interstellar material is incorporated in the ice mantles on dust grains. Therefore, the precursor material from which the hot core and warm core molecules must form in this evolutionary situation is mostly contained in the ice. But the ice, as we have seen, is dominated

Table 9.3 List of detected interstellar and circumstellar molecular species, of five atoms or larger (data from Agúndez and Wakelam 2013;[22] and Cologne Database for Molecular Spectroscopy – website[23]).

Number of atoms	Detected species
5	C_5, CNCHO, HNC_3, HC_2NC, HC_3N, C_4H^-, C_4H, HCOOH, NH_2CN, H_2C_2O, HNCNH, CH_2CN, H_2C_3, c-C_3H_2, l-C_3H_2, H_2COH^+, CH_3O, SiH_4, CH_4, C_4Si, NH_4^+, H_2NCO^+
6	C_5N^-, C_5N, C_5H, c-H_2C_3O, HC_2CHO, HC_3NH^+, HC_4N, H_2C_4, CH_3SH, NH_2CHO, NHCHCN, CH_2CNH, CH_3NC, CH_3CN, CH_3OH, C_2H_4, HC_4H^+
7	HC_5N, C_6H^-, C_6H, CH_2CHCN, CH_2CHOH, c-C_2H_4O, CH_3CHO, CH_3C_2H, CH_3NH_2
8	C_7H, C_6H_2, CH_2CCHCN, CH_3C_3N, $HCOOCH_3$, CH_2OHCHO, CH_3COOH, NH_2CH_2CN, CH_2CHCHO, CH_3CHNH, l-HC_6H^+
9	HC_7N, C_8H^-, C_8H, CH_3C_4H, CH_3CONH_2, CH_3CH_2CN, CH_3CH_2OH, CH_3OCH_3, CH_2CHCH_3, CH_3CH_2SH
10	CH_3C_5N, CH_3CH_2CHO, $(CH_2OH)_2$, CH_3COCH_3
>10	HC_9N, CH_3C_6H, CH_3COOCH_3, C_2H_5OCHO, $C_6H_6^+$, C_3H_7CN, $HC_{11}N$, C_{60}, C_{60}^+, C_{70}

by H_2O, CO, and CO_2, with traces of CH_4, NH_3, and XCN. Can there be a chemistry of the solid state which leads to the molecular complexity revealed by the observations of hot cores? The association of hot and warm cores with the cold very dense gas in which the molecular material is largely in the ices seems to imply that the ice material—somehow—enables the production of complex molecules in a hot core. Or is it simply that the desorbing ices trigger a rapid gas phase chemistry in the hot core?

For a few molecular species, their abundances in the ice are capable of providing their gas phase abundances in a hot core. For example, methanol is typically present in ices at a few percent of H_2O by number, and H_2O itself is about 10^{-4} relative to all hydrogen; thus, methanol in the ice is about 10^{-6} relative to hydrogen. It is observed at about this relative abundance in hot cores. In other words, methanol prepared in or on the ice by surface reactions (see Section 9.2, above) and retained in the ice is released to the gas when the ices sublimate when heated in a hot core. Unfortunately, we do not know whether this type of mechanism works for other molecular species, because the limit of detection in the solid state is poor. However, most hot core molecules are present at abundances relative to hydrogen of about 10^{-8}. They could have precursors in the ice at the level of 0.01% of H_2O, but these components would not be detectable with present instrumentation. The example of methanol is a powerful inducement to consider that other species are also formed in the ice and released to the gas. A large number of laboratory experiments also support this view; they indicate that simple ices under irradiation by ultraviolet radiation or by fast particles (representing cosmic rays) may create complex molecular species such as those detected in hot cores. Some of these experiments will be discussed in Section 9.3.2. Theoretical simulations of ice processing are discussed in Section 9.3.1, below.

The alternative to "cooking" the primitive ice by ultraviolet or fast particle irradiation, to prepare hot core molecules in the ice, is simply to assume that the ices formed in the prestellar phase by freeze-out and surface reaction are desorbed into the gas phase, and that chemical conversion to greater complexity occurs by reactions in the gas phase. This process is now considered to be less favourable than ice processing by irradiation because of the difficulty of creating a suitable chemistry in largely neutral saturated molecular gas (but see also Section 9.4, where grain explosions are discussed). However, this alternative approach does provide a relatively simple approach to testing some important aspects of hot core chemistry that are hard to incorporate in more complex simulation treatments.

For example, it seems obvious that the warming of dust by infrared radiation from the protostar cannot be instantaneous, and that the heating rate must depend on the distance from the protostar and the column density of the dust. Thus, at any particular distance, the temperature of an ice-mantled grain will decrease monotonically with distance so that grains further from the protostar will generally be cooler than those closer to the protostar. Laboratory studies by Collings, McCoustra, *et al.* (2004)[24] showed that the mixed ices that may be present in interstellar clouds had a complex desorption behaviour under slow warming, with different molecular species released to the gas phase at several distinct temperature bands. Carbon monoxide from pure solid CO was the first species and H_2O was the last species to enter the gas phase (see also Section 9.3.2). As this temperature-banded desorption occurs, the ice itself changes its chemical composition. Thus, the gas-phase chemistry and the ice chemistry during the warm-up phase of the protostar are both time- and space-dependent, as first noted by Viti & Williams (1999).[25] The sensitivity of the gas-phase chemistry to these subtleties of desorption can be explored through such models.

However, the chemical complexity that can be incorporated in such models is rather limited; it is usually assumed that hydrogenation of species occurs at the surface, so that CH_4, NH_3, H_2CO, and CH_3OH are present in the ice and are ejected at the appropriate temperature during the warm-up phase. While this relatively straightforward approach is viable and is useful when exploring astronomical influences on hot core chemistry, it does not address the problem of the origin of the relatively large species found in hot cores. For that purpose, one needs to use more complex simulations.

9.3.1 Simulations

This area of astrochemistry is still very much "work in progress". In this sub-section, therefore, we shall merely indicate the types of activities currently undertaken. An excellent discussion of present work on the simulation of hot core chemistry has been published recently (2013) by Robin Garrod and Susanna Widicus Weaver.[26]

The rate equation approach is the most widely used in the simulation of hot core chemistry. Rate equations have been the main method of exploring *pure gas-phase* chemistry in interstellar clouds. The methods are routinely used to describe time-dependent chemistry in static clouds or in regions that are dynamically evolving, such as gravitationally collapsing clouds, or in an interface, or suffering a dynamical or MHD shock. The rate equations express the rates of formation and loss of all the various species involved by their interactions in various types of reaction. The gas-phase schemes typically involve hundreds of species reacting in thousands of reactions. There has been a huge effort on the part of chemists, over nearly half a century, to provide reliable rate coefficients for these models of gas-phase chemistry, and a continuing effort in assessment and validation of the data (see, for example, Agúndez & Wakelam 2013[22]). This research programme has been very successful, but continues to be an immense task for gas-phase chemistry. However, the problem is even worse for surface reactions (as we saw for H_2 formation, Section 8.3).

The rate equation method for hot core chemistry couples the gas-phase chemistry with the surface chemistry. In principle, each gas phase species may collide with the surface, may possibly find a reaction partner and take part in reactions on or inside the surface, or simply become incorporated as part of the ice. If a reaction occurs, then the reaction product may be ejected or may be retained and be incorporated in the ice mantle. Mantle species are of course subject to photodesorption, photodissociation, and photoionization by the local radiation field. They may be ionized by fast particles. Therefore, ideally, we require information, for all species, on the species-surface potential and the sticking probabilities, on surface potentials for lateral motion and the mobilities by hopping or by quantum mechanical tunnelling, and on reaction probabilities and the energy budget, just as in the case for the case of H_2 formation (Section 8.3). We also require rates of radiative and collisional processes for molecules in or on the ice. These requirements represent a huge demand for data that are generally very poorly known and often very difficult to determine in a reliable way. Not surprisingly, simple expressions are often used to represent mobilities or sticking probabilities. These expressions contain implicitly a huge amount of uncertain data.

For the gas phase, the numbers of any species contained in an arbitrarily large volume can be large, resulting in small stochastic fluctuations in the numbers of species. This is the reason why the rate equations work so well to describe that part of the chemistry that is in the gas phase. However, as we saw also in Chapter 8, the populations of species on grain surfaces may be small enough, on the order of unity, that stochastic fluctuations become significant. For example, if species A and species B are on separate grains, there is no possibility that they can react to form a species AB, even though rate equations dealing with averages would predict that the species AB should form. Thus, rate equation methods may overestimate the formation rates of molecules by surface reactions when surface populations are low.

The straightforward alternative to the rate equation approach, as mentioned in Chapter 8, is to solve the chemical master equation for a full gas + grain chemical network using the Monte Carlo method. Vasyunin *et al.* (2009)[27]

have demonstrated that although this is in principle viable it is a very time-consuming indeed, and difficult to integrate with the rate equations that are used to solve the gas-phase chemistry. Therefore, given that rate equations can work well, there have been several attempts to modify the rate equations describing the surface chemistry in such a way as to deal with the stochastic problem outlined above.

Perhaps the most successful of these modifications to rate equations has been made by Garrod (2008),[28] who changed the actual form of the rate equation for the formation of molecule AB from the combination of A and B. He considered that on small grains (where the stochastic problems are likely to be prominent) the formation of molecule AB takes place through the arrival of A on a grain where B is already located (or *vice versa*), so that the reaction rate is controlled by the accretion rate of A (or B) on the grain. This is the so-called "accretion-limited" situation. Once A (or B) has arrived the reaction is assumed to take place. The canonical rate of formation of AB in the rate equation method is essentially

$$R_{prod}(AB) = k_{AB} \times \langle N(A) \rangle \times \langle N(B) \rangle \tag{9.1}$$

where k_{AB} is the rate coefficient for the surface reaction and $\langle N(A) \rangle$ is the expectation value of the number of A atoms on a grain. In the modified rate equation method, this expression is replaced by

$$R_{prod} = R_{acc}(B) \times \langle N(A) \rangle + R_{acc}(A) \times \langle N(B) \rangle \tag{9.2}$$

where $\langle N(A) \rangle$ and $\langle N(B) \rangle$ are both assumed to be much less than unity, and the reactions proceed with unit probability. If the modified production rate exceeds the canonical rate, then the canonical rate is used. Garrod showed that with a limited chemistry involving the production of H_2, H_2O, O_2, and CH_3OH the modified rate equations give solutions very close to master equation solutions, while the non-modified rate equations overestimate the formation rates by large factors, up to ~100. Results from a much larger network have been obtained by Garrod *et al.* (2009)[29] using the modified method and compared with the exact results obtained by Vasyunin *et al.* (2009)[27] using the Monte Carlo method. Results for several gas phase and ice species are shown in Figure 9.4.

Garrod *et al.* (2009)[29] also show "global agreement diagrams" for a huge range of parameter space. These indicate how much better the results for all species, computed using the modified equations, agree with the results of the Monte Carlo calculations than results from the non-modified equations.

The modified rate equation method appears to work well and provides a suitable balance between accuracy of results and speed of computation. The calculations demonstrate conclusively that non-modified rate equations may overestimate formation rates by large factors. Thus, a viable and economical approach to computation of molecular abundances in a gas + grain chemical network is available. However, it must not be forgotten that whichever method is used to compute the results, they are only as good as the basic data

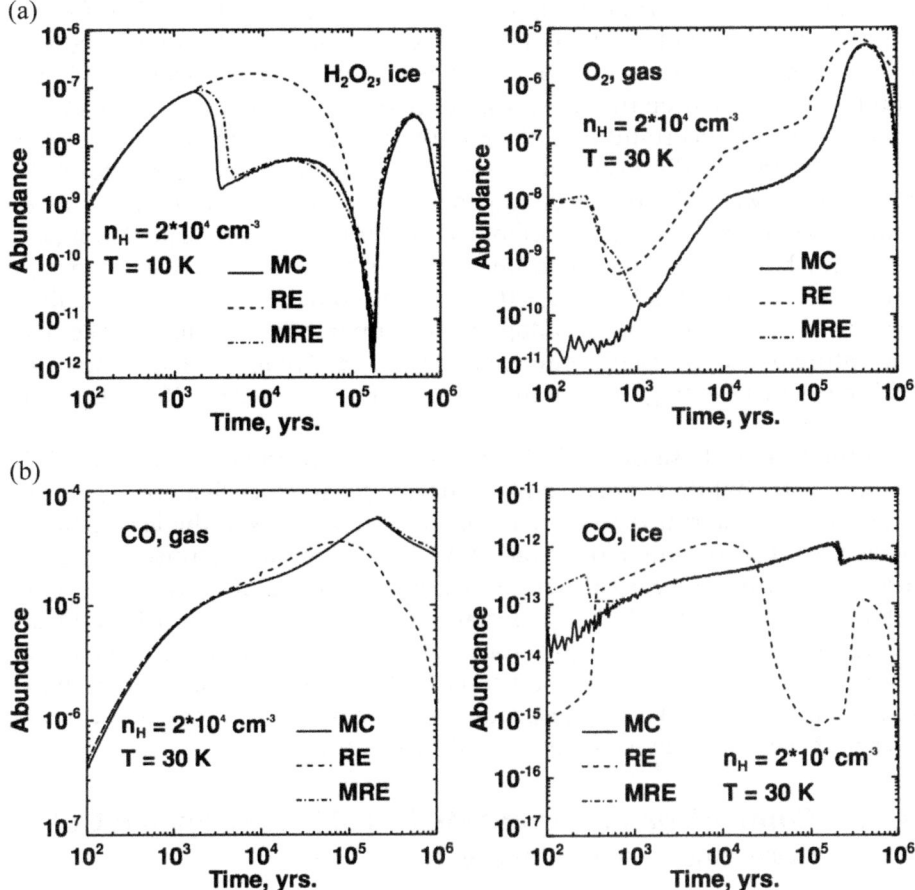

Figure 9.4 (a) Time-dependent abundances of representative species calculated with Rate Equations (RE), Modified Rate Equations (MRE), and Monte Carlo (MC) calculations. (b) Gas phase and grain surface abundances of CO. Reproduced with permission from Garrod *et al.* (2009).[29]

included in the models. As noted above, these data are at present very poorly known, and gross assumptions and simplifications are routinely made. The main uncertainties lie with several fundamental parameters describing the interaction between the species and the surface.

The first of these parameters is the binding energy between the species and the surface. While estimates of the binding energy for some species may be obtained from TPD experiments such as those of Collings *et al.* (2004),[24] data for radicals and atoms are poorly known. Secondly, in many conventional gas + grain models, surface reactions occur because species may diffuse over the surface to locate a reaction partner. Barriers to diffusion are essentially unknown, except in a few cases, and it is often assumed that the barriers are simply a substantial fraction of the binding energy. These barriers may be penetrated by thermal hopping or by quantum mechanical diffusion, but the probability of diffusion is extremely sensitive to the shape and size of the adopted barrier, so major uncertainties must be associated with any computed values. Even in the few cases where some experimental data exist, there are major uncertainties in those data because the surface may be poorly characterised. Thirdly, even when the species are in appropriate proximity, reactions between them may be impeded by activation barriers. The shapes and sizes of these barriers are unknown. As for the mobility calculation, the activation barriers may be penetrated quantum-mechanically, even at the low temperature of interstellar ices, ~10 K, but the computed reaction probabilities are highly sensitive to the adopted barrier parameters. Finally, of course, the nature of the ice itself is poorly characterized. Are reactions taking place within cavities existing in a porous ice, or is the ice in a dense form that has a low porosity?

In the face of these many difficulties that remain to be investigated, it is good to note that there are several important parameters that are capable of experimental determination: the response of molecules in the ices to fluxes of electromagnetic and particle radiation. One essential response for the gas + grain networks discussed here is the creation of radicals from the saturated molecules of which the ice is mainly composed. Because these radicals are mobile, they may undergo association reactions and produce the larger molecular species observed in hot cores and other locations. External fluxes of this kind may also lead to desorption of species from the ices. Some examples of these experiments are described in Section 9.3.3.

9.3.2   Comparison of Current Models of Pre-Stellar and Hot Core Phases with Observational Data

The gas + grain models discussed in the previous sub-section describe a lengthy process of chemical evolution from simple species to relative complexity, in an environment in which the gas number density may be increasing during this evolution by several orders of magnitude. First, there is chemistry in the gas phase and the initiation of deposition of gas phase species on to grain surfaces, creating ice mantles on dust grains. This deposition is accompanied by the chemical processing of those ices by continuous irradiation with electromagnetic and particle fluxes, and in current models the creation of radicals. Then, as a nearby newly-formed star begins to warm its immediate environment, radicals become mobile in the ices and radical–radical reactions occur, forming greater chemical complexity. Finally, the ice mantles are desorbed thermally and are mixed with the ambient gas.

This is an immensely complex chemical process. Yet it represents a minimal physical description of the behaviour of gas in a star-forming region. Ideally, such chemical descriptions would be based on proper hydrodynamical (HD)—or even MHD—models. The interaction of the protostar with its environment may be more complex than simply warming the gas on an appreciable timescale; it may involve interactions of jets and winds from the protostar, the creation of interfaces between those winds and the ambient gas, and the presence of HD or MHD shocks as the winds impinge on the gas. In the face of the rather simplistic physical description that—quite properly—has been adopted at this stage, it is perhaps unreasonable to expect any agreement between the results of computational models and observational data.

Perhaps the most remarkable feature of this programme of work, so far, is that there is a very substantial amount of *qualitative* agreement between theory (as explored by Garrod *et al.* 2008;[30] see also Section 9.5, below) and observational data. While there are many detailed uncertainties, it seems clear that chemical complexity created by the association of simple radicals that are released by photolysis from simple molecular species in the ice gives a range of product molecules that is very similar to those relatively large species detected in star-forming regions. Evidently, there is something fundamentally correct about the process of radical creation and addition, even allowing for the many uncertainties in the adopted parameters. This is a remarkable outcome.

There are also some encouraging *quantitative* agreements between theory and observations (such as for methanol), in spite of the many theoretical uncertainties. The models are perhaps more useful as aids in exploring the apparent sensitivity of the complex chemistry to age, to warm-up period, and to the spatial position of the distributions. This work is on-going.

9.3.3 Recent Advances in Laboratory Astrochemistry Related to Hot Cores

In the interstellar medium, ices of simple species can be "cooked" in various ways to produce more complex species. The "cooking" can be driven by cosmic rays (fast particles; mainly protons, α-particles, and electrons) and by the ultraviolet radiation field (which—if external stellar ultraviolet is totally excluded—may be the cosmic ray induced radiation field, see Section 1.5). The more complex species may be retained in the ice or released when the ices are warmed by the radiation of a newly forming star. All these processes can be replicated in laboratory experiments. We can learn from such experiments the likely products of such "cooking" and incorporate that information in our models. In this sub-section we shall highlight some recent laboratory work relevant to the drive to chemical complexity in interstellar molecular clouds.

9.3.3.1 Photodesorption of Ice Molecules

Carbon monoxide is, as we have seen, one of the most abundant components of interstellar ices. It is deposited directly from the gas phase on to H_2O ices, and is to some extent distinct from the H_2O ice. Therefore, it is important to explore the extent to which CO is retained on the ice, in the face of various desorbing mechanisms, including thermal desorption and photodesorption. Some CO in the ice may be converted to formaldehyde and methanol (see below), and some may remain unchanged in the ice. From an astrochemical point of view, it is important to understand the balance between these competing processes.

Muñoz Caro *et al.* (2010)[31] designed the InterStellar Astrochemistry Chamber (ISAC) for the study of solids and their response to thermal and ultraviolet irradiation. It operates at an exceptionally low pressure, down to 2.5×10^{-11} mbar. It has been used to provide new results on thermal and photodesorption of CO from CO ice. The ice thickness and temperature are closely monitored and detection of released CO molecules is made by QMS. In the thermal experiment, the CO ice is deposited at 7 K and warms at a rate of I K per minute. The signal is shown in Figure 9.5.

The peak rate of CO desorption occurs at a temperature of 28 K, consistent with earlier measurements. However, it is interesting that CO desorption actually begins in a much lower temperature range, 15–23 K. This is attributed to a release of CO induced by H_2, a low-level impurity in the system that becomes trapped in micropores in the CO ice. The authors note that this process may well be occurring in the interstellar medium, in which H_2 is abundant when CO is being deposited.

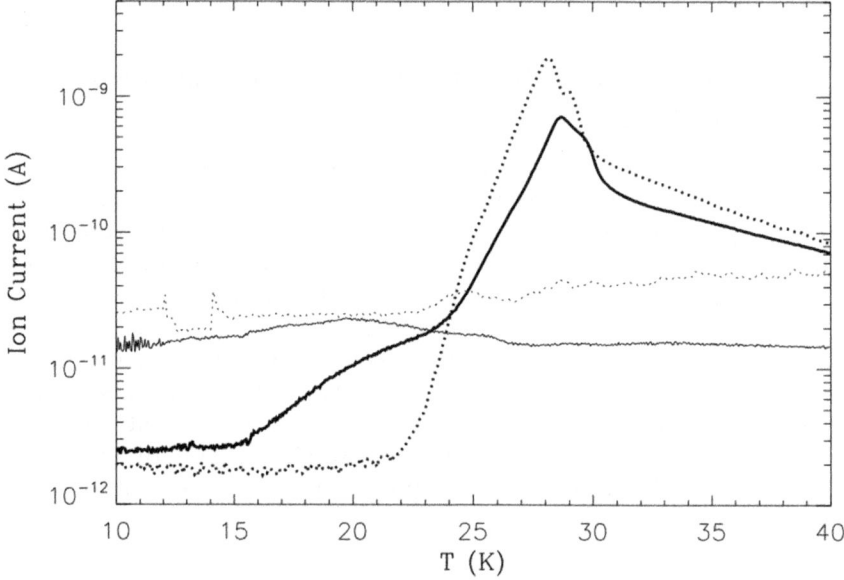

Figure 9.5   Thermal desorption of CO from ices deposited at two different temperatures, 7 K (thick solid trace) and 20 K (thick dotted trace). The desorption of H_2 is also shown; 7 K deposition (thin solid trace) and 20 K deposition (thin dotted trace). Reproduced with permission from Muñoz Caro *et al.* (2010).[31]

Photodesorption experiments were made using a hydrogen discharge lamp as the ultraviolet source. CO ices of various thicknesses were deposited, and photodesorption rates were determined for ices at various temperatures. The rates at 7 K, 8 K, and 15 K are 6.4×10^{-2}, 5.4×10^{-2}, and 3.5×10^{-2} molecules per ultraviolet photon, and the rate at 15 K is about one order of magnitude larger than previous determinations. The rates are independent of ice thickness for thicknesses greater than 5 monolayers, suggesting that it is only absorptions in the top five layers that lead to desorption. With these thermal desorption and photodesorption values, the computed abundance of CO in the gas phase in dense clouds in which the freeze-out rate is rapid is then consistent with observation.

Photodesorption efficiencies may also be obtained from theoretical work. Extensive classical molecular dynamics calculations to determine the photodesorption efficiency of water molecules from ice as a function of temperature have been made by Koning *et al.* (2013).[32] These calculations include both H and D isotopes in water, and have been used by Arasa *et al.* (2015)[33] to study fractionation in astronomical ices. The photodesorption of H_2O is fundamentally different from that of CO; in the case of CO, the molecule is excited by an ultraviolet photon but not dissociated, while in the case of water dissociation occurs and a variety of possible pathways opens up. These are the desorption of H, or of OH, or of both H and OH, or of H_2O, or the retention of both H and OH, or of H_2O. Also, the energetic H atom may interact with, and "kick out", a neighbouring H_2O molecule. Results are dependent on the ice temperature, and on the depth within the ice at which the photon is absorbed. It appears that absorptions in the top four monolayers of water ice contribute significantly to the efficiency. Desorption efficiencies are typically $\sim 10^{-3}$ molecules per ultraviolet photon.

9.3.3.2 *Hydrogenation of CO in and on Interstellar Ice*

Perhaps the most important interstellar surface chemistry (after the formation of H_2 and H_2O) is the formation of H_2CO (formaldehyde) and CH_3OH (methanol) by the hydrogenation of CO, because—as we shall see below—methanol in the ice is generally regarded as a very important component of the "feedstock" for the formation of large interstellar species such as those found in hot cores. While these two species can be formed in rather low abundance through gas phase reactions, the large amounts associated with the ices cannot arise simply by the freeze-out of these molecules from the gas phase. The species in the ices must, therefore, be formed by chemical processing of the ices themselves. The sequence of hydrogenation reactions on or in ices

$$CO \rightarrow HCO \rightarrow H_2CO \rightarrow CH_3O \rightarrow CH_3OH$$

has been studied in the laboratory by many authors. Watanabe *et al.* (2003)[34] noted that these chemical conversions in $H_2O + CO$ ice were sensitive to ice temperature and to the H atom flux, and that these reactions should properly be considered in competition with other hydrogenations such as H_2 and H_2O formation. They noted that the observed wide range of CH_3OH/H_2O in ices towards low mass and high mass protostars may be a consequence of these temperature effects. In a later paper,[35] they showed that the hydrogenation reactions must also compete with photolysis.

In their 2007 experiments, therefore, Watanabe *et al.*[35] created ice samples at UHV by deposition of gases on to a cold substrate. The sample ices were either an intimate H_2O/CO mixture (4:1) or a pure ice layer with one monolayer of CO on the surface. The ultraviolet source provides mainly lines of H_2 Lyman and Werner band emission, rather similar to the cosmic-ray generated radiation field expected to be present in dark clouds. The ultraviolet flux is measured by a photodiode and the H atom flux by a quadrupole mass spectrometer. The samples were irradiated for up to 100 minutes, corresponding to a fluence on the order of 10^{17} cm^{-2}. In a dark cloud, a fluence of this order would be achieved in about 10^5 years. The hydrogenation in the experiment was created by a cooled atomic H beam generated by microwave discharge. Irradiation by H atoms was also carried out for periods up to 100 minutes corresponding to an interstellar exposure of about 10^5 years. Experiments were carried out for irradiation solely by ultraviolet, solely by H atoms, and by co-irradiation (ultraviolet + H). The results in all cases showed the decline in CO with time, as the abundances of product molecules CO_2, H_2CO, and CH_3OH grew; see Figure 9.6. The effective chemical network adopted by Watanabe *et al.* to account for these results is shown in Figure 9.7.

The results, together with some observational data, are shown in Table 9.4. They show encouraging agreement with observational data and confirm that hydrogenation of H_2O–CO ices does lead to H_2CO and CH_3OH. Photolysis is important for larger ultraviolet fluxes. Hydrogenation occurs preferentially on capped ices, *i.e.* on surface CO rather than CO embedded in the H_2O ice. The species CO_2 and HCOOH tend to form for higher ultraviolet fluxes (>10^4 cm^{-2} s^{-1}). The chemical composition and physical structure of the ice will significantly affect the chemical evolution under irradiation.

9.3.3.3   Ultraviolet Irradiation of Methanol-Rich Ice and the Formation of Large Molecules

It is generally accepted that interstellar ices form sequentially, so that H_2O forms first and CO ice appears on top (see earlier in this chapter, Section 9.1). As the previous discussion in this section has shown, CO may be hydrogenated to methanol. Therefore, methanol in interstellar ice may be relatively pure or mixed with CO. Can methanol provide a source of radicals that may drive a solid-state chemistry in the ice, with products molecules similar to those seen in hot cores?

In a very extensive laboratory investigation, Öberg *et al.* (2009)[36] studied the chemistry induced by ultraviolet irradiation of CH_3OH ice under conditions that are appropriate for dense interstellar clouds. They have shown

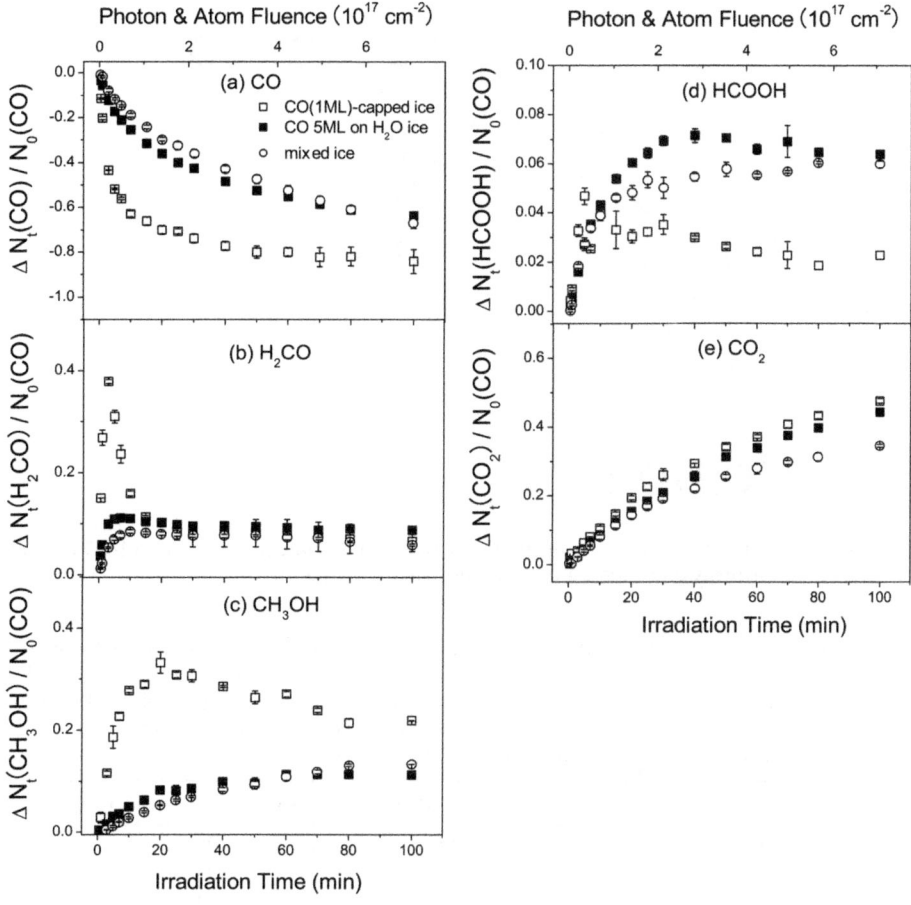

Figure 9.6 Time dependence of (a) CO, (b) H_2CO, (c) CH_3OH, (d) HCOOH, and (e) CO_2 released by the co-irradiation of various ice samples (as indicated) with UV and H atoms at 15 K. Reproduced with permission from Watanabe *et al.* (2007).[35]

that a wide range of relatively large species of the kind found in hot cores arise in these experiments, and they have been able to estimate formation rates from their experiments. The ices are either pure methanol, or are mixtures of methanol–carbon monoxide or methanol–methane. The ices are grown under UHV to thicknesses in a range from 3 to 66 monolayers, and their temperatures are accurately controlled in the range 20–200 K. In the experiments, the ices are irradiated by ultraviolet from a broadband hydrogen microwave-discharge lamp which has peak intensity at Lyman α (121 nm) with a range 115–170 nm. The apparatus includes a Fourier transform infrared spectrometer in reflection-absorption mode (RAIRS), operating in a range that includes the vibrational bands of many molecules. RAIRS identifies complex molecules formed during the irradiation and also

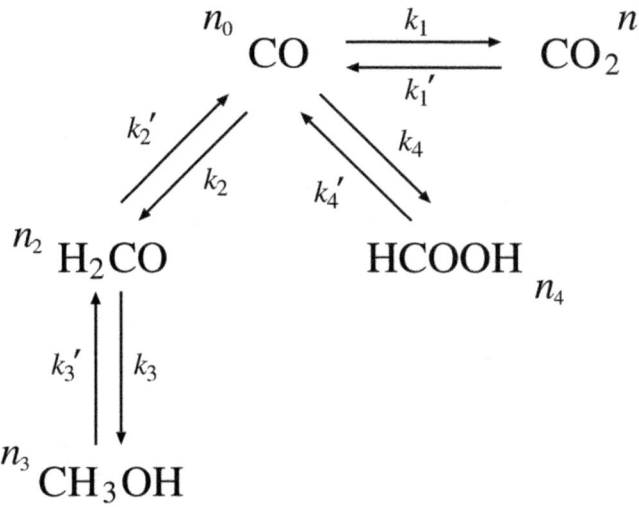

Figure 9.7   Effective reaction network adopted by Watanabe *et al.* (2007)[35] for photolysis of ices. Reproduced with permission.

Table 9.4   Molecular abundances relative to CO for lines of sight towards W33A (a high mass protostar), Elias 29 (a low mass protostar), and Elias 16 (a field star) compared to data from experiments in which ultraviolet + H atom co-irradiation on mixed H_2O + CO ice and on H_2O ice capped with one monolayer of CO (data from Watanabe *et al.* 2007[35]).

Molecule	W33A	Elias 29	Elias 16	Co-irradiation, mix	Co-irradiation, cap
CO	100	100	100	100	100
CO_2	163	394	97	123	298
H_2CO	38	36	—	24	41
CH_3OH	207	312	12	61	137
HCOOH	46	1.8	—	18	14

measures chemical changes in the ice during the experiments. The ice composition is also determined using TPD, in which the evaporated molecules are detected by a QMS.

The apparatus is used initially to determine the methanol photolysis cross section and the methanol photodesorption yields. The first of these determines how much of the methanol is converted into radicals that drive the subsequent chemistry. The photodissociation cross section is found to be $2.6 \pm 0.9 \times 10^{-18}$ cm^2 at 20 K. It increases with temperature. The second determines how much of the methanol is lost to the gas phase and unavailable for ice processing. The photodesorption yield is $2.1 \pm 1.0 \times 10^{-3}$ molecules per ultraviolet photon at 20 K, and increases with temperature. For the fluences adopted in the experiments, typically more than half of the methanol is destroyed. With these data, accurate formation rates for other species can be established.

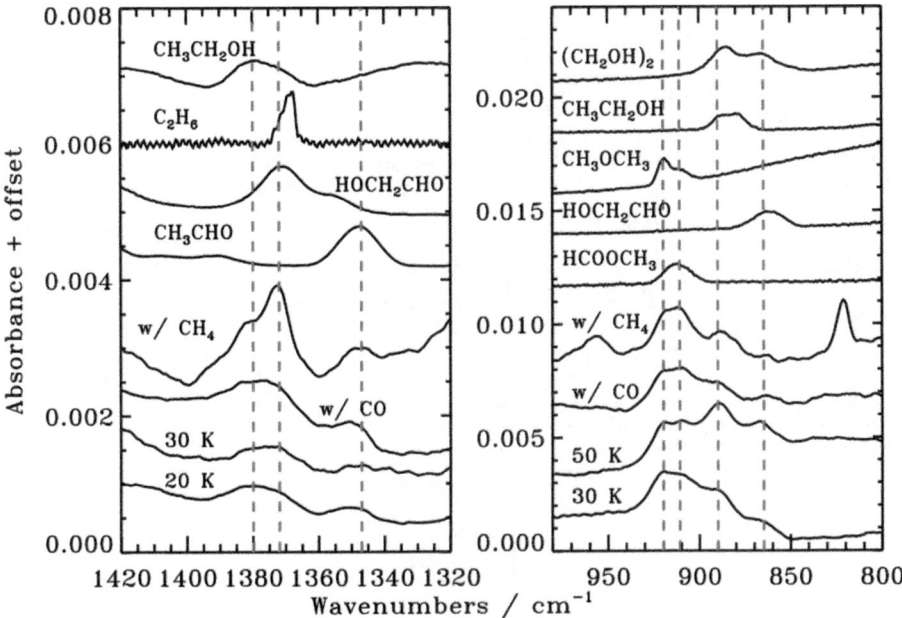

Figure 9.8 Left panel: spectra of UV-irradiated pure CH_3OH ice at 20 and 30 K, and CH_3OH–CO 1:1 and CH_3OH–CH_4 1:2 ice mixtures at 30 K after the same fluence of 2.4×10^{17} cm^{-2}, together with pure ice spectra of the complex species that absorb in this region. Right panel: the complex absorption pattern between 980 and 820 cm^{-1} in pure ices at 30 and 50 K and mixed ices at 30 K, together with possible carriers of these bands. In both panels, CH_3OH bands have been subtracted from the spectra for visibility and the features have been scaled to the initial CH_3OH abundance in each experiment to facilitate comparison. Red dashed lines: to guide the eye between band positions in the irradiated ices and in pure complex ice spectra. Reproduced with permission from Öberg *et al.* (2009).[36]

A large number of complex organics are detected as a result of ultraviolet irradiation; for an example, see Figure 9.8.

The identified species include CO, CO_2, CH_4, HCO, H_2CO, CH_2OH, CH_3CHO, CH_3OCH_3, CH_3CH_2OH, $(CH_2OH)_2$, and various CHO and COOH containing species. The dependence on temperature is small, suggesting that radicals diffuse and react quickly. Following irradiation, the ices are warmed from 20 K to 70 K. During this warming additional chemistry occurs, indicating that diffusion of radicals through the ice is occurring, with barriers to diffusion in the following order: H, CH_3, HCO, CH_3O, and CH_2OH. In general, the product composition depends on the fluence and the temperature, but not on the flux nor on the ice thickness. From the distributions of related products, Öberg *et al.* are able to deduce that the CH_3OH photodissociation branching ratio is [CH_2OH + H]:[CH_3O + H]: [CH_3 + OH] of 5:1:<1.

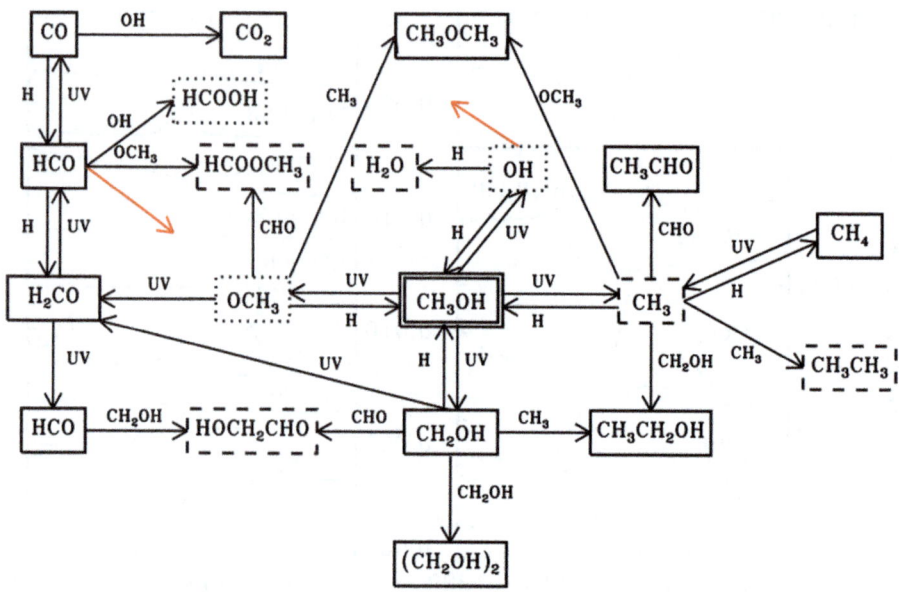

Figure 9.9   Reaction scheme proposed by Öberg *et al.* (2009)[36] to form the observed products following the irradiation of pure CH_3OH ice (solid boxes), of CO–CH_3OH or CH_4–CH_3OH ice mixtures (dashed boxes) and products only constrained with upper limits (dotted boxes). All reactions are reversible. Red arrows indicate that the radicals take part in reactions outside this scheme. Reproduced with permission.

Öberg *et al.* propose a reaction scheme to form the products identified in their experiments of ultraviolet irradiation of ices. This scheme is shown in Figure 9.9.

Evidently, methanol may be a key species in the formation of complex interstellar molecules. This is also emphasized in the following section.

9.3.3.4   Irradiation of Methanol Ice by X-Rays, by Energetic Electrons, and by Heavy Nuclei

9.3.3.4.1   X-Ray Irradiation. Young solar-type stars can be powerful sources of X-rays, with X-ray intensities some thousands of times greater than that of our middle-aged Sun. Is it possible that X-rays could initiate chemical changes in interstellar ices, similar to the transformations brought about by ultraviolet radiation? A series of experiments to investigate this idea have been carried out by Chen *et al.* (2013).[37] These authors selected methanol ice as a simple representation of interstellar ice, and irradiated it with X-rays that were either monochromatic (photon energy either 300 eV or 550 eV) or broadband (250–1200 eV). In their experiments, the methanol ice is created on a cold substrate at a temperature of 14 K in UHV chamber in which the typical pressure is $(1$–$3) \times 10^{-10}$ mbar. After the X-ray irradiation, a Fourier

Table 9.5 Products of X-ray irradiation of methanol ice assigned by infrared spectroscopy (data from Chen *et al.* 2013[37]).

Infrared band (cm^{-1})	Broad band	300 eV	550 eV	Assignment
2340	y	y	y	CO_2
2134	y	y	y	CO
1842	y		y	HCO
1746	y		y	$HCOCH_2OH + HCOOH + CH_3COOH$
1726	y	y	y	$HCOOCH_3 + H_2CO + HCOOH$
1690			y	$HCOCH_2OH$
1498	y		y	H_2CO
1304	y	y	y	CH_4
1249	y		y	H_2CO
1193	y	y	y	CH_2OH
1161	y	y	y	$CH_3OCH_3 + HCOOCH_3$
1091	y		y	$(CH_2OH)_2 + CH_3OCH_3 + CH_3CH_2OH$
1066	y		y	$HCOCH_2OH$
1025				CH_3OH
920	y		y	CH_3OCH_3

transform infrared (FTIR) spectrometer is used to examine changes in the chemical nature of the ice, and QMS monitors the sample purity during deposition and measures photodesorption in the warm-phase that follows the irradiation.

The effect of X-ray irradiation is different from that of ultraviolet irradiation. Photolysis by ultraviolet is a single event, whereas absorption of X-rays leads to a multistep process. X-rays ionize or excite core electrons in atoms in the sample, and the resulting energy cascade promotes a series of subsequent ionization and excitations that generate a population of secondary electrons. The experiments of Chen *et al.* sought to determine whether this radically different process produces chemical changes that are also different from those created in ultraviolet irradiation. The products of X-ray irradiation of methanol ice found by Chen *et al.* in their experiments are shown in Table 9.5.

It is interesting to note that there are differences in the chemistry between the two monochromatic experiments; these may be explained by the fact that 300 eV photons tend to break C–H bonds, so generating the CH_2OH radicals, while 550 eV photons tend to break C–O bonds. Within any experiment the amount of product molecules increases with energy absorbed. However, monochromatic irradiation is much more efficient in creating new species than broadband. The main result is that the list of products generated by X-ray irradiation—CO, CO_2, HCO, CH_2OH, H_2CO, CH_4, $HCOOH$, $HCOCH_2OH$, CH_3COOH, CH_3OCH_3, $HCOOCH_3$, and $(CH_2OH)_2$—is remarkably similar to the list of products obtained from ultraviolet irradiation. Evidently, the complexity of the energy input from X-rays does not generate widely differing product species.

9.3.3.4.2 Energetic Electrons. Ices in interstellar clouds are exposed to cosmic rays, *i.e.* a flux of fast ions and electrons with energies ranging from less than one MeV to more than 10^{20} eV. The ions are mostly protons and helium nuclei, with (as we discuss below) smaller abundances of heavier nuclei. The distribution in energy is very strongly skewed towards the lower energies in this range, and the most effective energies at which the ions interact with target material is in the MeV range. In the interaction of MeV particles with atoms, secondary electrons are generated, with energies up to a few keV. Therefore, Bennett *et al.* (2007)[38] and Bennett & Kaiser (2007)[39] simulate the interaction of cosmic rays with ices by irradiating solid methanol with 5 keV electrons and exploring the chemical consequences.

The methanol ices are deposited at 11 K on a silver mono-crystal. A FTIR spectrometer monitors chemical changes in the sample under irradiation by fast electrons, and during the subsequent warming after the irradiation is complete. Spectra of the methanol taken before and after irradiation, and during the subsequent warm-up to 113 K are shown in Figure 9.10.

With some theoretical guidance, Bennett *et al.* interpret these results in the following way: dissociation of methanol occurs into the channels $CH_2OH + H$, $CH_3O + H$, $H_2CO + H_2$, and $CH_4 + O$. In the subsequent warm-up the radicals generated within the ice matrix become mobile and begin to associate. The new products are then methyl formate, (CH_3OCHO), glycolaldehyde (CH_2OHCHO), and ethylene glycol ($CH_2OH)_2$. The list of detected products is therefore similar (if not quite as extensive) to those resulting from ultraviolet and X-ray irradiation. Evidently, keV electrons are effective in generating radicals that may under appropriate circumstances (warming to ~100 K) associate to form molecules of greater complexity than was originally the case.

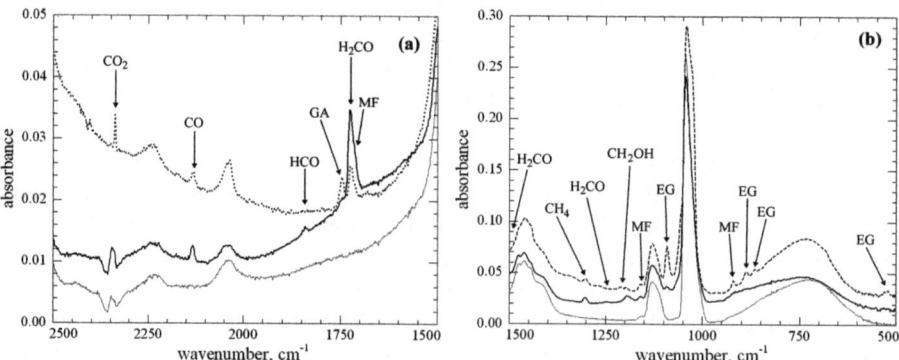

Figure 9.10 Comparison of the infrared spectra of the methane frost at 11 K before irradiation by fast electrons (grey line), after irradiation (solid black line), and during the warm-up at 113 K (dashed line) at regions (a) between 2500–1500 cm^{-1} and (b) 1500–500 cm^{-1}. GA: glycolaldehyde, MF: methyl formate, EG: ethylene glycol. Reproduced with permission from Bennett *et al.* (2007).[38]

9.3.3.4.3 Heavy Cosmic Ray Particles. As mentioned above, cosmic rays include not only the relatively abundant ions but also minor components of heavier nuclei, in relative abundances that are very approximately like the elemental relative abundances measured in the Sun. Therefore, oxygen ions, for example, are present in cosmic rays with abundances about one thousandth of that of protons. de Barros *et al.* (2011)[40] have examined the chemical effects induced by a variety of heavy ions on methanol ices. The ions are taken from the medium energy beamline of the heavy ion accelerator Grand Accélérateur National d'Ions Lourd (GANIL) at Caen, France, and include $^{16}O^{5+}$ at 16 MeV, $^{16}O^{7+}$ at 220 MeV, $^{65}Zn^{20+}$ at 606 MeV, and $^{86}Kr^{31+}$ at 774 MeV.

Methanol molecules were deposited in a CsI substrate at 15 K and the methanol ice sample was irradiated with particular heavy ions, up to a value of the fluence in the range $1.0–2.5 \times 10^{13}$ ions cm^{-2}. The ice was interrogated with an FTIR spectrometer, and products CO, CO_2, HCO, H_2CO, CH_2OH, CH_4, $HCOOCH_3$, CH_3COOH, and C_2H_5OH were detected. Carbon monoxide, formaldehyde and methane were the most abundant of these products. The radicals CH_2OH and HCO were fragile. The radiochemical yields were found to be much greater for the Zn and Kr beams than for beams of oxygen, helium, or hydrogen nuclei, and it is suggested that iron nuclei known to be present in cosmic rays may be important in promoting chemical changes in methanol ice.

The range of products obtained by heavy ion irradiation is similar to—if not quite as extensive—the ranges of products obtained from experiments with other forms of excitation.

9.4 Sulfur – A Special Case

Interstellar sulfur is an interesting case that is still not fully resolved and requires some discussion. In the diffuse interstellar medium through which starlight passes fairly readily, atomic sulfur—which has an ionization potential of 10.4 eV, lower than that of atomic hydrogen—is mainly ionized. The amount of sulfur combined into gaseous molecules such as SH or H_2S is tiny. In fact, relative to hydrogen, the amount of sulfur, as S^+, in diffuse regions of interstellar space is close to the relative abundance of sulfur in the Sun, often regarded as a standard cosmic abundance (see Table 1.1). Evidently, almost all the sulfur that we might expect to find in diffuse regions is present as S^+.

However, when observations are made of sulfur species in dense clouds through which starlight does not readily pass, the situation is very different. Certainly, sulfur species are detected in the gas phase in dense clouds, for example in the forms of SO, SO_2, H_2S, *etc.*, and to a modest extent in ice mantles on dust grains as OCS; however, the total amount of sulfur in these forms is relatively tiny compared to the amount implied by the cosmic abundance of sulfur. In dense interstellar clouds, the amount of this element in all its detected forms appears to be *depleted* by a factor of up to one thousand compared to the cosmic abundance. Where has 99.0–99.9% of the sulfur gone? Why are diffuse clouds and dense clouds so different in this respect? Sulfur is chemically similar to oxygen, yet evidently these elements behave differently in interstellar clouds. We think we understand how the oxygen budget is distributed in both diffuse and dense interstellar clouds, but obviously there is a sink of sulfur in dense clouds that is currently undetected.

Since millimetre- and submillimetre-wave band searches for molecular lines emitted from gaseous species are very sensitive and have not revealed a large reservoir of sulfur, it is generally assumed that large amounts of sulfur must be locked in the dust in some form that has yet to be detected. Ruffle *et al.* (1997)[41] pointed out that since atomic sulfur has an ionization potential less than that of carbon, it should remain ionized to greater depths within dense clouds so that the electrostatic attraction between S^+ ions and the (normally) negatively charged grains could enhance the freeze-out of sulfur onto grain surfaces. One might expect that—like oxygen atoms being hydrogenated to form H_2O ice—the sulfur would end as H_2S molecules on the grain surfaces. However, no detections of hydrogen sulfide in interstellar ices have yet been made, and the chemical conversion of H_2S into some other form— as yet undetected—needs to be considered.

Such experiments have been carried out using the *InterStellar Astrochemistry Chamber*[31] (see Section 9.3.3.1, above) to study the ultraviolet irradiation of ices of H_2S, H_2S/H_2O, H_2S/CO, and H_2S/CH_3OH ices.[42,43] Infrared spectroscopy and quadrupole mass spectrometry were used to monitor the gaseous and solid-state products. In the case of H_2S/H_2O ice, products included species containing two sulfur atoms (not yet detected in the interstellar medium), H_2S_2 and S_2, and the existence of polymeric sulfur (up to S_8) was inferred. Such polymers would be solids at relevant temperatures in interstellar clouds. For other ice combinations, sulfur-bearing species detected included H_2S_2, HS_2, CS_2, and OCS, and a complex organic refractory residue of sulfur chains. It appears likely that a significant fraction of available sulfur is present in the interstellar medium in these products and that polymeric sulfur may account for the "missing" sulfur. The experiments also suggest that species containing two sulfur atoms may be relatively abundant in the interstellar medium.

9.5   Exploding Ices

In 1973, J Mayo Greenberg[44] noted that the timescale for interstellar molecules to adhere to interstellar grains in dense interstellar clouds was short, yet molecules clearly existed in the gas phase. He proposed that explosions in ices in the interstellar medium could return material from ices to the gas phase, and that these explosions could be driven by chemical energy, released sporadically when the growing population of radicals in the ices spontaneously associated. Explosions, therefore, could be an alternative to other forms of desorption. Later, d'Hendecourt *et al.* (1982)[45] reported the results of laboratory experiments in which explosions were seen to occur in laboratory analogues of interstellar ices, under ultraviolet irradiation and modest warming to a temperature of 27 K. Grain explosions have also been considered in connection with H_2 formation (see Section 8.5). Rawlings and co-workers in 2013 [46,47] have modelled the exploding ices idea and have further proposed that a three-body chemistry can occur in the very high density but very short-lived gas created when the solid ice is transformed rapidly to gas in the explosions. They evaluated the mechanism for the conditions of dense interstellar clouds in which a weak ultraviolet field maintained by the flux of cosmic ray particles is sufficient to allow a population of radicals to be created in the ices. The mechanism would work equally well in dense cores around pre-stellar objects. In fact, one of the main predictions made by Rawlings *et al.* is that some of the "hot core type of molecules" should also be found in cool dense interstellar clouds, not associated with star-forming regions.

Given that the ice composition in dense interstellar clouds is known, then the radicals created by the ultraviolet radiation field can be identified. For a standard molecular cloud ice composition, the radicals should include OH, CH_3, CH_2, CH, NH_2, NH, CHO, CH_3O, and CH_2OH. The fraction of the ice converted to radicals depends on the strength of the radiation field and the duration of exposure to it. Rawlings *et al.* estimate that this fraction should be in the range 10^{-2}–1%.

It is possible to write down a set of reactions involving these species in three body reactions with a dominant partner, expected to be H_2O in this explosion. The set of reactions proposed for these three body reactions is shown in Table 9.6.

Obviously, there are similarities in this radical–radical chemistry to the chemistry proposed by Garrod *et al.* for surface and solid state chemistry in ices. However, an important difference is that Rawlings *et al.* propose that the reactions take place in the gas phase, in a very high density (possibly as great as $\sim 10^{20}$ cm^{-3}) and very short-lived (~ 10 ns) phase, *via* three-body reactions. Such reactions are often very fast, occurring on every collision, and although few of these reactions have been studied in the laboratory (as is certainly also the case for the surface and solid state reactions proposed by Garrod *et al.*) it seems likely that most of the three-body reactions will be fast.

Even without computing molecular abundances arising from this explosions model, the predicted range of species shown in Table 9.6 is striking. Almost all the detected complex organic molecules (COMs) should arise from the post-explosion chemistry. Of course, the explosions should be considered as a process operating alongside other mechanisms.

To compute molecular abundances arising from the explosions model, one needs to make a number of assumptions about three-body rate coefficients and about the explosion parameters. The results for an optimistic case are shown in Table 9.7.

These abundances relative to H_2O in the exploding gas are then mixed with the ambient gas on some assumed periodic event related to the ice deposition timescale. On the basis of the explosions model, a computation of

Table 9.6 Products from radical association reactions (data from Rawlings et al. 2013 [46]).

			Products			
No.	Reactant 1	Reactant 2	Molecule	Radical	Name	Detected in ISM?
1	OH	OH	H_2O_2		Hydrogen peroxide	
2	OH	CH$_3$	CH$_3$OH		Methanol	Yes
3	OH	CH$_2$		CH$_2$OH		
4	OH	CH	H$_2$CO		Formaldehyde	Yes
5	OH	NH$_2$	NH$_2$OH		Hydroxylamine	
6	OH	NH		NHOH		
7	OH	CHO	HCOOH		Formic acid	Yes
8	OH	CH$_3$O	CH$_3$OOH		Methyl hydroperoxide	
9	OH	CH$_2$OH	CH$_2$(OH)$_2$			
10	CH$_3$	CH$_3$	C$_2$H$_6$		Ethane	
11	CH$_3$	CH$_2$		CH$_3$CH$_2$		
12	CH$_3$	CH		CH$_3$CH		
13	CH$_3$	NH$_2$	CH$_3$NH$_2$		Methylamine	Yes
14	CH$_3$	NH		CH$_3$NH		
15	CH$_3$	CHO	CH$_3$CHO		Acetaldehyde	Yes
16	CH$_3$	CH$_3$O	CH$_3$OCH$_3$		Dimethyl ether	Yes
17	CH$_3$	CH$_2$OH	C$_2$H$_5$OH		Ethanol	Yes
18	CH$_2$	CH$_2$	C$_2$H$_4$		Ethylene	Yes
19	CH$_2$	CH		CH$_2$CH		
20	CH$_2$	NH$_2$		CH$_2$NH$_2$		
21	CH$_2$	NH	CH$_2$NH		Methylenimine	Yes
22	CH$_2$	CHO		CH$_2$CHO		
23	CH$_2$	CH$_3$O		CH$_3$OCH$_2$		
24	CH$_2$	CH$_2$OH		CH$_2$CH$_2$OH		
25	CH	CH	C$_2$H$_2$		Acetylene	
26	CH	NH$_2$		CHNH$_2$		
27	CH	NH		CHNH		
28	CH	CHO		CHCHO		

(continued)

No.	Reactant 1	Reactant 2	Product	Name	Detected
29	CH	CH_3O	CH_3OCH		
30	CH	CH_2OH	$CHCH_2OH$		
31	NH_2	NH_2	N_2H_4	Hydrazine	Yes
32	NH_2	NH	N_2H_3		
33	NH_2	CHO	NH_2CHO	Formamide	Yes
34	NH_2	CH_3O	CH_3ONH_2	Methoxyamine	
35	NH_2	CH_2OH	CH_2OHNH_2		
36	NH	NH	N_2H_2	Diazine	
37	NH	CHO	$NHCHO$		
38	NH	CH_3O	CH_3ONH		
39	NH	CH_2OH	CH_2OHNH		
40	CHO	CHO	$(CHO)_2$		
41	CHO	CH_3O	CH_3OCHO	Methyl formate	Yes
42	CHO	CH_2OH	CH_2OHCHO	Glycolaldehyde	Yes
43	CH_3O	CH_3O	$(CH_3O)_2$		
44	CH_3O	CH_2OH	CH_3OCH_2OH	Methoxymethanol	
45	CH_2OH	CH_2OH	$(CH_2OH)_2$	Ethyleneglycol	Yes
46	OH	CH_2OH	CH_2OHOH		
47	OH	$NHOH$	$NHOHOH$		
48	OH	CH_3CH_2	CH_3CH_2OH	Ethanol	Yes
49	OH	CH_3NH	CH_3NHOH	Methylhydroxylamine	
50	OH	CH_2CH	CH_2CHOH	Vinylalcohol	Yes
51	OH	CH_2NH_2	CH_2NH_2OH		
52	OH	CH_2CHO	CH_2OHCHO	Glycolaldehyde	Yes
53	OH	CH_3OCH_2	CH_3OCH_2OH	Methoxymethanol	
54	OH	CH_2CH_2OH	$(CH_2OH)_2$	Ethyleneglycol	Yes
55	OH	$CHNH$	$CHOHNH$		
56	OH	N_2H_3	N_2H_3OH	Hydrazinehydroxyl	
57	OH	$NHCHO$	$NHOHCHO$		
58	OH	CH_3ONH	CH_3ONHOH		
59	OH	CH_2OHNH	$CH_2OHNHOH$		
60	CH_3	CH_2OH	CH_3CH_2OH	Ethanol	Yes
61	CH_3	$NHOH$	CH_3NHOH	Methylhydroxylamine	
62	CH_3	CH_3CH_2	$CH_3CH_2CH_3$	Propane	

Table 9.6 (continued)

| | | | Products | | | |
No	Reactant 1	Reactant 2	Molecule	Radical	Name	Detected in ISM?
63	CH_3	CH_3NH	CH_3NHCH_3		Dimethylamine	
64	CH_3	CH_2CH	CH_3CHCH_2		Propylene	Yes
65	CH_3	CH_2NH_2	$CH_3CH_2NH_2$		Ethylamine	Yes
66	CH_3	CH_2CHO	CH_3CH_2CHO		Propanal	
67	CH_3	CH_3OCH_2	$CH_3CH_2OCH_3$		Ethylmethylether	
68	CH_3	CH_2CH_2OH	$CH_3CH_2CH_2OH$		Propanol	
69	CH_3	$CHNH$	CH_3CHNH		Ethanimine	
70	CH_3	N_2H_3	$CH_3N_2H_3$		Methylhydrazine	
71	CH_3	$NHCHO$	CH_3NHCHO		Methylformamide	
72	CH_3	CH_3ONH	CH_3ONHCH_3		Dimethylhydroxylamine	
73	CH_3	CH_2OHNH	CH_3NHCH_2OH		Methylaminomethanol	
74	CH_2	CH_3CH	CH_3CHCH_2		Propylene	Yes
75	CH_2	$CHNH_2$	CH_2CHNH_2		Vinylamine	
76	CH_2	CH_3OCH	CH_3OCHCH_2		Methoxyethene	
77	CH_2	$CHCH_2OH$	CH_2CHCH_2OH		Allylalcohol	
78	CH_2	$CHCHO$	CH_2CHCHO		Propenal	Yes
79	NH_2	CH_2OH	NH_2CH_2OH		Aminomethanol lab ices	
80	NH_2	$NHOH$	NH_2NHOH		Hydroxylhydrazine	
81	NH_2	CH_3CH_2	$CH_3CH_2NH_2$		Ethylamine	
82	NH_2	CH_3NH	CH_3NHNH_2			
83	NH_2	CH_2CH	CH_2CHNH_2		Related to aspargine	
84	NH_2	CH_2NH_2	$CH_2(NH_2)_2$			
85	NH_2	CH_2CHO	CH_2CHONH_2			
86	NH_2	CH_3OCH_2	$CH_3OCH_2NH_2$		N-Methoxymethanamine	
87	NH_2	CH_2CH_2OH	$CH_2CH_2OHNH_2$		Monoethanolamine	
88	NH_2	$CHNH$	$CHNH_2NH$			
89	NH_2	N_2H_3	NH_2NHNH_2		Triazine	
90	NH_2	$NHCHO$	NH_2NHCHO		Formichydrazide	

(continued)

91	NH$_2$	CH$_3$ONH	CH$_3$ONHNH$_2$		
92	NH$_2$	CH$_2$OHNH	CH$_2$OHNHNH$_2$		
93	NH	CH$_3$CH	CH$_3$CHNH	Ethanimine	Yes
94	NH	CHNH$_2$	CHNH$_2$NH		
95	NH	CH$_3$OCH	CH$_3$OCHNH		
96	NH	CHCH$_2$OH	CH$_2$OHCHNH		
97	NH	CHCHO	CHCHONH		
98	CHO	CH$_2$OH	CH$_2$OHCHO	Glycolaldehyde	Yes
99	CHO	NHOH	CHONHOH		
100	CHO	CH$_3$CH$_2$	CH$_3$CH$_2$CHO	Propanal	Yes
101	CHO	CH$_3$NH	CH$_3$NHCHO	Methylformamide	
102	CHO	CH$_2$CH	CH$_2$CHCHO	Acrolein	
103	CHO	CH$_2$NH$_2$	CH$_2$NH$_2$CHO		
104	CHO	CH$_2$CHO	CH$_2$(CHO)$_2$		
105	CHO	CH$_3$OCH$_2$	CH$_3$OCH$_2$CHO	Methoxyacetaldehyde	
106	CHO	CH$_2$CH$_2$OH	CH$_2$CH$_2$OHCHO	Betahydroxyaldehyde	
107	CHO	CHNH	CHCHONH		
108	CHO	N$_2$H$_3$	N$_2$H$_3$CHO	Formylhydrazine	
109	CHO	NHCHO	NH(CHO)$_2$		
110	CHO	CH$_3$ONH	CH$_3$ONHCHO	N-Methoxyformamide	
111	CHO	CH$_2$OHNH	CH$_2$OHNHCHO		
112	CH$_3$O	CH$_2$OH	CH$_3$OCH$_2$OH	Methoxymethanol	
113	CH$_3$O	NHOH	CH$_3$ONHOH		
114	CH$_3$O	CH$_3$CH$_2$	CH$_3$CH$_2$CH$_3$O	Ethylmethylether	
115	CH$_3$O	CH$_3$NH	CH$_3$OCH$_3$NH	N-Methoxymethanamine	
116	CH$_3$O	CH$_2$CH	CH$_2$CHOCH$_3$	Methoxyethene	
117	CH$_3$O	CH$_2$NH$_2$	CH$_3$OCH$_2$NH$_2$		
118	CH$_3$O	CH$_2$CHO	CH$_3$OCH$_2$CHO	2-Methoxyacetaldehyde	
119	CH$_3$O	CH$_3$OCH$_2$	CH$_2$(CH$_3$O)$_2$	Dimethoxymethane	
120	CH$_3$O	CH$_2$CH$_2$OH	CH$_3$OCH$_2$CH$_2$OH	2-Methoxyethanol	
121	CH$_3$O	CHNH	CH$_3$OCHNH		
122	CH$_3$O	N$_2$H$_3$	CH$_3$ON$_2$H$_3$	Methoxyhydrazine	
123	CH$_3$O	NHCHO	CH$_3$ONHCHO		
124	CH$_3$O	CH$_3$ONH	CH$_3$ONHCH$_3$O		

Table 9.6 *(continued)*

| No | Reactant 1 | Reactant 2 | Products | | | Detected in ISM? |
			Molecule	Radical	Name	
125	CH_3O	CH_2OHNH	CH_3OCH_2OHNH			
126	CH_2OH	CH_2OH	$(CH_2OH)_2$		Ethylene glycol	Yes
127	CH_2OH	$NHOH$	$CH_2OHNHOH$			
128	CH_2OH	CH_3CH_2	$CH_3CH_2CH_2OH$		Propanol	
129	CH_2OH	CH_3NH	CH_3NHCH_2OH		Methylaminomethanol ices	
130	CH_2OH	CH_2CH	CH_2CHCH_2OH		Allyl alcohol	
131	CH_2OH	CH_2NH_2	$CH_2NH_2CH_2OH$		Ethanolamine	
132	CH_2OH	CH_2CHO	CH_2CHOCH_2OH			
133	CH_2OH	CH_3OCH_2	$CH_3OCH_2CH_2OH$		2-Methoxyethanol	
134	CH_2OH	CH_2CH_2OH	$CH_2(CH_2OH)_2$			
135	CH_2OH	$CHNH$	$CH_2OHCHNH$			
136	CH_2OH	N_2H_3	$CH_2OHN_2H_3$			
137	CH_2OH	$NHCHO$	$CH_2OHNHCHO$			
138	CH_2OH	CH_3ONH	CH_3ONHCH_2OH			
139	CH_2OH	CH_2OHNH	$(CH_2OH)_2NH$			

Table 9.7 Predicted molecular abundances relative to H_2O of various species after explosions of the ice mantles (data from Rawlings *et al.* 2013[47]).

Species	Abundance	Species	Abundance	Species	Abundance
CH_3OH	1.3×10^{-4}	CH_2CHOH	3.2×10^{-6}	CH_2NH	1.9×10^{-6}
HCOOH	1.0×10^{-4}	$C_2H_5CH_2OH$	7.8×10^{-7}	CH_3OCHO	1.2×10^{-6}
CH_3CHO	7.2×10^{-6}	CH_3CHNH	7.6×10^{-7}	$(CH_2OH)_2$	1.2×10^{-5}
C_2H_5OH	6.3×10^{-6}	H_2CO	9.3×10^{-5}	C_2H_5CHO	9.5×10^{-7}
NH_2CHO	1.9×10^{-6}	CH_3NH_2	2.4×10^{-6}	CH_3CHCH_2	1.9×10^{-6}
CH_2OHCHO	1.3×10^{-5}	CH_3OCH_3	1.5×10^{-6}		

the abundance of propylene, CH_3CHCH_2, a molecule detected in the Taurus Molecular Cloud but apparently having no viable gas phase route, gives a value in the range 10^{-10}–10^{-9} relative to molecular hydrogen. This is in the range observed. Thus, the explosions model has the potential to be an important contributor to the abundance of COMs in the interstellar medium.

9.6 Conclusions

It must be re-emphasized that this area of research—determining the origin of the interstellar COMs—is very much work in progress and that definitive answers to all questions cannot be given at present. Nevertheless, considerable progress has been made.

It now seems inescapable that the processing of interstellar ices by electromagnetic and particle irradiation must be responsible for the remarkable chemical complexity that is found in denser regions of interstellar space. Those are the regions where interstellar ices form and for which purely gas phase chemical networks have been shown to fail to produce COMs in adequate quantities. But experiments show that the ices can readily be processed by available energy sources, firstly to provide radicals which, secondly, may be sufficiently mobile to create by radical association (either in the ice matrix or in explosions) the observed COMs.

This general scenario for COMs production is now widely accepted. The relative importance of different energy sources has received some attention, and will undoubtedly be the focus of further research. However, it is interesting that all the energy sources discussed here derive ultimately from cosmic rays – even the ultraviolet field in dark clouds arises indirectly from this source. So, ultimately, the balance between the various energy sources may not matter all that much. However, cosmic ray rates vary within the Milky Way galaxy and between galaxies. Therefore, the abundance of COMs may be related to the local cosmic ray rate. That would be a remarkable outcome: COMs as cosmic ray tracers!

The field is advancing rapidly, both in experimental work and in theoretical modelling. The modelling faces an almost overwhelming task of characterising many aspects of radical–surface interactions; these data are hard to measure and difficult to estimate. Even in the suggested model of ice explosions, where many of the surface data are not required, large numbers of three body association rates are needed. However, this is a problem of lesser difficulty and smaller in scope than the problem that faced gas phase interstellar chemistry in the 1970s. Although experiments are making huge advances in this field, models are still required to allow astrophysicists to explore the COMs formation in different locations within the Milky Way and especially in external galaxies. Therefore, some way around the data difficulty needs to be found.

Further Reading

1. E. F. van Dishoeck, E. Herbst and D. A. Neufeld, *Chem. Rev.*, 2013, **113**, 9043.
2. F. C. Gillett and W. J. Forrest, *Astrophys. J.*, 1973, **179**, 483.
3. M. Accolla, E. Congiu, *et al.*, *Phys. Chem. Chem. Phys.*, 2011, **13**, 8037.
4. A. C. Cheung, D. M. Rank, C. H. Townes, D. D. Thornton and W. J. Welch, *Nature*, 1969, **221**, 626.
5. J. E. Chiar, Y. J. Pendleton, *et al.*, *Astrophys. J.*, 2011, **731**, 9.
6. A. P. Jones and D. A. Williams, *Mon. Not. R. Astron. Soc.*, 1984, **209**, 955.
7. A. G. G. M. Tielens and W. Hagen, *Astron. Astrophys.*, 1982, **114**, 245.
8. J. F. Roberts, J. M. C. Rawlings, S. Viti and D. A. Williams, *Mon. Not. R. Astron. Soc.*, 2007, **382**, 733.
9. R. T. Garrod, D. A. Williams and J. M. C. R. Rawlings, *Mon. Not. R. Astron. Soc.*, 2006, **373**, 577.
10. H. M. Cuppen, S. Ioppolo, C. Romanzin and H. Linnartz, *Phys. Chem. Chem. Phys.*, 2010, **12**, 12077.
11. T. Lamberts, H. M. Cuppen, S. Ioppolo and H. Linnartz, *Phys. Chem. Chem. Phys.*, 2013, **15**, 8287.
12. J. He and G. Vidali, *Astrophys. J.*, 2014, **788**, 50.
13. F. Dulieu, E. Congiu, *et al.*, *Nat. Sci. Rep.*, 2013, **3**, 1338.
14. Q. Chang and E. Herbst, *Astrophys. J.*, 2012, **759**, 147.
15. A. C. A. Boogert, J. E. Chiar, C. Knez, K. I. Öberg, L. G. Mundy, Y. J. Pendleton, A. G. G. M. Tielens and E. F. van Dishoeck, *Astrophys. J.*, 2013, **777**, 73.
16. O. Morata, J. M. Girart and R. Estalella, *Astron. Astrophys.*, 2005, **435**, 113.
17. K. I. Öberg, A. C. A. Boogert, K. M. Pontoppidan, S. van den Broek, E. F. van Dishoeck, S. Bottinelli, G. A. Blake and N. J. Evan II, *Astrophys. J.*, 2011, **740**, 109.
18. Y.-J. Chen, K.-J. Chuang, G. M. Muñoz Caro, M. Nuevo, C.-C. Chu, T.-S. Yih, W.-H. Ip and C.-Y. R. Wu, *Astrophys. J.*, 2014, **781**, 15.
19. A. C. A. Boogert, *et al.*, *Astrophys. J.*, 2008, **678**, 985.
20. C. J. Lintott, S. Viti, D. A. Williams, J. M. C. Rawlings and I. Ferreras, *Mon. Not. R. Astron. Soc.*, 2005, **360**, 1527.
21. L. A. Zapata, J. Schmid-Burgk and K. M. Menten, *Astron. Astrophys.*, 2011, **529**, A24.
22. M. Agúndez and V. Wakelam, *Chem. Rev.*, 2013, **113**, 8710.

23. https://www.astro.uni-koeln.de/cdms.
24. M. P. Collings, M. A. Anderson, R. Chen, J. W. Dever, S. Viti, D. A. Williams and M. R. S. McCoustra, *Mon. Not. R. Astron. Soc.*, 2004, **354**, 1133.
25. S. Viti and D. A. Williams, *Mon. Not. R. Astron. Soc.*, 1999, **305**, 755.
26. R. T. Garrod and S. L. Widicus Weaver, *Chem. Rev.*, 2013, **113**, 8939.
27. A. I. Vasyunin, D. A. Semenov, D. S. Wiebe and T. Henning, *Astrophys. J.*, 2009, **691**, 1459.
28. R. T. Garrod, *Astron. Astrophys.*, 2008, **491**, 239.
29. R. T. Garrod, A. I. Vasyunin, D. A. Semenov, D. S. Weibe and Th. Henning, *Astrophys. J.*, 2009, **700**, L43.
30. R. T. Garrod, S. Widicus Weaver and E. Herbst, *Astrophys. J.*, 2008, **682**, 283.
31. G. M. Muñoz Caro, A. Jiménez-Escobar, J. Á. Martín-Gago, C. Rogero, C. Atienza, S. Puertas, J. M. Sobrado and J. Torres-Rodondo, *Astron. Astrophys.*, 2010, **522**, A108.
32. J. Koning, G.-J. Kroes and C. Arara, *J. Chem. Phys*, 2013, **138**, 104701.
33. C. Arasa, J. Koning, G.-J. Kroes, C. Walsh and E. F. van Dishoeck, *Astron. Astrophys.*, 2015, **575**, A121.
34. N. Watanabe, T. Shiraki and A. Kouchi, *Astrophys. J.*, 2003, **588**, L121.
35. N. Watanabe, O. Mouri, A. Nagaoka, T. Chigai, A. Kouchi and V. Pironello, *Astrophys, J.*, 2007, **668**, 1001.
36. K. I. Öberg, R. T. Garrod, E. F. van Dishoeck and H. Linnartz, *Astron. Astrophys.*, 2009, **504**, 891.
37. Y.-J. Chen, A. Ciaravella, G. M. Muñoz Caro, C. Cecchi-Pestellini, A. Jiménez-Escobar, K.-J. Juang and T.-S. Yih, *Astrophys. J.*, 2013, **778**, 162.
38. C. J. Bennett, S.-H. Chen, B.-J. Sun, A. H. H. Chang and R. I. Kaiser, *Astrophys. J.*, 2007, **660**, 1588.
39. C. J. Bennett and R. I. Kaiser, *Astrophys. J.*, 2007, **661**, 899.
40. A. L. F. de Barros, A. Domaracka, D. P. P. Andrade, P. Boduch, H. Rothard and E. F. da Silveira, *Mon. Not. R. Astron. Soc.*, 2011, **418**, 1363.
41. D. P. Ruffle, T. W. Hartquist, J. M. C. Rawlings and D. A. Williams, *Astron. Astrophys.*, 1997, **334**, 678.
42. A. Jiménez-Escobar and G. M. Muñoz Caro, *Astron. Astrophys.*, 2011, **536**, A91.
43. A. Jiménez-Escobar, G. M. Muñoz Caro and Y.-J. Chen, *Mon. Not. R. Astron. Soc.*, 2014, **443**, 343.
44. J. M. Greenberg, in *Molecules in the Galactic Environment*, ed. M. A. Gordon and L.E. Snyder, John Wiley and Sons, 1973, p. 93.
45. L. B. d'Hendecourt, L. J. Allamandola, F. Baas and J. M. Greenberg, *Astron. Astrophys.*, 1982, **109**, L12.
46. J. M. C. Rawlings, D. A. Williams, S. Viti, C. Cecchi-Pestellini and W. W. Duley, *Mon. Not. R. Astron. Soc.*, 2013, **430**, 264.
47. J. M. C. Rawlings, D. A. Williams, S. Viti and C. Cecchi-Pestellini, *Mon. Not. R. Astron. Soc.*, 2013, **436**, L59.

Section IV
Roles of Dust in the Universe

CHAPTER 10

The Roles of Dust in the Formation of Stars and Planets

10.1 Introduction

Star formation and planet formation are among the topics of greatest interest in modern astronomy. The conversion of interstellar cloud gas at a number density of a few hundred H atoms per cm^3 and a temperature of ~10 K into a star like the Sun requires an increase in number density by more than twenty orders of magnitude and in central temperature by a factor of about a million. Somehow, these immense changes must be brought about by the single force of gravity, opposed by thermal, turbulent, and magnetic pressures, and by dynamical outflows with eroding interfaces, and by angular momentum considerations. Fortunately, star formation in the Milky Way galaxy is an on-going phenomenon, and with modern astronomical instrumentation we can observe the process in many locations and in various states of evolution.

The formation of a star is a process that is important not only for the star itself and its immediate environment; the formation also affects its parent galaxy as a whole. It is through star formation that a galaxy evolves, converting its reservoir of interstellar gas to stars, flooding interstellar space with infrared, optical, ultraviolet and X-radiation, restructuring and ionizing the environment around the stars, enhancing the content of heavy atoms and of dust in interstellar space through stellar winds and explosions, and—through supernovae explosions—maintaining a low density million degree background gas as the stage on which a galaxy evolves.

Star formation and planet formation are both very active fields and the current understanding in them is well represented in many reviews and other articles (see, for example, in Further Reading, articles by Shu,

The Chemistry of Cosmic Dust
By David A. Williams and Cesare Cecchi-Pestellini
© David A. Williams and Cesare Cecchi-Pestellini 2016
Published by the Royal Society of Chemistry, www.rsc.org

Adams, & Lizano 1987;[1] Kennicutt 1998;[2] Evans 1999;[3] McKee & Ostriker 2007;[4] Bergin & Tafalla 2007;[5] Zinnecker & Yorke 2007;[6] Draine 2011;[7] Ward-Thompson & Whitworth 2011;[8] Kennicutt & Evans 2012;[9] and Aikawa 2013 [10] for star formation; and articles by Bergin 2009;[11] Williams & Cieza 2011;[12] Kley & Nelson 2012;[13] Henning & Semenov 2013,[14] and Testi *et al.* 2014 [15] for topics related to planet formation).

However, our purpose in this book, which is dedicated to the chemistry of dust, is much more restricted than in these wide-ranging reviews. We ask:

What are the specific roles that dust grains have in star and planet formation in the nearby Universe?

Having identified these roles we may then address the supplementary question:

If the physical and chemical nature of dust varies from place to place within a galaxy or between one galaxy and another, do these variations influence star and planet formation in these situations?

The initial stages of star formation occur at rather low temperatures, and it is in these early stages that interstellar molecules and dust play important roles. Once the temperature in the star-forming region begins to rise significantly, then their influence is limited. However, it is during the early low temperature stages that the path to success or failure in star formation is determined. In Section 10.2 we shall describe the roles of dust in star formation in an environment that is similar to that of the Milky Way galaxy. Star formation in the distant Universe is different from that in the local Universe, and cannot be directly observed: in particular, the first stars in the Universe are long dead. We shall make some remarks about the formation of the first stars in the Universe in Chapter 11. The major difference between current star formation in the local Universe and formation of the first stars in the very distant Universe is the presence of dust in the former and its absence in the latter.

The first discoveries of planets orbiting normal stars outside the Solar System—*i.e.* exoplanets—were made about two decades ago and since then about 1800 exoplanets have been identified (up to 2014). Around a quarter of these may be in multiple planetary systems. The Kepler Space Telescope in 2014 identified 715 newly verified exoplanets in orbits around 305 stars. These planets have masses between those of the Earth and Neptune (*i.e.* between 1 and 17 times an Earth mass). The Kepler mission also identified many more candidate exoplanets. If the sample examined so far is typical, then it is possible that there may be more than 10 billion exoplanets of sizes comparable that of the Earth, in orbits that are in "habitable zones" around their central stars. Not surprisingly, this extraordinary period of discovery has stimulated much theoretical work about planet formation. The formation of a planetary system is clearly directly associated with the formation of the central star and associated circumstellar disk, but there are also "free-floating" planets, *i.e.* planets not associated with any particular star. Presumably, these have been ejected from the planetary system during the planet-forming phase.

Planet formation occurs as a result of the aggregation of interstellar dust grains into larger and larger bodies. This process is believed to occur in dense disks of gas and dust that arise naturally as part of the star-forming process, so that the two processes, star formation and planet formation, are closely linked. We shall see that the roles of dust grains are crucial in at least the early stages of this sequence of events. An image of a dusty protoplanetary disk around a star appears later in this chapter (see Figure 10.6); planet formation may occur in a protoplanetary disk such as this.

In Section 10.3 we will describe the roles of dust in planet formation in an environment similar to that of the Milky Way galaxy. We will describe the growth of particles from single grains to planetesimals—*i.e.* bodies that are large enough to have developed significant self-gravity—and then the continued growth to planet-sized objects. Finally, in Section 10.4 we shall ask whether the evident variations in dust properties in the local Universe are enough to affect the formation of stars and planets.

10.2   An Overall Picture of Star and Planet Formation

Although we shall not discuss all the details of the star and planet formation processes, it is useful to have a crude picture of star and planet formation in mind as the basis for our discussions of the roles of dust. The general aspects of this picture are well established through observations and theoretical studies, although many details remain to be worked out.

Star formation begins in cold, dark, dense molecular clouds. It does not occur in the general interstellar medium, where radiation from hot stars pours large amounts of energy into the gas. This implies that potential star-forming regions must be shielded from the interstellar radiation field. Evidently, this is the first important role of dust in star-forming regions.

If a cold, dark, dense interstellar cloud at a given density and temperature has sufficient mass, it may become unstable to gravitational collapse. In other words, the kinetic pressure, magnetic effects, the internal turbulence of the cloud and its angular momentum may no longer be able to support the cloud against gravity and the cloud will begin to collapse. If we ignore for a moment all pressures inhibiting collapse, and imagine a collapse occurring solely under gravity without resistance, then this situation is termed *free-fall*, and the free-fall time (the time required for all mass to fall to the central point) is the minimum time necessary for the collapse to occur. It is given by

$$t_{ff} = (3\pi/32G\rho_0)^{1/2} \approx 10^8/n_H^{1/2} \text{ years} \qquad (10.1)$$

where G is the gravitational constant and ρ_0 is the mass density in the core; n_H is the total number density of H atoms (as H and as H_2) in the gas (for this and other relevant timescales, see Banerji *et al.* 2009[16]). Therefore, the minimum time for a core of number density ~10^4 H_2 cm^{-3} to collapse to high density is one million years. Of course, a real collapse is much more complex than free-fall. Parcels of gas at the edge of the cloud and within it begin to fall towards the centre of gravitational attraction but in doing so they collide with other cloud material. The collisions generate heat from this release of gravitational potential energy. If this heat can be radiated away from the system and the temperature maintained at a low level, then it is possible that the collapse may be able to continue, depending on the nature of the various supporting pressures. Otherwise, the heat is retained, driving up the temperature and the pressure so that the collapse is terminated; such a cloud may be in a quasi-static state, *i.e.* on the point of collapsing but being unable to do so until some external changes to the support mechanisms occur. However, if the collapse of the cloud is able to continue, then the higher densities achieved make it possible for individual parts of the cloud to become unstable; therefore, the collapsing cloud begins to fragment into an association of smaller, denser clumps of gas.

The collapse and further fragmentation of the cloud can continue until gas densities become very high compared to typical densities in interstellar clouds if the gravitational potential energy released as heat can be radiated away through some appropriate mechanism, and if the magnetic and other supports of the gas can be controlled. Associations of dense clumps within a Milky Way molecular cloud can be observed using telescope arrays to obtain sufficient angular resolution. Some of the observed clumps appear to be gravitationally unstable. An example of a clumpy cloud in which some clumps may continue to collapse and form stars is shown in Figure 10.1.[17]

Eventually, the collapsing and fragmenting objects become very compact and dense, and consequently each fragment of the cloud becomes tightly bound by gravity. In this state its ability to radiate is much reduced. Central temperatures are able to rise significantly without affecting the stability of the structure, if gravity is strong enough. The fragment has become a roughly spherical core with radius on the order of 10^4 Earth–Sun distances (Astronomical Units, AU), with gas infalling from all directions. After about 0.1 My or less its rotation causes it to develop a disk within the continuing infall, and the protostar generates a fast jet or outflow of hot gas along the symmetry axis of the disk. Over a period of about a million years the outflow begins to erode the interface at the outflow/infall boundary, the opening angle of the outflow increases, and the rate of mass growth by infall decreases. The system appears at this stage to be a protostar with a well-established outflow, a weaker infall, and a disk. Eventually, the infall ceases (possibly because the outflow opens completely and cuts off the infall), the outflow weakens and the disk is largely eroded, although a remnant disk remains in orbit around the star. The disk—what we may now call a protoplanetary disk—may be around 100 AU in diameter at this

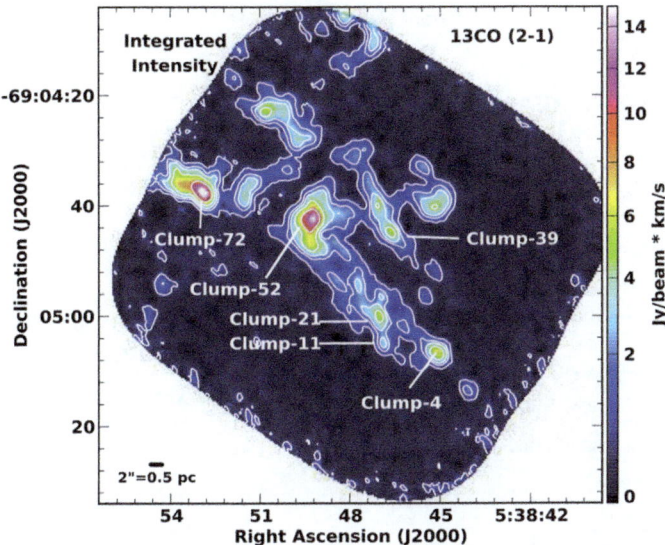

Figure 10.1   The clumpy structure of a molecular cloud revealed by molecular line observations. Reproduced with permission from Indebetouw *et al.* (2013).[17]

stage. The star/disk system may be in this phase for some millions of years, during which planet formation may occur. Eventually, the star remains, possibly with a planetary system generated from the remnant disk. A sketch illustrating these stages in the evolution of a star and its planetary system is shown in Figure 10.2.[18,19] It is worth noting our Solar System is embedded within the Öpik-Oort Cloud which has a diameter of around 10^5 AU. It contains a very large number of icy planetesimals, and is a source of long-period comets for the Solar System. These planetesimals may have originated much closer to the Sun, but have been ejected by interactions with newly-formed planets. The equivalent may also be found in other planetary systems.

The rough description given above applies to the formation of stars of low and moderate mass (up to around 8 solar masses). However, it does not apply accurately to stars of greater mass, which form in massive dense cores within infrared dark clouds. The more massive a star is, the more rapidly it evolves. Sufficiently massive stars have already begun hydrogen-burning in their interiors before the accretion of mass by intense infall from the massive parent core has been completed. The influence of the hydrogen-burning star on its environment is drastic and so the later stages of the sketch in Figure 10.4 do not apply to the formation of massive stars. Nevertheless, the sketch indicates that star and planet formation should be closely related phenomena. We now consider the roles of dust in star formation in Section 10.3 and of dust in planet formation in Section 10.4.

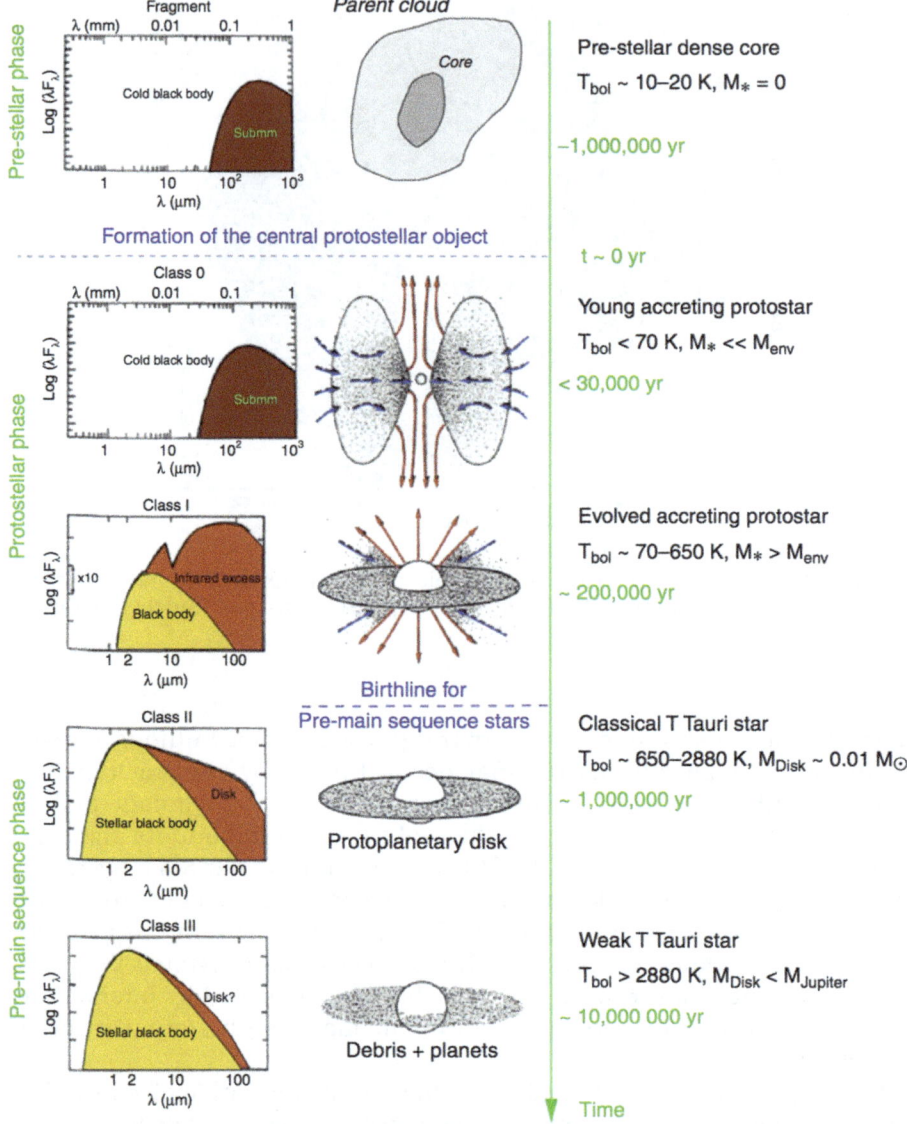

Figure 10.2   A sketch illustrating the evolution of a star and its planetary system. The later stages of the sketch do not apply to evolution of massive stars. Reproduced with permission from André.[18,19]

10.3   The Roles of Dust in Star Formation

10.3.1   Coolants in Collapsing Cores

Evidently, from the earlier discussion, the gravitationally collapsing core must be able to rid itself of heat generated during the collapse, otherwise the gas will heat up and consequently the increased pressure will terminate the collapse. The only effective cooling process for an isolated dense core is by radiation, and any specific cooling mechanisms must be able to operate at the temperature of cold dense cores, *i.e.* around 10 K. There are two mechanisms that can do this: emission from dust and emission from molecules.

Cooling by dust emission: In a typical spiral galaxy, about one third of the energy radiated by stars is absorbed by dust and re-emitted at long wavelengths. For example, continuous emission from dust grains heated by optical and ultraviolet starlight to temperatures of about 10 K covers a wide range of submillimetre wavelengths and typically peaks at wavelengths ~300 μm. There are also band emissions from vibrational modes of PAH structures in the range 1–20 μm. Evidently, radiation from dust grains can be an important coolant for a dense core which absorbs all the external starlight that falls upon it or which generates heat from gravitational collapse.

One can calculate the spectral energy distribution of the re-emitted radiation, for a given dust grain model and for a given interstellar radiation field. Important aspects of the model include the physical and chemical nature of the grains, their composition, and their size distribution. The spectrum is particularly sensitive to the size distribution of the dust grains. While large grains have temperatures that vary rather little and so are close to their mean temperatures, small grains—say, less than about 0.01 μm in radius—have temperatures that fluctuate widely (up to around 50 K), because the arrival of a single ultraviolet photon is sufficient to cause a stochastic fluctuation in the instantaneous grain temperature to a high value. Therefore, the averaged spectral energy distribution from a dense core may depend heavily on the dust grain size distribution.

Cooling by molecular line emission: Simple and relatively abundant gas-phase interstellar molecules such as CO and OH (and their isotopologues) can be effective radiators at the temperatures of cold dark interstellar clouds, through emission in their rotational spectra. For example, CO has a rotational constant of 1.9225 cm^{-1} (equivalent to 2.766 K) so that the 1–0 rotational transition occurs at 2.60 mm. Thus, low-lying rotational levels of CO molecules can readily be collisionally-populated in gas with temperatures as low as ~10 K and will radiate as the populations relax (the $J = 1$ level is 5.5 K above $J = 0$). If the emitted photons can escape from the core, then they contribute to cooling of the core. However, molecular hydrogen (the most abundant interstellar species) is useless as a coolant at these low temperatures, as the molecule has a rotational constant of 59.335 cm^{-1} (equivalent to 85.37 K) so that the first permitted quadrupole rotational transition 2–0 at 28 μm can only be excited in much warmer gas (the $J = 2$ level is 512 K above $J = 0$).

The chemistry of dense interstellar gas has been extensively studied over many years, and the ability of the more abundant molecular species to cool the gas through emission in rotational transitions has been well established. Thus, it is straightforward to determine the overall cooling rate for a dense interstellar core *via* such processes, when proper allowance is made for the transfer of the cooling radiation out of the core. For example, the core may become optically thick in 1–0 radiation at 2.60 mm from the common isotopologue ^{12}C^{16}O so that its effectiveness as a coolant is diminished.

A minor isotopologue, such as $^{13}C^{16}O$ or $^{12}C^{18}O$, radiating at slightly different wavelengths, may then be more effective than the more abundant version of the molecule. All these procedures are well established in the astrophysical literature.

However, the abundances of the coolant molecules are not well determined. As we described in Chapter 9, molecules tend to freeze-out on the surfaces of dust grains, forming molecular ices, and this process will certainly occur in cold dense cores where the freeze-out will be rapid at high densities. The freeze-out timescale in gas at a temperature of 10 K in which the total number density (cm^{-3}) of H atoms (as both H and H_2) is n_H, and for a dust-to-gas ratio equal to that in the Milky Way galaxy is

$$t_{fo} \approx 10^9/n_H \text{ years} \qquad (10.2)$$

which for a clump with number density $>10^4$ cm^{-3} is short compared to the free-fall time, t_{ff}. Indeed, in some circumstances, the freeze-out may be essentially complete so that the usual molecular line tracers of cold, dense gas are entirely missing. Figure 10.3 compares molecular line emission spectra from two dense cores, MM1 and MM2, detected by their submillimetre dust emission in a region (G11.92-0.61) where a cluster of protostars appears to be forming. MM1 appears to be a conventional object, where a massive protostar is warming the nearby dense gas not incorporated into the star; it has a rich spectrum. However, the submillimetre spectrum of MM2 is entirely featureless.

Cyganowski *et al.* (2014)[20] estimate that MM2 is cold (dust temperature 17–19 K), massive (>30 solar masses), compact (radius <1000 AU), and extremely dense (molecular hydrogen number density $>10^9$ cm^{-3}). Its luminosity from continuum dust emission is 5–7 solar luminosities. It appears that MM2 is a massive starless core, in which star formation has not yet occurred but may do so in the future. However, MM2 has none of the usual tracers of star-forming objects. Its temperature appears to be controlled by dust emission rather than molecular line emission, and freeze-out of gas phase tracers on to dust grains appears to swamp any desorbing mechanisms – possibly due to the extremely high inferred H_2 number density.

However, a complete absence of emission from gaseous molecular tracers is unusual. Dense cores usually show both solid-state features due to ices and spectral lines due to the gaseous molecules, in varying proportions. Evidently, non-thermal desorption processes can be effective in maintaining a population of gaseous molecules that may act as coolants of dense cores.

10.3.2   Non-Thermal Desorption Mechanisms

Non-thermal desorption mechanisms occur on or in dust grains and may, as we have seen in the previous chapter, contribute to maintaining a population of cooling molecules in the gas phase of a potentially collapsing core. Therefore, these dust-related mechanisms have an astronomical

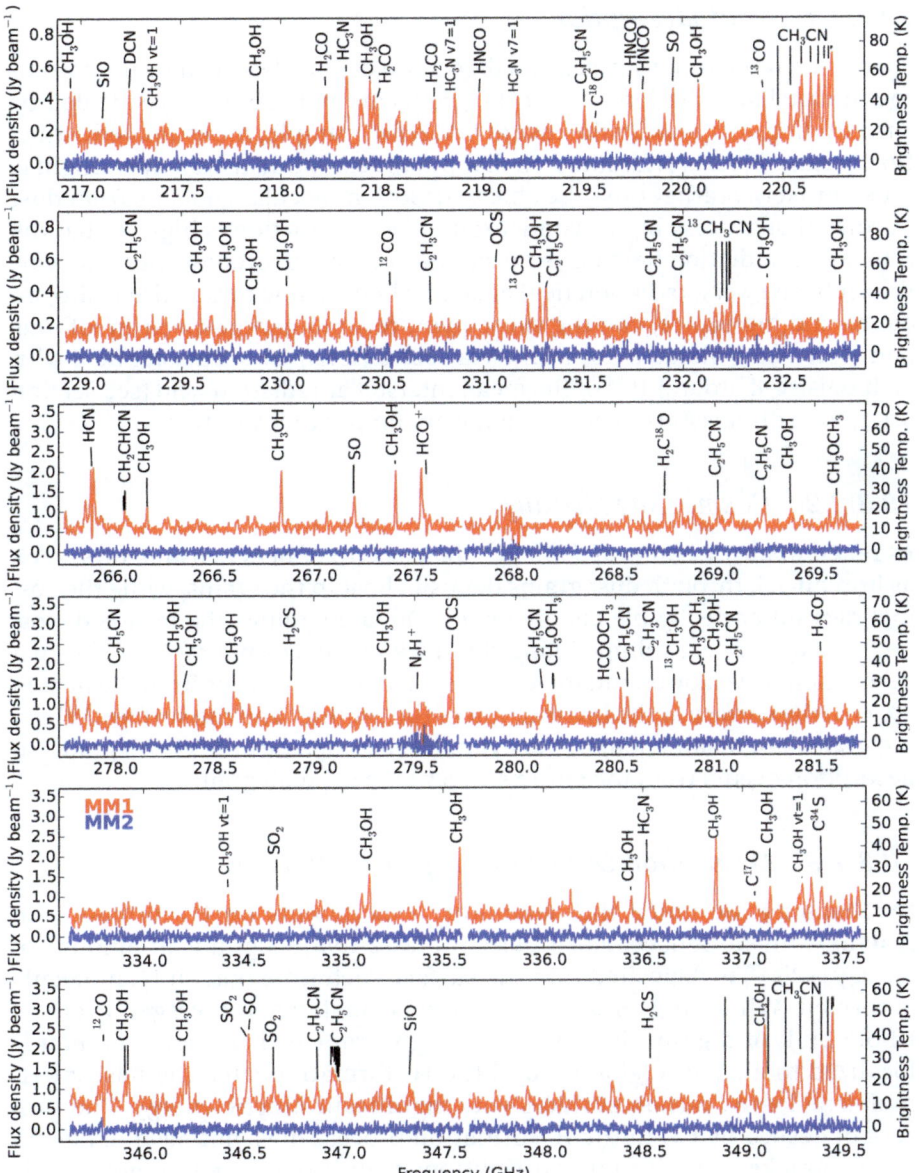

Figure 10.3   Continuum-subtracted spectra towards MM1 (red) and MM2 (blue) continuum peaks. MM1 spectra are offset for clarity. Reproduced with permission from Cyganowski *et al.* (2014).[20]

significance: differences in the physics and chemistry of dust grains may affect the population of cooling molecules available in dense interstellar gas, and therefore alter the propensity for star formation in that region of space. Details of various desorption mechanisms can be found in Roberts *et al.* (2007).[21]

10.3.2.1   Photodesorption

Species on or near the surface of a dust grain can be desorbed by photons of the interstellar radiation field, if the grains are not too heavily shielded by surrounding dusty gas. In the case of MM2, for example, the estimated extinction in this object is enormous, so that photodesorption driven by starlight is most unlikely. Starlight may be able to desorb molecular coolants from dust in cores that are less highly condensed. Where extinction is high so that the interstellar radiation field plays no significant role, photodesorption may be driven by the weak radiation field created when cosmic rays ionize hydrogen atoms and molecules, and the subsequent recombination spectrum of ions and electrons generates (for Milky Way conditions) an ultraviolet field with an intensity of around 10^{-4} of the mean interstellar radiation field (see Section 9.1.1, for a discussion of some recent work on photodesorption).

10.3.2.2   Cosmic Ray Heating

As noted first by Léger *et al.* (1985),[22] the passage of heavy cosmic rays, such as iron nuclei, through dust grains deposits heat in the grains, while the cosmic rays suffer negligible energy losses. On large grains, the heat is deposited in a cylindrical volume along the track of the cosmic ray, and thermal desorption of weakly bound molecules such as CO can occur from the heated areas of the surface at the intersection of the cylinder and the surface. Where the grains are small enough, the cylinder encompasses the entire grain, so desorption occurs from the entire surface area of small grains.

10.3.2.3   Desorption Driven by Surface Chemistry

In Chapter 8 we noted that laboratory experiments on H_2 formation in reactions on carbonaceous surfaces indicate that a significant amount of energy (~2 eV) is deposited into the surface each time that an H_2 molecule is formed. While this amount of energy is small when averaged over the whole bulk of a grain, it is, however, significant if that energy is mainly localized to a small region around the H_2 formation site. The nascent H_2 molecule leaves carbon surfaces with significant kinetic energy. The process of desorbing CO in the H_2 formation process does not need to be efficient to make a significant contribution: even if only one CO molecule is released from the surface during the formation of one hundred H_2 molecules through surface reactions, this can amount to a significant CO desorption rate.

10.3.2.4   Grain Explosions

As we described in Chapters 8 and 9, chemically-driven explosions on dust grains may also act as non-thermal desorption mechanisms. The expected temperature rise in explosions could be—as indicated in experiments—as high as a thousand Kelvin. If such a process operates, then desorption would not be limited to weakly bound species such as CO.

10.3.3   Magnetic Support of Dense Cores

We discussed in Chapter 2 the implications of the observational result that visual starlight is weakly linearly polarized in the interstellar medium, and we noted that this polarization may be created by partially aligned large dust grains. The interaction of the local magnetic field with large rotating grains generates a dissipation of one component of angular momentum of grains, leading to their partial alignment. If so, then we may obtain from polarization data some information about the magnitude and orientation of the local magnetic field wherever the light of background stars is not too heavily extinguished in a molecular cloud. There are, of course, more direct ways of studying the magnetic fields in molecular clouds: they may also be probed using the Zeeman effect on the 21 cm line of atomic hydrogen, on the OH Λ-doubling lines at 1612, 1665, 1667, and 1720 MHz, and on the CN 1–0 rotational transition at 2.6 mm. Alternatively, one may use maps of the observed polarization of far infrared or submillimetre emission from dust in the cloud. Figure 10.4 (ref. 23) shows the results of polarimetry measurements at 350 μm of two small regions in a Giant Molecular Cloud of the Milky Way galaxy, NGC 6334.

The polarization of a few percent is large and suggests that the grains must be well-aligned by the local magnetic field. Also, the field must be relatively uniform along the line of sight. Detailed studies of these and other regions have shown that magnetic energies within molecular clouds and cores are at least as great as the local turbulent kinetic energies. Evidently, the magnetic field could be capable of affecting cloud dynamics. In particular, if suitable mechanisms exist, the magnetic field may be sufficient to resist the gravitational collapse of a core in a molecular cloud. How could it do this?

As is well known, the magnetic field interacts directly only with charged particles that have a velocity component transverse to the field direction, causing them to gyrate around the field lines, while motion along the field lines is unaffected. Thus, the field lines and the charged particles are indeed connected dynamically. This connection also extends to the neutrals in the gas, because collisions between charged particles and neutrals ensure that there is also a drag between an ion flow and the neutrals. Since the free movement of the charged particles transverse to the field direction is constrained, then neutral material mixed with the charged particles also feels the same constraint, through collisions. Neutrals can therefore be prevented from moving freely under gravity, at least transverse to the field lines, where the magnetic field resists the collapse of the charged component of partially ionized gas. Evidently, the resistance is not spatially uniform: for a strong field permeating a partially ionized cloud under gravity, collapse may be possible along the field lines while resisted in the orthogonal direction.

Where the ionization fraction is reasonably high, the field and the charged particles are essentially locked together. However, if the ionization fraction is

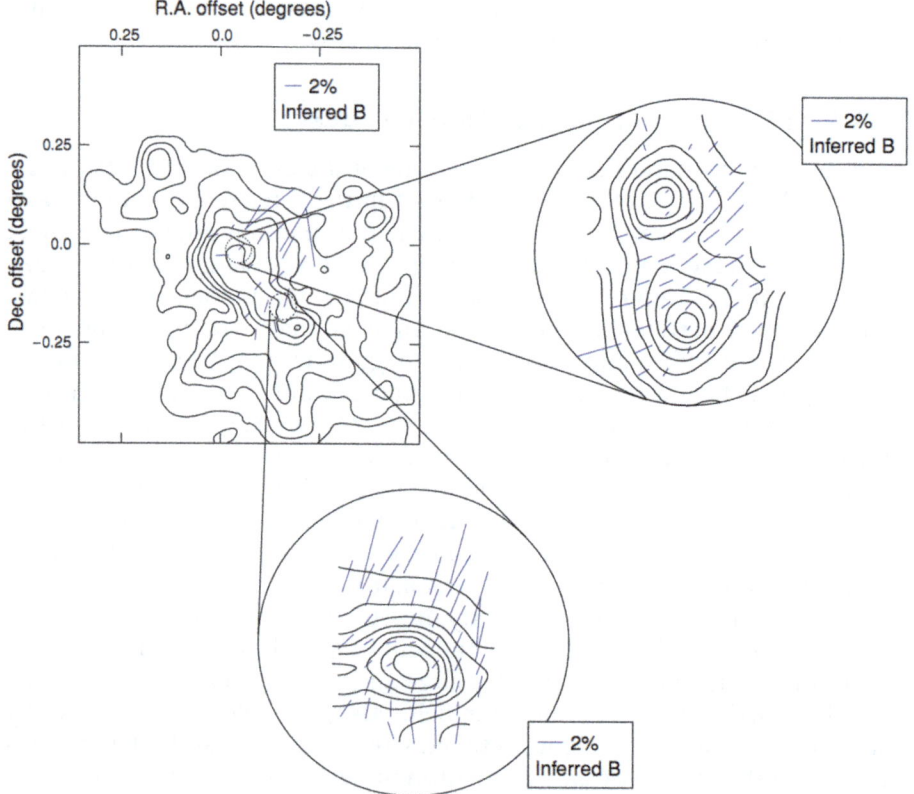

Figure 10.4 Submillimetre polarimetry of the giant molecular cloud NGC 6334. The map at upper left shows 450 µm polarization measurements superposed on 100 µm intensity contours. Expanded views of two small regions of NGC 6334 show 350 µm polarization measurements. Blue bars indicate the inferred magnetic field direction, and the length of the bars is proportional to the polarization. Reproduced with permission from Novak *et al.* (2009).[23]

sufficiently low, the field and the charged particles may begin to separate in a process known as *ambipolar diffusion*. This term means that the magnetic field in a dense core may begin to drift out of the core, so that support against gravitational collapse is weakened. If ambipolar diffusion is rapid compared to the collapse timescale, then magnetic support against gravity may not be significant. An approximate timescale for ambipolar diffusion, t_{AD}, in dense cores can be written

$$t_{AD} \sim 10^6 \, (x_i/10^{-8}) \text{ years} \tag{10.3}$$

where $x_i = n_i/n_H$ is the fractional ionization (a dimensionless quantity), and n_i is the number density of ions in the gas. If x_i is indeed on the order of 10^{-8}, or larger, then ambipolar diffusion is relatively slow compared to the minimum collapse timescale, t_{ff}, and magnetic support must be considered in any study of gravitational collapse. However, if x_i is small compared to 10^{-8}, then magnetic support may be negligible because the magnetic field may drift rapidly out of the core. So the question of whether magnetic support can play a role reduces to the value of the fractional ionization in the core, and here—at last—is a situation in which there is a possible role for dust grains.

In dense interstellar gas in which starlight is excluded, the fractional ionization is established as cosmic rays ionize hydrogen molecules to form H_2^+ which—in rapid reaction with another H_2 molecule—forms H_3^+. This is an important molecule in gas-phase chemistry because of its ready propensity to donate protons to almost any other species. In particular, it can donate a proton to the abundant molecule CO to form HCO^+ (a detected and abundant interstellar ion) for which an effective loss route is in dissociative recombination with electrons, a very efficient process

$$HCO^+ + e^- \rightarrow H + CO.$$

If CO is indeed abundant in dense gas, then HCO^+ may be the most abundant ion, and—for physical parameters appropriate for dense cores in the Milky Way galaxy—the fractional ionization is computed to be $\sim 10^{-8}$ and so the ambipolar diffusion timescale is on the order of a million years, comparable with the collapse time.

However, this conclusion depends on the abundance of CO in the gas, which is determined by surface area of dust grains per unit volume and by desorption processes. The ionization balance is also affected by the accumulation of charge on grains. In some circumstances, the charge resides mainly on grains rather than in a minor component of ionized gas. Grain charge is also determined by the surface area per unit volume and electrical properties of the grain material. The ambipolar diffusion timescale must be examined carefully in each case. Comprehensive modelling of the influence of dust grains on potential magnetic support has yet to be fully explored.

10.4 The Roles of Dust in Planet Formation

10.4.1 Description of Protoplanetary Disks, Gas and Dust: Current Ideas of Planet Formation

We indicated schematically in Figure 10.2 how a dense star-forming core develops a disk from which a planetary system may form. To elucidate the roles of dust in planet formation, we now need to consider in some more detail how the disk may evolve from its initial massive gas-rich form to a low mass "debris disk" in which planet formation actually occurs. The disks as formed initially are massive (possibly containing a significant fraction of a solar mass) but gravitationally unstable. They evolve quickly to a less massive and more stable form in which they exist for timescales on the order of some millions of years. We shall regard these stable disks as the initial states of the evolutionary process, referred to above, whose final states are debris disks. These initial states are sites of an abundant gas-phase chemistry that reflects the local physical conditions. Therefore, molecular line studies, together with infrared and submillimetre thermal emissions from the various dust components that they contain, reveal the mass, size, and detailed structure of these initial disks in some detail. The sketch in Figure 10.5 indicates the variations in physical conditions and summarizes the various processes occurring in the initial states.

These disks are warm and dense, and may have masses on the order of five Jupiter masses (1 Jupiter mass $\approx 10^{-3}$ solar masses). They are commonly flared, so that far ultraviolet and extreme ultraviolet and X-rays from the central star can reach the outer layers, raising the temperature in them to $\sim 10^3$ K or higher and generating a characteristic type of gas-phase chemistry (so-called PDR chemistry) in the outer layers. However, the bulk of the material, deeper inside the disk, is well-shielded and unaffected by these radiation sources; this interior material has lower temperatures (roughly $\sim 10^2$ K) and higher number densities ($n_H \sim 10^9$ cm^{-3} and greater). These conditions are such that water molecules (for example) remain in the gas phase for extensive regions of the disk (where they take up nearly all oxygen not already in carbon monoxide), apart from in the very densest (possibly as high as $n_H \sim 10^{14}$ cm^{-3}) and coolest ($\leq 10^2$ K) regions in the mid-plane of the disk, where ice is deposited on grain surfaces, outside a critical radius (sometimes called the "snow line"). Spectroscopy reveals that this cold mid-plane is generally devoid of gas phase molecular species.

Close to the star is a region in which material is being transferred on to the star. This accretion zone is important in determining the final mass of the forming star. In the bulk of the disk, various transport mechanisms are operating. Viscous transport carries material in the plane of the disk towards the star, to feed the continuing accretion of material on to the star. Dust settling, for large dust grains, increases the number density of large grains in the central plane of the disk, while small dust grains (say, less than about 0.1 µm) are swept along in the turbulent gas. Grain growth is therefore crucial to the evolution of the disk. It occurs primarily by collisions between dust grains, rather than by accumulation of molecules in mixed ices on the surface of dust grains (sometimes referred to as growth by condensation). Condensation occurs on grains of all sizes to the same extent, but since small grains provide much of the surface area per unit volume, they take up most of the molecular material that forms the ices, while large grains undergo relatively little change in volume. Grain collisions, however, may cause growth or may shatter weakly-bound large grains into smaller entities. These processes, and the influence of the emerging star, cause the disk to evolve, and the end-point of this evolution is the debris disk, in which planet formation occurs, see Figure 10.6.

Figure 10.5 indicates how the disk loses mass by accretion on to the star and by photo-evaporation. Grains grow and settle into the mid-plane, and also provide less extinction for ultraviolet. The disk mass decreases and the supply of mass to the central star also decreases, but the outer disk is no

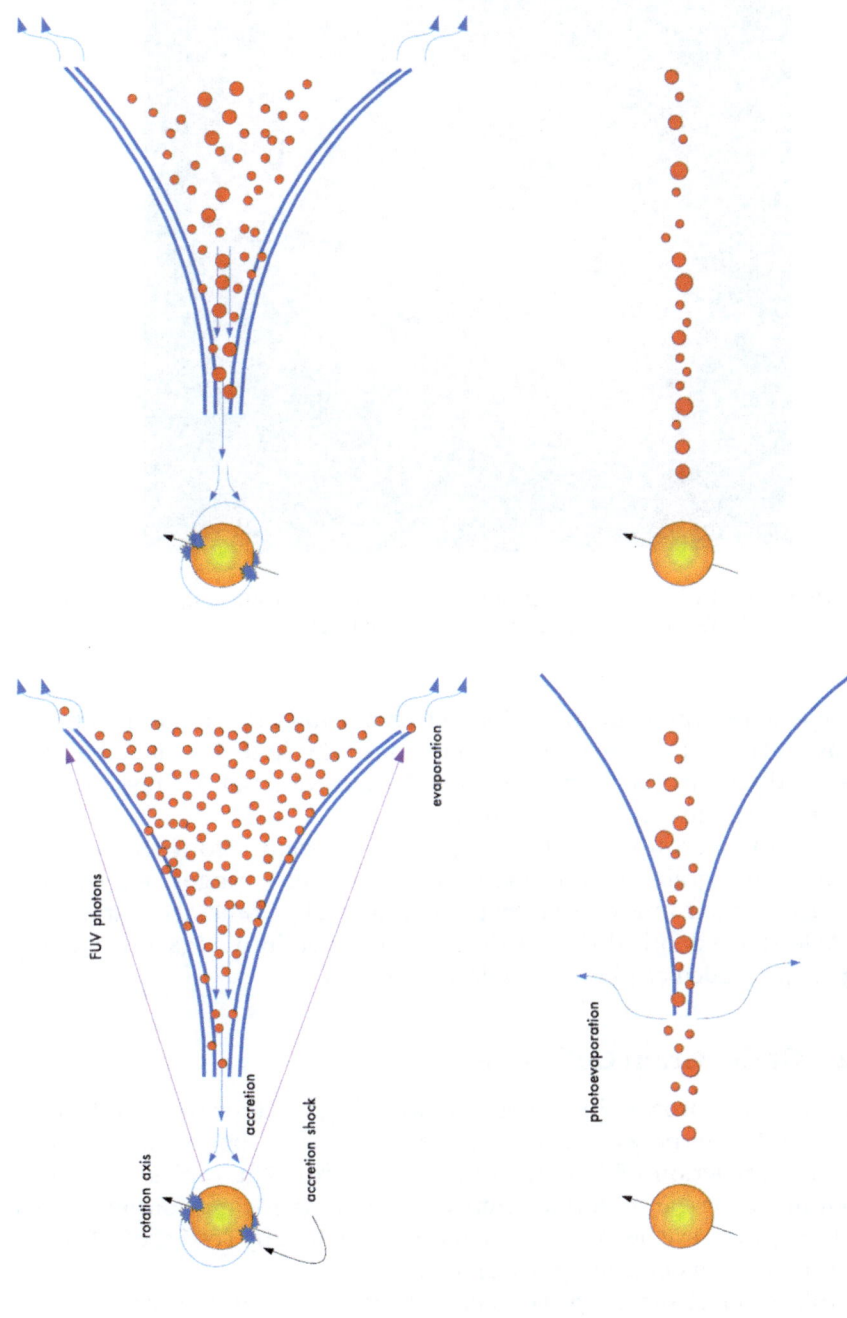

Figure 10.5 Schematic diagram indicating the evolution of a protoplanetary disk. Based on Williams & Cieza (2011).[12]

Figure 10.6   ALMA image of the protoplanetary disk surrounding the young star HL Tauri. Credit: ALMA (ESO/NAOJ/NRAO).

longer able to re-supply the inner disk with mass and the accretion on to the star ceases. Then an inner hole develops; once the hole is established the disk is rapidly destroyed from the inside by ultraviolet and X-rays, leaving only the larger grains and planetesimals.

Observations certainly confirm that growth occurs from the original submicron-sized grains to millimetre and centimetre ranges. Figure 10.6 shows a false colour image taken by the Atacama Large Millimeter Array telescope (ALMA) of a debris disk. The debris is corralled into rings, presumably by large objects such as planetesimals and planets.

10.4.2   Grain–Grain Collisions

The fundamental process in planet formation is grain growth by collisions. Evidently, collisions between grains may lead to sticking, bouncing, fragmentation, or compaction of loosely-formed structures. The topics have been and continue to be the subject of intense laboratory work and computational simulations, and have been comprehensively reviewed by Testi *et al.* (2014).[15] This is very much a developing research area.

The role of electrostatic forces between charged grains is unclear and is often ignored. A crude comparison of the likely kinetic energy of grains with the energy of electrostatic repulsion between typical grains with single charges suggests that it is safe to ignore these effects, unless the grain charge is large (say, 30 electrons). The interactions between grains are then simply rather weak van der Waals forces and the collisions are essentially kinetic. Velocities between grains of different sizes are due to random motions in the gas, and to drift velocities; all these velocities are sensitive to grain size so grain–grain collisions may be expected to occur.

Experiments suggest that there is a critical velocity above which sticking does not occur. This critical velocity depends sensitively on many factors, including grain size, grain material, grain morphology, and the mass ratio of the colliding grains. For micron-sized silica grains, the experimentally determined critical velocity is ~1 m s^{-1}, while the figure for ice grains may be ~30 m s^{-1}. Smaller grains may have much larger critical velocities.

For collisions well below the critical velocity, sticking occurs at the first contact. Numerical simulations indicate that early grain growth in these circumstances leads to very fluffy grains. As the mass of the aggregate increases, then compaction begins to play a role. This leads to the formation of more recognisably solid structures, although with varying degrees of porosity. The increasing mass of aggregates and the accompanying increase in compaction leads to a reduction in the sticking probability and an increase in the likelihood of bouncing and possibly of erosion. This suggests that there is a maximum aggregate size that can be formed in direct sticking collisions. Such a barrier to further growth may be in the millimetre to centimetre regime.

However, there is another mechanism that has been observed to occur in laboratory experiments. For velocities close to the fragmentation barrier, collisions between a large aggregate and a small aggregate are found to lead to a large mass transfer from the small to the large aggregate. Numerical simulations suggest that this mass transfer process may allow growth of aggregates to planetesimal sizes.

10.5 Consequences of Grain Properties on the Formation of Stars and Planets

The answers for the specific roles of dust in star formation in the local Universe are clear from the preceding discussion. (1) Dust is an essential component in shielding star-forming regions from the interstellar radiation field created by stars in the visible and ultraviolet. (2) Continuum emission from dust is an effective coolant and contributes significantly to cooling of regions undergoing gravitational collapse. (3) Dust affects the abundance of small molecules, which are also effective coolants of collapsing gas, by providing sinks—through freeze-out—for these molecules, and also alternative sources through various desorption mechanisms that release molecules back from icy mantles into the gas phase. (4) Dust grains may, in very dense regions, become significant carriers of negative charge and thereby may influence ambipolar diffusion and magnetic support.

The second question posed in Section 1.1.1 asks what might be the effect on star formation if—as is very likely—the physical and chemical nature of dust varies somewhat from place to place within a galaxy or between one galaxy and another. Let us consider each role of dust in turn.

(1) To provide adequate extinction for a star-forming region, dust must be present in sufficient quantities in the gas, and must provide extinction in the ultraviolet and far ultraviolet. Thus, the dust-to-gas ratio must not depart much from the Milky Way value, and the dust grain size distribution must have a major population of small grains. A region of space in which a dust population with the canonical dust-to-gas mass ratio of 0.01 is present, but without a major component of small grains (say, less than 0.1 μm) will not be adequately shielded from the interstellar radiation field – regardless of grain composition. Such a region will be less favoured for star formation.

(2) Dust grains of various size ranges and compositions are effective radiators and can be important coolants. The cooling rate will be proportional to the dust-to-gas ratio, and will have a dependence on the size distribution: small grains are more effective coolants at elevated temperatures.

(3) The roles of dust grains as sinks and sources of interstellar molecules are not very sensitive to the chemical nature of the bare dust grains. The gas-to-dust ratio combined with the grain size distribution determines the available cross section per unit volume; this is an important parameter in both freeze-out and desorption.

(4) If conditions are such that dust grains become significant carriers of charge, it is likely that the charge level is very low, ambipolar diffusion is relatively fast and so magnetic support is minimal. However, if the cosmic ray ionization rate is enhanced by, for example, the presence of nearby massive stars then magnetic support may be significant.

(5) The role of dust in planet formation is a fundamental one: dust grains provide the solid material from which planets are formed.

The initial steps in this grand process are the interactions of pairs of dust grains, their growth by sticking in random associations, or otherwise. There appear to be significant differences in the critical velocities above which sticking will not occur, ice being favoured over bare silicates. In fact, dust grains are almost certainly coated with ice at the stage where growth is likely to occur. Growth into fractal structures, compaction, and fragmentation appear to form dust readily to solids of millimetre or centimetre ranges. Mass transfer in binary collisions appears to be necessary for aggregates to grow beyond this size, and become planetesimals. None of these processes is particularly dependent on the nature of the original dust or its size distribution. The debris disk mass must be related to the initial dust-to-gas ratio.

Further Reading

1. F. H. Shu, F. C. Adams and S. Lizano, *Annu. Rev. Astron. Astrophys.*, 1987, **25**, 23.
2. R. C. Kennicutt, Jr., *Annu. Rev. Astron. Astrophys.*, 1998, **36**, 189.
3. N. J. Evans II, *Annu. Rev. Astron. Astrophys.*, 1999, **37**, 311.
4. C. F. McKee and C. E. Ostriker, *Annu. Rev. Astron. Astrophys.*, 2007, **45**, 565.
5. E. Bergin and M. Tafalla, *Annu. Rev. Astron. Astrophys.*, 2007, **45**, 339.
6. H. Zinnecker and H. W. Yorke, *Annu. Rev. Astron. Astrophys.*, 2007, **45**, 481.
7. B. T. Draine, *Physics of the Interstellar and Intergalactic Medium*, Princeton University Press, Princeton and Oxford, 2011.
8. D. Ward-Thompson and A. P. Whitworth, *An Introduction to Star Formation*, Cambridge University Press, Cambridge, 2011.
9. R. C. Kennicutt, Jr. and N. J. Evans II, *Annu. Rev. Astron. Astrophys.*, 2012, **50**, 531.
10. Y. Aikawa, *Chem. Rev.*, 2013, **113**, 8961.
11. E. Bergin, in *Physical Processes in Circumstellar Disks around Young Stars*, ed. P. J. V. Garcia, University of Chicago Press, Chicago, 2009, p. 55.
12. J. P. Williams and L. A. Cieza, *Annu. Rev. Astron. Astrophys.*, 2011, **49**, 67.
13. W. Kley and R. P. Nelson, *Annu. Rev. Astron. Astrophys.*, 2012, **50**, 211.
14. T. Henning and D. Semenov, *Chem. Rev.*, 2013, **113**, 9016.
15. L. Testi, T. Birnstiel, *et al.*, in *Protostars and Planets VI*, ed. H. Beuther, R. Klessen, C. Dullemond and Th. Henning, 2014, p. 339.
16. M. Banerji, S. Viti, D. A. Williams and J. M. C. Rawlings, *Astrophys. J.*, 2009, **692**, 289.
17. R. Indebetouw, C. Brogan, *et al.*, *Astrophys. J.*, 2013, **774**, 73.
18. P. André, in *Star Formation and the Physics of Young Stars*, ed. J. Bouvier and J.-P. Zahn, EDP Sciences, 2002, p. 1.
19. P. André. in *Encyclopedia of Astrobiology*, ed. G. Muriel, Springer, 2011, p. 1549.
20. C. J. Cyganowski, C. L. Brogan, *et al.*, *Astrophys. J.*, 2014, **796**, L2.
21. J. F. Roberts, J. M. C. Rawlings, S. Viti and D. A. Williams, *Mon. Not. R. Astron. Soc.*, 2007, **382**, 733.
22. A. Léger, M. Jura and A. Omont, *Astron. Astrophys.*, 1985, **144**, 147.
23. G. Novak, J. L. Dotson and H. Li, *Astrophys. J.*, 2009, **695**, 1362.

CHAPTER 11

Dust in the Far Distant Universe

11.1 Introduction

Almost everything we have said about interstellar dust so far in this book refers to dust in the Milky Way and in neighbouring galaxies; *i.e.* what we might call the local or nearby Universe. Within the Milky Way, and between the Milky Way and relatively nearby galaxies, significant variations in physical conditions certainly occur. Radiation fields, cosmic ray fluxes, and even relative elemental abundances may differ from galaxy to galaxy and even from region to region within a galaxy. But it is clear from observations that dust is present and plays various roles in many interstellar locations and in all nearby galaxies. It is also clear that while the nature of the dust may vary from one location to another, its properties can generally be described by models that originated from our studies of the Milky Way (see Chapter 3). In spite of variations in physical conditions, we still expect interstellar dust in the local Universe to be mostly "soot and sand" (to use the crude shorthand description given by Hartquist and Williams in 1995[1]). Therefore, the chemical processes on dust surfaces that we have described in Chapters 8 and 9 will—with some comprehensible variations between locations—usually apply throughout the local Universe, and the molecular products of those chemical processes exert their crucial astronomical roles as described in Chapter 10.

But what can we say about dust in countless galaxies that are very distant from us, so distant that their light has been travelling to us for an appreciable fraction of the age of the Universe? Is dust present in galaxies very near the edge of the observable Universe? If so, when did dust first appear in the Universe? What was its nature? What was its function in those very early times?

The Chemistry of Cosmic Dust
By David A. Williams and Cesare Cecchi-Pestellini
© David A. Williams and Cesare Cecchi-Pestellini 2016
Published by the Royal Society of Chemistry, www.rsc.org

In the Big Bang cosmology, the Universe began at an exceptionally high temperature in which only dark matter of unknown nature, together with a very minor component of familiar matter in the form of subatomic particles existed, along with electromagnetic radiation. In the Universal expansion and cooling that followed, atoms of hydrogen and helium and their isotopes, and a trace of lithium, were formed. There were no more massive elements at this stage in the Universe. Dust as we understand it in the local Universe could not have been present at this early stage.

The first stars in the Universe, the Population III stars, were formed as normal (*i.e.* baryonic) matter sank into dark matter "minihaloes" of approximately one million solar masses; the baryonic matter collapsed under its own weight in the initial stages of star formation. The only available coolants effective at low temperature in this primordial gas were trace amounts of molecular hydrogen and deuterium hydride – a much more restricted situation than in star formation in the local Universe (see Chapter 10). Collisions excited populations into rotational levels of H_2 and HD, and emission from these levels provided the cooling that permitted the collapse to continue. Eventually, the gas density increased during the continuing collapse that three-body collisions were able to convert all the atomic hydrogen to molecules, and the cooling rate increased and the collapse continued. Ultimately, a cluster of Population III stars was formed in each event. The mass of individual stars was probably in the range of a few tens of solar masses. Their evolution would have been extremely rapid. More detailed information on the formation of Population III stars may be found in the 2013 review by Volker Bromm.[2]

Population III stars changed the Universe dramatically. Their radiation re-ionized the baryonic matter. Their supernovae populated space with the first heavy elements such as carbon nitrogen, oxygen, silicon and iron, *etc.*, and dust was at last a possible constituent of the Universe. Accumulation of minihaloes and the death of the Population III stars led to the formation of the first galaxies in which a new generation of conventional stars, the Population II stars, was formed.

Astronomers often use the concept of red shift, z, to specify evolutionary age in the Universe. Red shift refers to the increase in wavelength of radiation received on Earth from a distant source; the increase arises because the Universe is expanding as a result of the Big Bang. The great advantage of using the red shift is that it is a readily and directly measurable quantity. A red shift of zero means that the radiation comes from a nearby region and there is no increase in wavelength, while a positive value means the source of radiation is very distant such that the expansion velocity is great. The light travel time is also long and so the radiation must have been emitted when the object was young. Using current data from cosmological observations, the age of the Universe since the Big Bang is believed to be around 13.8 billion years. The earliest star formation (of Population III stars) is believed to have occurred at 180–100 million years post-Big Bang ($z \sim 20$–30), while a major burst of conventional star formation (of Populations II and I stars) is believed to have occurred at 480–420 million years ($z \sim 10$–11) post-Big Bang.

In this cosmology, heavy elements and dust should, therefore, have begun to accumulate in gas from the epoch in which Population III stars exploded in supernovae. Observations in the far-infrared and submillimetre wavebands of extremely luminous and very distant galaxies confirm that dust grains (probably composed of the familiar heavy elements) must have had appreciable abundances back to remarkably early times after the Big Bang. Almost all of the detected radiation from high red shift matter is simply thermal emission from dust grains heated by ultraviolet emission from massive stars. An example of such a galaxy is HFLS3, a galaxy so distant that the infrared and sub-mm radiation we receive left that galaxy just 890 million years after the Big Bang (corresponding to a red shift z of 6.3). It is a galaxy in which the star formation rate appears to be an astonishing 3000 solar masses per year (for comparison, the mean star formation rate in the Milky Way is about one solar mass per year). It is worth noting, however, that recent and very sensitive observations by the Atacama Large Millimeter Array (ALMA) astronomical facility of some huge starburst galaxies suggest that these are, in fact, multiple galaxies that were previously unresolved.[3,4]

Nevertheless, considered as a single object, HFLS3 contains about a billion solar masses of dust, has a dust-to-gas mass ratio of 1% and a dust-to-stellar mass of 4%, similar to values for nearby starburst galaxies. The abundant dust clouds in which star formation occurred shrouded the newly-formed massive stars, absorbed the intense stellar ultraviolet and converted almost all of it ultraviolet to infrared and sub-mm radiation. Evidently, large amounts of dust had formed very early in the evolving Universe. Did this dust have roles other than simply transmuting ultraviolet to infrared? What was the nature of this dust?

11.2   Chemical Questions

The example of HFLS3 (see above) and of other high red shift objects shows that dust was present in some locations at remarkably early times; abundant dust was probably present in some objects by 500 million years post-Big Bang. However, the picture is far from uniform: an actively star-forming object, Himiko, at a red shift $z = 6.62$ (840 million years post-Big Bang) has a low dust content and seems to be composed of near-primordial interstellar gas. Since we know that dust plays crucial chemical and physical roles in the contemporary Universe, we may ask similar questions about dust in the high red shift Universe. In particular, from the astrochemical perspective of this book, some interesting questions are:

(1) When did dust grains first become widely available in the Universe?
(2) Why is the dust distribution not uniform with z?
(3) How did the dust-to-gas ratio for these first grains vary in time?
(4) What were the composition and the size distribution of the first grains?
(5) Were these first grains effective in promoting chemistry at high red shifts?

We shall begin with a survey of the astronomical observations relevant to dust formation at high red shift. We shall discover, however, that this very active subject is still in its infancy, so that definitive answers to most of these questions are extremely difficult to find, given the present state of astronomical data. On the positive side, progress is rapid.

11.3 Observations

Fortuitously, it turns out to be easier to observe high red shift galaxies in the far-infrared, submillimetre, and millimetre than in other wavebands, and—for objects at red shifts of ~10 or greater—the received flux densities in the millimetre waveband should exceed those in all other bands. Thus, wavebands in the sub-mm and mm are ideal for studies of very distant objects. This means that, unlike observations of dust-related phenomena in the local Universe, much of the information about high red shift dust is generally obtained from long wavelength data (*i.e.* far-infrared, sub-mm, and mm wavebands) rather than data from the infrared, visible and ultraviolet (as discussed at length in Chapter 5, for dust in the Milky Way). Unfortunately, observational spectra at these long wavelengths tend to suffer from a lack of identifying features and spectral bands that enable reliable fits to be made, so the task of identifying the nature of grains is more difficult in the high-z Universe than in the local Universe.

Observational studies of very low metallicity objects (*i.e.* objects having low abundances of the common elements C, N, O, S, Si, Mg, Fe, *etc.*, relative to hydrogen) in the *local* Universe have been made and used as analogues of high red shift objects. One might expect that objects with low metallicity also have low dust-to-gas ratios. For example, this plausible view is supported by studies of the local low-metallicity dwarf galaxy I Zw 18;[5] it is found to have a low dust-to-gas ratio. If this object is regarded as an analogue of the very distant object Himiko (referred to in the previous section), then Himiko should have a dust-to-gas ratio that is only 1% of that of the Milky Way; in other words, the low metallicity of Himiko implies a low dust-to-gas ratio. However, this simplistic view of the link between metallicity and dust is contradicted by studies of another very low metallicity local galaxy SBS 0335-052 which—rather surprisingly—has dust-to-gas ratios close to those for the Milky Way[6] in spite of its low metallicity. Evidently, the connection between metallicity and dust is not a simple one.

11.4 Dust Emission as a Probe of Star Formation

In principle, the far-infrared and sub-mm dust emission from a distant galaxy can be regarded as a direct tracer of the star formation rate in that object, since almost all the emission comes from dust grains heated by stellar ultraviolet. For astronomers, the star formation rate is fundamental to understanding the evolution of galaxies, so an observational approach to determine the rate of formation of totally obscured stars is highly desirable. However, the task is not straightforward and to perform it accurately it would be necessary to know (*inter alia*) the optical properties of the dust (which depend on the chemical and physical composition of the dust material) and the mass spectrum of stars in a distant galaxy (since this spectrum determines the ultraviolet spectral energy distribution of the radiation heating the dust grains). However, the dust composition is essentially unknown, although one can make assumptions based on properties of dust in the local Universe; from the astrochemical point of view, progress on this topic would be highly desirable. Similarly, the ensemble of stars heating the dust cannot be directly viewed, because by definition the stars are totally obscured by their natal dust clouds. The nature of stellar ensembles in high red shift galaxies is precisely the kind of information that astronomers would like to discover. So these direct approaches linking dust emission to star formation rate have major difficulties associated with them. As a result, astronomers have developed empirical methods (see for example Kennicutt 1998[7]) to infer galactic-scale star formation rates. Such methods have been successful, but the complexities of dust heating and emission are ignored in them. The current star formation rate in a high-z galaxy derived from the total far-infrared luminosity in this approach depends on the star formation history of that galaxy (see reviews by Kennicutt and Evans 2012;[8] Lutz 2014[9]).

From the perspective of the astrochemical question listed in Section 11.2, we shall ignore these astronomical difficulties and consider the problems of formation of dust and its nature in high red shift galaxies.

11.5   Formation of Dust at High Red Shifts

The intense far-infrared emission from dust grains is not only a signature of massive star formation at early times in the evolution of the Universe, it is also indisputable evidence that very large amounts of dust were formed rapidly very early on in many objects. In some galaxies, about 100 million solar masses of dust are apparently present and this dust must have been formed within about 500 million years after the Big Bang. What are the processes in which such active dust formation occurs? Do these processes give insight into the nature of the dust and therefore of its chemical roles?

We discussed in Chapter 5 the formation of dust in several environments of the Milky Way and in Chapter 6 the response of Milky Way dust to electromagnetic radiation and its destruction in shocks. The main sources of dust in the Milky Way were described in Chapter 6 to be in the cooling outflows from asymptotic giant branch (AGB) stars, and in the ejecta of supernovae. Both of these dust-forming events occur at or near the end of the lives of these objects. Of course, supernovae can be both source and sink of dust grains, so the overall contribution of supernovae to interstellar dust needs to be rather carefully considered. For dust on the Milky Way galactic-scale, AGB stars and supernovae appear to be the only significant sources. Is it possible that these processes also operate similarly in the distant Universe as in the local Universe? This assumption would appear to be a good starting position.

On this basis, Gall *et al.* (2011)[10] have made an extensive and detailed investigation of the contributions from AGB stars and supernovae to dust formation in high-z galaxies. However, they note that there is (at least) one striking difference from the situation in the local Universe: in high-z galaxies, there must be sufficient time for a star to reach the end of its life (*i.e.* its dust-forming phase) at that particular red shift. Low mass stars evolve slowly and may not have reached the ends of their lives at the epoch implied by the red shift. If so, they cannot contribute to dust formation. Gall *et al.* (2011)[10] found that a star formed at red shift $z = 10$ could be a source of dust at $z = 6$ (at which red shift many submillimetre sources are detected) only if its mass is ≥ 3 solar masses. Stars with greater mass evolve much more quickly than stars with lower mass. If a star has less than this critical value of 3 solar masses, then its lifetime is longer than the age of the Universe at this red shift. Gall *et al.* also showed that the metallicity with which a star is born made little difference to this conclusion.

Gall *et al.* therefore considered the dust production in stars in the range 3–40 solar masses. At the lower end of this range, 3–8 solar masses, stars evolve into AGB stars, while more massive stars end their lives as supernovae. Making assumptions based on observational data and theoretical models concerning dust formation in AGB stars and in supernovae, and on the mass distribution of stars in the high-z galaxy, they estimated the total dust mass formed in high-z galaxies. There are, of course, many uncertainties in these highly complex models, particularly in the dust-forming efficiencies, in dust destruction, and in the mass distribution of stars in the high-z galaxy.

The computational results appear to predict larger dust masses than detected. This may be due to poorly understood mechanisms of grain destruction. On the other hand, observational estimates of dust mass generated in supernovae appear to be increasing. In Chapter 5, we referred to the detection in 2011 of massive amounts of dust in supernova SN1987A in the Large Magellanic Cloud, a galaxy neighbouring the Milky Way.[11] More recently, Gall *et al.* (2014)[12] reported the growth of dust mass in dense gas surrounding the core-collapse supernova CCSN 2010jl in the nearby galaxy UGC 05189. The extinction properties of the grains in SN 2010ij show the grains to be large, *i.e.* of diameter well in excess of 1 μm, making them able to withstand destruction in shocks from the explosion (see Chapter 6). In early observations of SN 2010ij, the dust mass was $\sim 10^{-4}$ solar masses, but this increased ten times on a rapid timescale of months. If dust production continued at the same rate, CCSN 2010ij could attain a similar dust mass to that in SN1987A (possibly as much as 0.7 of a solar mass).

Thus, while uncertainties remain both for observations and modelling, the picture is an encouraging one. It seems possible that the dust formation processes operating in the Milky Way and the local Universe may also act in high-z objects. If so, the nature of dust at high red shifts may be similar to local dust.

11.6 Observations and Modelling of Extinction at High Red Shifts

As we have seen in Chapter 3, interstellar extinction (and other) observations provide useful (if not conclusive) constraints on models of interstellar dust in the Milky Way and nearby galaxies. Ideally, one would like to carry out similar comparisons between observations and modelling for high-z dust. It is, however, much more difficult to obtain reliable observational data for extinction in high-z galaxies. It is possible to obtain extinction measurements for an entire galaxy, taking account of absorption, emission, and scattering from both gas and dust. These studies are highly dependent on assumed distributions of stars, gas and dust. Direct measurements of extinction can in principle be made if the line-of-sight to a background source—such as a quasar or a gamma-ray burst afterglow—passes through a high-z galaxy. However, these measurements are difficult because the spectrum of the background radiation may not be well-determined. Various ways of dealing with these problems have been developed by astronomers, but they need not concern us here.

In the local Universe, a huge variety of interstellar extinction curves (ISECs) exist. We showed a sample of four ISECs for the Milky Way and its near neighbours, the Large and Small Magellanic Clouds (LMC and SMC), in Chapter 2 (see Figure 2.2). These curves differ in the strength of the 217.5 nm "bump", and in the behaviour in the far-ultraviolet. In general, the bump is strong in the Milky Way and weak (*e.g.* LMC) or even absent (*e.g.* SMC) elsewhere. The ISEC for the SMC has a particularly strong rise in the far-ultraviolet.

The SMC-type of ISEC is found widely. Richards *et al.* (2003)[13] found this type of ISEC in some quasars with red shifts in the range 0–2.2. Gallerani *et al.* (2010)[14] examined the optical and near-infrared spectra of 33 quasars with red shifts in the range 3.9–6.4 and determined that seven of these spectra were subject to significant extinction. The inferred ISECs are shown in Figure 11.1. Note that the normalization is at a wavelength of 300 nm, rather than in the visual region. These ISECs are clearly similar to the SMC ISEC: they have no detectable bump, and they rise into the far-ultraviolet, if not quite as strongly as the SMC ISEC.

Hjorth *et al.* (2013)[15] have determined the presence of ultraviolet extinction in a sample of 21 quasars, all at a red shift of about 6 (see Figure 11.2). Although there is considerable variation, the mean extinction curve shows no bump and is consistent with the SMC ISEC.

The importance of these ISECs is that they offer the opportunity to learn about the composition of the dust that generates the extinction in these galaxies. The simplest approach is simply to vary the nature and abundance of the various components assumed to be contributing to extinction until the computed ISEC matches the observed. Here, we use the unified model in which the various components are linked through the physical properties of the region (see Chapter 3 and Cecchi-Pestellini *et al.* 2014[16]). Such calculations have predictive power. The SMC-type of curve is

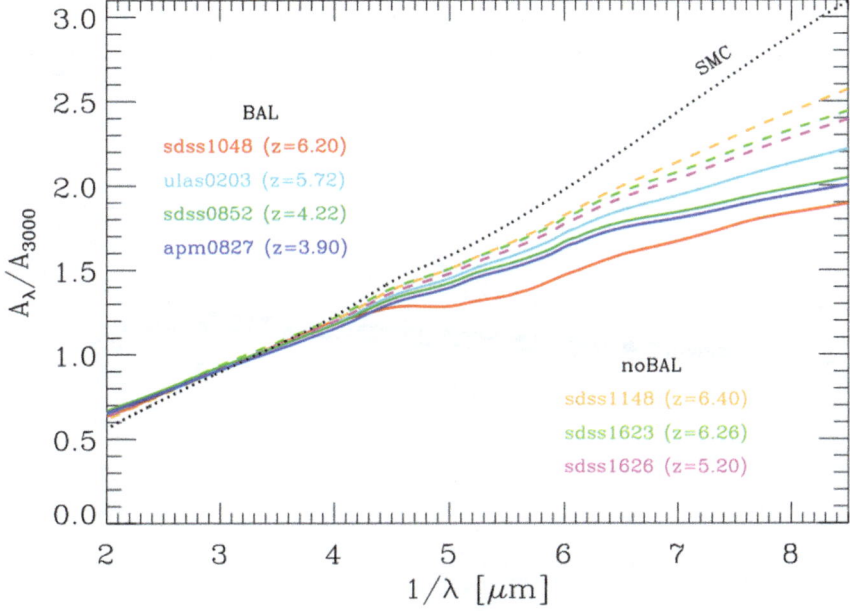

Figure 11.1   Inferred interstellar extinction curves for seven high red shift quasars, with red shifts, z, in the range 3.9–6.4. Four quasars show broad
absorption lines (BAL) and may be younger objects. Also shown is
the extinction curve for the Small Magellanic Cloud, a neighbouring
galaxy to the Milky Way. Reproduced with permission from Gallerani
et al. (2010).[14]

readily obtained (see the fits shown in Figure 11.2 for SMC, and the Milky
Way galaxy). A fit to an even more extreme type of ISEC obtained for the
gamma-ray burst afterglow GRB 080605 was shown in Figure 2.2.

Evidently, it is possible to account for ISECs of a variety of shapes obtained
from objects at red shifts up to at least 6 by a model description based on
Milky Way data. This supports the view that dust is essentially the same
material throughout the Universe – which is the conclusion from the discussion of dust formation in the previous section. The dust may respond to the
local physical conditions and take on different optical properties, but essentially it is the same material.

11.7   Conclusions: What do We Know About Dust at High Red Shifts?

Interstellar dust at high red shifts is poorly constrained by observations,
but the limited evidence that exists supports the view that it is formed in
processes that are familiar from studies of the local Universe. Extinction
data suggest that the nature of dust and its optical properties are also

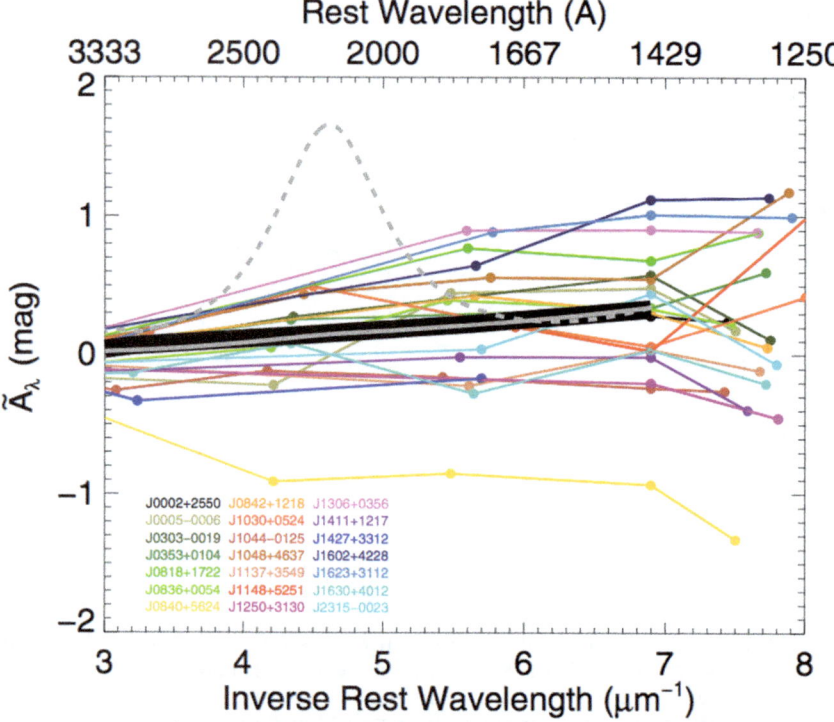

Figure 11.2   Extinction curves for 21 quasars with redshifts $z \sim 6$. The thick black curve is the median and the grey curve close to the median is a representative curve for the Small Magellanic Cloud, a neighbouring galaxy to the Milky Way. The dashed curve is characteristic of the Milky Way extinction in this spectral region. Reproduced with permission from Hjorth *et al.* (2013).[15]

similar to those of the local Universe. The subject of high-*z* dust is relatively new, but if these inferences are supported by further work then from the chemical point of view the issues are straightforward: we may assume that surface chemistry operates in the high-*z* Universe more or less as it does in the local Universe.

As regards the questions raised in Section 11.2, at least some preliminary answers may be given. Dust has generally been available in the Universe since about 500 million years after the Big Bang, or possibly earlier. However, dust formation depends on a number of factors, including the mass distribution of stars: low mass stars will not contribute to dust at $z \sim 6$. Therefore, the dust-to-gas ratio is affected by the mass distribution of stars, as well as by the availability of heavy elements. A stellar mass distribution that favours low mass stars will have a low dust-to-gas ratio, regardless of the available elemental abundances. The time-dependence of the dust-to-gas ratio is unclear; however, it appears that if dust formation is possible then it occurs rapidly. The composition of high-*z* dust may be similar to that in the local

Universe, and possibly this may also be true for the size distribution of dust grains. It seems that surface chemistry in the high red shift Universe is likely to contribute to the overall chemistry in much the same way as in the local Universe.

Further Reading

1. T. W. Hartquist and D. A. Williams, *The Chemically Controlled Cosmos*, Cambridge University Press, Cambridge, 1995.
2. V. Brom, *Rep. Prog. Phys.*, 2013, **11**, 112901.
3. Y. D. Hezaveh, D. P. Marrone, *et al.*, *Astrophys. J.*, 2013, **767**, 132.
4. C. M. Casey, D. Narayanan and A. Cooray, *Phys. Rep.*, 2014, **541**, 45.
5. D. B. Fisher, A. D. Bolatto, *et al.*, *Nature*, 2014, **505**, 186.
6. L. K. Hunt, E. Palazzi, *et al.*, *Astron. Astrophys.*, 2014, **565**, 112.
7. R. C. Kennicutt, *Annu. Rev. Astron. Astrophys.*, 1998, **36**, 189.
8. R. C. Kennicutt and N. J. Evans, *Annu. Rev. Astron. Astrophys.*, 2012, **50**, 531.
9. D. Lutz, *Annu. Rev. Astron. Astrophys.*, 2014, **52**, 373.
10. C. Gall, J. Hjorth and A. C. Anderson, *Astron. Astrophys. Rev.*, 2011, **19**, 43.
11. M. Matsuura, E. Dwek, *et al.*, *Science*, 2011, **333**, 1258.
12. C. Gall, J. Hjorth, *et al.*, *Nature*, 2014, **511**, 326.
13. G. T. Richards, P. B. Hall, *et al.*, *Astron. J.*, 2003, **126**, 1131.
14. S. Gallerani, R. Maiolino, *et al.*, *Astron. Astrophys.*, 2010, **523**, A85.
15. J. Hjorth, P. M. Vreeswijk, C. Gall and D. Watson, *Astrophys. J.*, 2013, **768**, 173.
16. C. Cecchi-Pestellini, S. Viti and D. A. Williams, *Astrophys. J.*, 2014, **788**, 100.

CHAPTER 12

Dust Chemistry and Astrobiology

12.1 Introduction

Only a very small fraction of the organic compounds in nature are found in planets or comets and other condensed objects. By far the larger quantity, more than 99.9% by mass, reside in the enormous molecular clouds in interstellar space of the Milky Way and other spiral galaxies. Abiotic organic chemistry, as observed in molecular clouds, offers a glimpse of the chemical evolution preceding the onset of life on our own planet, and allows us to evaluate the possibility that—during the evolution from a molecular cloud to a planetary system—COMs (*i.e.* complex organic molecules) are formed, transformed and preserved until they are incorporated into comets and meteorites. Indeed, as we have seen in Section 7.3, complex chiral organics, such as amino acids, have been found in space. Moreover, analyses of cosmic debris show that some amino acids present an excess of the L-conformation enantiomer (*e.g.* Engel & Nagy 1982;[1] Cronin & Pizzarello 1997[2]), in straightforward similarity with terrestrial biomolecular homochirality. This coincidence is too striking to be fortuitous; it points out that products of routine cosmic chemistry contributed to the early Earth organic pool and facilitated prebiotic molecular evolution. This chapter is concerned with the development of chemical complexity in astronomy, and with the introduction of chirality into abiological molecules.

Familiar prebiotic organic compounds in space coexist with exotic organic species. Ion-molecule reactions, driven by cosmic-ray ionisation, are important in interstellar chemical synthesis because they are very fast and typically proceed with no activation barrier, and therefore proceed at temperatures of only a few Kelvin. The chemistry initiated by such reaction routes can

The Chemistry of Cosmic Dust
By David A. Williams and Cesare Cecchi-Pestellini
© David A. Williams and Cesare Cecchi-Pestellini 2016
Published by the Royal Society of Chemistry, www.rsc.org

produce molecules of modest complexity, with low to moderate hydrogen saturation. However, as we have emphasized throughout this book, cold gas-phase chemistry fails to produce molecules at the next level of complexity in large partly hydrogen-saturated species, *i.e.* structures containing three or more H atoms with no carbon–carbon multiple bonds. Examples of such species observed in space are ethyl cyanide (CH_3CH_2CN), methyl formate ($HCOOCH_3$), dimethyl ether ($(CH_3)_2O$), acetone ($(CH_3)_2CO$), glycolaldehyde ($HCOCH_2OH$) and acetic acid (CH_3COOH). Many of these molecules are also of biological interest. For example, glycolaldehyde is the simplest of the monosaccharide sugars, while acetic acid shares the C–C–O backbone with glycine, the simplest amino acid, from which it differs by an amino group, NH_2. The recently discovered iso-propyl cyanide ($i\text{-}C_3H_7CN$) in the giant gas cloud Sagittarius B2, a region of ongoing star formation close to the centre of the galaxy, is the first detection of a carbon-bearing molecule with a branched structure in interstellar space,[3] and is a good omen for future detections of amino acids, for which this branched structure is a key characteristic.

How are molecules at this level of complexity formed in interstellar space? Condensation chemistry can generate large molecules, but will produce mainly carbon clusters or hydrocarbons with one and two hydrogen atoms when the product complexity reaches the level of three or four carbon atoms. In Chapter 8 we concluded that H_2 formation on interstellar dust analogue surfaces is well constrained by experiment and theory, and that the implications for H_2 formation in the interstellar medium are consistent with the observational requirements. Evidently, as we have stressed throughout this book, the surfaces of dust grains are chemically reactive. We have also shown in Chapter 9 that a large variety of molecular species can be formed in ices in the laboratory: alcohols, quinones, esters, and even amino acids. Thus, dust appears to be the vehicle that creates increasing chemical complexity in space. The inferred gap in mass between the largest interstellar molecules and the smallest grains is now small, suggesting that some of the smaller grains are in fact large, mainly organic, molecules with specific but unknown structures. We shall discuss the chemistry of prebiological molecules in Sections 12.3 and 12.4.

Do COMs survive the processes of star and planet formation, as described in Chapter 10? The formation of a planet is a violent event, so the intricate chemical history of the gas from which the planet forms may be obliterated, requiring chemical evolution to be continuously restarted. On the other hand, the chemical mechanisms that generate biomolecules in space could be transferred to newly formed planets during a bombardment phase by the dust grains aggregates, comets, asteroids, and meteorites, so there is a potential connection between prebiotic organic chemistry and the chemistry of the interstellar medium. These exogenous products could, of course, have been complemented by substances arising on Earth by reactions of the kind Stanley Miller and others have envisaged and explored.

Cosmic dust may have contributed to the emergence of life on Earth many times during the long evolutionary pathway that eventually gave rise to our planetary system, in the following ways: (i) contributing precursors for pre-biotic chemistry in larger bodies, (ii) serving as building blocks from which future comets, asteroids, and other celestial bodies may originate, (iii) inducing the formation of the Earth itself and—more generally—planets, (iv) delivering COMs to the early Earth and Mars during the late heavy bombardment 4.5 billion years ago, (v) providing a stable and reducing environment for the ingredients needed to start life early and quickly, and (vi) contributing to chiral selection in space. While there is considerable agreement regarding the points from (i) to (iv), the last two issues are rather speculative and no general (or even partial) consensus on the production and selection of chiral organic molecules in protoplanetary disks has been reached.

Chirality is important for theories that link extraterrestrial organic matter with the origin of life, because terrestrial biology preferentially generates L amino acids whereas abiotic reactions produce racemic mixtures. Two theories provide different explanations for the origin of homochirality: one (biotic) explains it as a consequence of evolution, the other (abiotic) as a prerequisite of the evolution. It is a necessary consequence of the abiotic theory that the amino acids were already formed in interstellar space so that preferential destruction of D-amino acids could take place. We shall discuss these points in the last two sections.

12.2 Dust Chemistry and the Formation of Prebiotic Materials

To understand how life may begin on a habitable planet such as the Earth, it is essential to know what organic compounds were likely to have been available. The key questions to be addressed are: how are simple organic compounds assembled into more complex molecular systems? And what are the essential processes and pathways by which complex systems can develop those basic properties, such as homochirality, critical to the origin of life? Carbon-bearing meteorites provide the only natural sample of chemical evolution large enough for direct laboratory analyses. They are material records of abiotic chemistry that preceded vital processes on Earth, and may reveal whether it is realistic to assume that these or similar materials, i.e. either by direct delivery or analogy of formation, might have helped or even induced molecular evolution toward biogenesis. With a lesser degree of complexity organics have been detected in other small bodies of the Solar System such as comets[4] and asteroids.[5]

About 85% of meteorites that presently fall on Earth are chondrites, conglomerates with an elemental composition similar to that of the Sun if the loss of the most volatile elements is taken into account. Chondrites are fragments of ancient asteroids that have remained relatively unprocessed since the formation of the Solar System 4.6 billion years ago. Of all the chondrites, it is the carbonaceous chondrites that are considered the most primitive and these objects represent a window on to the earliest Solar

System. These objects may contain up 5% carbon, much of it organic. The carbon in carbonaceous chondrites is present in several forms: as silicon carbide; as graphite; as nanometre-sized diamonds, originating as condensation products of stellar outflows as well as from catastrophic supernovae explosions that occurred long before the birth of the Sun (see Section 7.2); and as younger carbonate minerals. All of these forms of carbon are less abundant than organic matter. A large body of detailed chemical analyses on carbonaceous chondrites found that the majority of organic matter (about 70–90%) is present as highly cross-linked aromatic network or macromolecular material, with average elemental abundance $C_{100}H_{46}N_{10}O_{15}S_{4.5}$. There have also been identified smaller free molecules such as hydrocarbons, carboxylic acids, amino acids, hydroxy acids, sulfonic acids, phosphonic acids, sugars, poly-hydroxyl compounds and many other species that are of interest to the origins of life. A list of the classes of organic compounds found in the Murchison meteorite is reported in Table 12.1.[6] Among them, amino acids have attracted great attention because they are the building blocks for proteins. The majority of the more than 80 different amino acids identified in carbonaceous meteorites are nonexistent, or rare, in terrestrial proteins.

One would expect to detect similar molecular and chiral signatures in the chemical content of comets. Knowledge of cometary ices is obtained by remote observations of volatile species, stored as ice in the cometary nucleus and released as gas into the coma. With the exception of C_2H_6 and S_2, the known cometary molecules comprise a subset of those identified in the interstellar medium. The chemical inventory in comets is qualitatively and, in many cases, quantitatively consistent with that determined for astronomical ices as measured by infrared spectroscopy toward embedded protostars and background stars (see Table 12.2).[7] So far, only the achiral simplest amino acid glycine has been tentatively identified and only in a single object, during a re-analysis of samples returned to Earth by NASA's STARDUST mission, which flew by Comet Wild 2 in 2004 to capture particles in the proximity of the 5-kilometre object.[8]

The isotopic composition is a good indicator of the chemical synthetic history of the various species, because the mass differences between isotopomers result in energy differences in their bond formation, the larger the energy difference between isotope bonds and the lower the temperature, the greater the potential for heavy isotope enrichment. Cold molecular clouds are therefore regions where molecules exhibiting substantial fractionation of isotopes of H, C, and N can form. The extreme sensitivity of these fractionation reactions to the gas temperature provides a strong evidence of an interstellar origin of the enhanced isotopic fractionation found in primitive Solar System material, comets and meteorites.

However, there are isotopic as well as molecular trends within the Murchison organic suite, for example, or in comets, that reveal significant formative distinctions between individual compounds that cannot be accounted for by the simplistic assumption of a straightforward interstellar heritage.

Table 12.1 Organic compounds in the Murchison meteorite.

Class	Example	3D structure
Carboxylic acids	Acetic acid	
Amino acids	Alanine	
Hydroxy acids	Lactic acid	
Ketoacids	Pyruvic acid	
Dicarboxylic acids	Succinic acid	
Sugar alcohols and acids	Glyceric acid	
Aldehydes and ketones	Acetaldehyde	
Amines and amides	Ethyl amine	
Pyridine carb. acids	Nicotinic acid	
Purines and pyrimidines	Adenine	
Aliphatic hydrocarbons	Propane	
Aromatic hydrocarbons	Naphthalene	
Polar hydrocarbons	Isoquinoline	

Very little is known of the sequence of chemical events that would have taken place in a prestellar core. However, the whole process would have been characterized by several stages of temperature, pressure, with corresponding chemical regimes, following the initial collapse of the presolar portion of the interstellar medium. This wealth of physical conditions would have produced a highly non-statistical isotopic distribution. We have to keep in

Table 12.2 Representative abundances (in % with respect to water) of molecules observed in comets and interstellar sources. All comets originate in the Oort Cloud or in the scattered disk, with the exception of 73P/SW3 that is a Kuiper Belt object; L134N is a cold dark interstellar cloud, IRAS 16293 a binary solar-type protostar, and Sgr B2(N) a hot prestellar core. Data from Mumma & Charnley (2011).

Molecule	Comets					Interstellar sources		
	Halley	Hyakutake	Hale–Bopp	2001 A2	73P/SW3	L134N	IRAS 16293	Sgr B2(N)
CO	3.5–11	14–30	12–23	4	0.5	26 400	8000	1000
CO$_2$	3–4		6					
CH$_4$	<0.8	0.8	1.5	1.2	<0.25			
C$_2$H$_2$	0.3	0.2–0.5	0.1–0.3	0.5	<0.04			
C$_2$H$_6$	0.4	0.6	0.6	1.7	0.14			
CH$_3$OH	1.8	2	2.4	2.8–3.9	0.22	1	10	3–200
H$_2$CO	4	1	1.1	0.24	0.14	6.6	2	0.5
HCOOH			0.09			0.1	<0.03	0.1
HCOOCH$_3$			0.08				0.09	1
CH$_3$CHO			0.02			0.2	<0.07	0.2
(CH$_2$OH)$_2$			0.25					0.1
CH$_3$OCH$_3$			<0.5				0.16	0.3
NH$_2$CHO			0.02				<0.001	0.2
NH$_3$	1.5	0.5	0.7		<0.3	0.08	6	
HCN	0.1	0.1–0.2	0.25	0.1–0.6	0.25	1.3	0.2	0.4
HNCO		0.07	0.1				0.3	0.6
HNC		0.01	0.04	0.01	<0.0013	2	0.01	0.1
CH$_3$CN		0.01	0.02	0.03	0.03		0.25	30
HC$_3$N			0.02			0.06	0.03	5
H$_2$S	0.4	0.8	1.5	1.15	0.25	0.3	3	
CS		0.1	0.1	0.07	0.11	0.3	0.6	
OCS		0.1	0.4			0.7	8	2
SO			0.3			6.6	8	
SO$_2$			0.2			0.1	3	30
CS$_2$	0.2	0.1	0.2					
H$_2$CS			0.02			0.2	0.02	20
NS			0.02				0.02	
S$_2$		0.01						

mind that the complexity of molecules in life is far greater than that seen in meteorites, and that the obstacles inherent to purely endogenous life origins would be the same as those faced by any developing organic system irrespective of the source of starting material. The minor bodies of the Solar System therefore probably present just a tiny and well hidden sample of the prebiotic potential of cosmic synthetic processes.

All the matter in the Solar System bodies was once in the interstellar medium: how much of this matter survived initial incorporation into the protosolar nebula and later the impact of a hot, energetic nebular chemistry? Transformations of organic compounds, or their synthesis from inorganics, occur in response to thermodynamic drivers, modulated by the kinetic properties of individual reactions, which in turn are entirely determined by the types and the ways the energy is deposited in the mixture, and by the physical environments in which the synthesis is taking place that may channel and select among competing chemical outcomes.

The search for amino acids (in particular glycine) in the interstellar medium has been attempted for almost 40 years without a positive result. Despite these failures, it was proposed amino acids would be formed in the interstellar medium by energetic processing of ices on dust grain surfaces. The ices would be evaporated, releasing the amino acids into the gas phase. Indeed, the potential precursors of these prebiotic molecules are abundant in the icy mantles of interstellar dust particles. The interstellar ice mantles formed on dust grains in cold molecular clouds grow by the accretion of atoms and molecules from the gas. The simple hydrides form by H atom additions to heavy atoms or molecules. Hydrogenation and oxidation of condensed CO molecules produces CO_2, H_2CO, and CH_3OH. Extension of these mechanisms to addition of C, O, and N atoms shows that, even at 10 K, a rich chemistry is possible in which various multiply bonded molecules are formed and can be subsequently hydrogenated (*e.g.* Hidaka *et al.* 2009[9]). In addition to chemical reactions on the ice surface, driven by accretion of reactive species, ices can also undergo secondary processing due to damage caused by photons and energetic particles (see Chapter 9). $HCOOCH_3$ (methyl formate) cannot be formed *via* atom addition reactions or in gas-phase ion–molecule chemistry involving methanol and formaldehyde, but—as we saw in Chapter 9—is readily produced by irradiation of for example, binary mixtures of methanol and carbon monoxide with ultraviolet radiation, particles and X-rays. In these experiments, radicals such as HCO, CH_2OH, OH, CH_3, and COOH and CH_3O can associate in radical–radical reactions in grain mantles to produce a variety of organics including glycolaldehyde, and ethylene glycol. The degree of complexity in the interstellar ice analogues reaches the level of amino acids. Using a range of mixtures containing H_2O, CH_3OH, NH_3, CO, and CO_2 two research groups demonstrated simultaneously the formation, induced by ultraviolet photolysis, of a variety of amino acids under simulated interstellar conditions.[10,11] These findings have been confirmed by subsequent experiments (*e.g.* Nuevo *et al.* 2006[12]). Bernstein *et al.* (2002)[11] detected glycine, alanine, and serine as products after acid hydrolysis of the residue. Muñoz Caro *et al.* (2002)[10] identified 16 amino acids in the hydrolysed and derivatized organic residue. Recent experiments of Meinert *et al.* (2012)[13] provided a list of 26 amino acids, including proteinogenic and non-proteinogenic amino acids, and diamino acids. All amino acids were detected as racemates, as no chiral influence was introduced into the system. Subsequent experiments produced other prebiotic species such as purine and pyrimidine compounds, and urea. Most of these species are identified in meteorites (see Burton *et al.* 2012[14] for a comprehensive overview of the organic composition of carbonaceous chondrites). In carbonaceous meteorites, more than 80 species of amino acid have been unambiguously identified, with these species containing from two to nine carbon atoms (see Table 12.3).

Table 12.3 Amino acids detected in meteorites (tentative identifications in italics). Data from Burton *et al.* (2012).

2	3	4	5	6	7	9
Glycine	Alanine	Threonine	Valine	Leucine	2-Amino-2,3,3-trimethyl-butanoic acid	Phenylalanine
	β-Alanine	*allo*-Threonine	Norvaline	Isoleucine	2-Amino-2-ethyl-3-methyl-butanoic acid	Tyrosine
	Serine	α-Aminobutyric acid	Isovaline	*allo*-Isoleucine	2-Amino-2-ethylpentanoic acid	
	Sarcosine	β-Aminobutyric acid	3-Aminopentanoic acid	Norleucine	2-Amino-3-ethylpentanoic acid	
		β-Aminoisobutyric acid	*allo*-3-Amino-2-methylbutanoic acid	Pseudoleucine	2-Amino-2,3-dimethylpentanoic acid	
		N-Ethylglycine	3-Amino-2,2-dimethylpropanoic acid	Cycloleucine	2-Amino-2,4-dimethylpentanoic acid	
		N,N-Dimethylglycine	3-Amino-2,3-dimethylpropanoic acid	2-Methylnorvaline	2-Amino-3,3-dimethylpentanoic acid	
		N-Methylalanine	3-Amino-2-ethylpropanoic acid	Pipecolic acid	2-Amino-3,4-dimethylpentanoic acid	
		N-Methyl-β-alanine	4-Aminopentanoic acid	2-Amino-2-ethylbutanoic acid	2-Amino-4,4-dimethylpentanoic acid	
		Aspartic acid	4-Amino-2-methylbutanoic acid	2-Amino-2,3-dimethylbutanoic acid	*allo-2-Amino-2,3-dimethylpentanoic acid*	
		2,3-Diaminobutanoic acid	4-Amino-3-methylbutanoic acid	*3-Amino-2-ethylbutanoic acid*	*allo-2-Amino-3,4-dimethylpentanoic acid*	

(continued)

Table 12.3 *(continued)*

2	3	4	5	6	7	9
		2,4-Diaminobutanoic acid	5-Aminopentanoic acid	*3-Amino-2,3-dimethylbutanoic acid*	2-Amino-2-methylhexanoic acid	
		3,3'-Diaminoisobutanoic acid	Glutamic acid	*3-Methylamine-pentanoic acid*	2-Amino-3-methylhexanoic acid	
			2-Methylaspartic acid	*4-Aminohexanoic acid*	*allo-2-Amino-3-methylhexanoic acid*	
			3-Methylaspartic acid	*4-Amino-3,3-dimethylbutanoic acid*	2-Amino-4-methylhexanoic acid	
			allo-3-Methylaspartic acid	*4-Amino-2-methylpentanoic acid*	*allo-2-Amino-4-methylhexanoic acid*	
			N-Methylaspartic acid	*4-Amino-3-methylpentanoic acid*	2-Amino-5-methylhexanoic acid	
			4,4'-Diaminoisopentanoic acid	*4-Amino-4-methylpentanoic acid*	2-Aminoheptanoic acid	
				6-Aminohexanoic acid	α-Aminopimelic acid	
				α-Aminoadipic acid	1-Aminocyclohexanecarboxylic acid	
				β-Aminoadipic acid 2-Methylglutamic acid		

The exact formation pathway of amino acids in interstellar ices is unknown. All types of amino acids discovered in meteorites correlate well with the presence of liquid water. α-Amino acids may form *via* reactions of aldehydes and ketones with ammonia and HCN (Strecker-cyanohydrin synthesis). These reactions are generally accepted as responsible for both α-amino and α-hydroxy acids in Murchison (and other aqueously altered meteorites), because HCN and ammonia are abundant in space, as C=O groups are in carbonaceous meteorites. Amino acids have also been found in meteorites that had experienced high temperatures. This suggests that the production of amino acids occurred through high-temperature processes (rather than the "cold" Strecker synthesis) as their parent asteroids gradually cooled down.[15] Such a process may involve gas containing hydrogen, carbon monoxide, and nitrogen, reacting in a Fischer–Tropsch scheme. This reaction is commonly used in industry to produce fuels (*i.e.* complex hydrocarbons) by the catalytic hydrogenation of carbon monoxide. As discussed above, amino acids may also have formed outside meteorite parent bodies, on ice surfaces. β-Amino acids are believed to have formed by Michael addition (one of the most useful methods for the formation of C–C bonds) of ammonia to α- and β-unsaturated nitriles. Like the Strecker-cyanohydrin reactions, Michael addition requires liquid water. The origins of γ- and δ-amino acids are poorly understood. One possibility is that they are decarboxylation products of α-amino dicarboxylic acids. Other suggestions involve the hydrolysis of lactams (cyclic amides) that have been identified in meteorites. Possible formation pathways for amino acids in meteorites are shown in Figure 12.1.

As we saw in Chapter 9, Rawlings *et al.* (2014)[16] proposed an alternative scenario, based on gas-phase reactions that occur in high density interstellar gas following catastrophic ice mantle sublimation. Laboratory experiments involving the slow warming of chemically-mixed ices demonstrate that desorption occurs in several distinct and narrow temperature bands, of which the most important in this context is the co-desorption band. This band is where the major component of the ice, H_2O, desorbs and carries with it all other species. The work of Rawlings *et al.* relies on other laboratory evidence showing that the catastrophic recombination of hydrogen atoms and other accumulated radicals in a solid may suddenly raise the temperature of the solid to ~1000 K. For abrupt temperature excursions of this magnitude, any ices that are present will be completely converted to gas and sublimated explosively. The ice material sublimated into the gas phase is then free to expand into a vacuum at some fraction of the sound speed, inducing for an amount of time of the order of nanoseconds multi-body chemical reactions. Within this scheme glycine is formed by just one single-stage reaction

$$CH_2COOH + NH_2 + H_2O \rightarrow NH_2CH_2COOH + H_2O$$

resulting in predicted concentrations in agreement with derived upper limits in the interstellar medium.

Independently of the mechanism of synthesis, once formed, amino acids would need to survive the exposure to violent cosmic processing in the

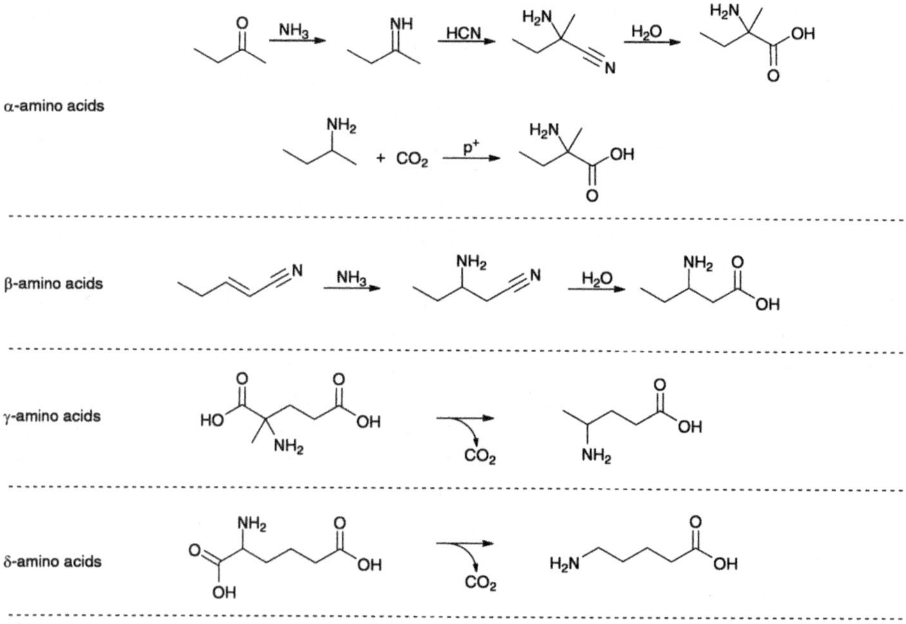

Figure 12.1   Possible formation pathways for amino acids in meteorites. Reproduced with permission from Burton *et al.* (2012).[14]

interstellar medium, or in protoplanetary disks. In particular, the photostability of amino acids in interstellar gas, as of many other molecular species, is very low, and therefore they do not survive in diffuse interstellar gas. This does not imply the lack of amino acids in space, but instead requires amino acids to be incorporated into dark and protected environments such as hot molecular cores, the interior of comets, asteroids, meteorites, and interplanetary dust particles.

The existence of volatile ices in comets is consistent with the chemistry associated with the formation and processing of interstellar ices, but it may be difficult to extend such an analogy to a higher degree of molecular complexity, as seen in meteorites. The role of interstellar processes, in particular those involving dust particles, if relevant, must be then considered from a different perspective.

12.3   The Potential of Ice Cavity Gas-Phase Chemistry for Chemical Complexity

All the early speculations about the origins of life on Earth were based on Solar System processes. Of these speculations, the more influential were the suggestions of possible Miller–Urey type syntheses in a reducing planet atmosphere, following production and recombination of radicals, and of catalytic, Fisher–Tropsch type processes in the early stages of the solar nebula.[17] In 1957 Miller[18] showed that formaldehyde and hydrogen cyanide were key intermediates in the synthesis of glycine. This led Oró and his co-workers (1960)[19] to study the products of a solution of ammonium cyanide (NH_4CN) in water, discovering that NH_4CN was converted in adenine, one of the four bases of DNA. Such (and other similar) discoveries determined the direction of research on prebiotic chemistry for many years. However, our understanding of the atmosphere on early Earth has changed since 1953, and it is now believed that the atmosphere consisted mostly of carbon dioxide, nitrogen, and water. Under such conditions, prebiotic molecules are produced only in trace amounts.

However, in the complex process leading to the formation of circumstellar disks around young stars from prestellar dense molecular clouds, physical conditions similar to those envisaged in the early prebiotic experiments may arise. Circumstellar disks are an inevitable consequence of angular momentum conservation during the formation of a star through gravitational collapse (Chapter 10). Initially, disks rapidly funnel material onto the star but, as the surrounding molecular core is used up or otherwise disperses, the accretion rate decreases, and only a small amount of the original material persists in the disk. That these disks can be considered protoplanetary is apparent not only in the geometry of the Solar System but also in the high detection rate of exoplanets. Dust dominates the opacity of protoplanetary disks and is the raw material for planetesimals. In the diffuse interstellar medium, dust is mainly composed of silicates, carbons and polycyclic aromatic hydrocarbons (Chapter 3). During the passage through cold dark clouds to protoplanetary disks, molecules freeze out from the gas phase onto dust grain surfaces, producing icy mantles (Chapter 9). In the high densities of the disks, these ice-coated dust grains collisionally agglomerate, assembling in loosely packed structures (Chapter 10) with much of their internal volume being vacuum and trapped ices. The timescale for the accretion of volatiles on grains is much faster than that of grain aggregation, implying that the latter process occurs when dust grains have already accreted ice mantles. Interstitial voids must occur even in highly organized, densely packed structures. For example, dense packing of like spheres in a face-centred cubic lattice leaves 26% of space unoccupied.[20] It is this internal volume that leads to a completely different chemical scenario with respect to that on the surface of grains. As discussed in Chapters 8 and 9, surface chemistry on individual grains proceeds in a sequence of steps: collision of a reactant with the surface, accommodation and sticking, reaction at the site of impact or delayed reaction with other grain adsorbates, and finally desorption. The time delay between the adsorption and reaction steps may be substantial if the reaction products are non-volatile, but can be short for prompt reactions. The products of surface reactions are either retained on dust or desorbed to the ambient gas. Duley (2000)[21] noted that the interior of dust aggregates offers a different intermediate possibility: the re-accretion of reaction products by other components of the aggregate. This redeposition may occur on the surface of other dust particles or on components of an ice mantle. As desorbed products can be in an energetic state, these secondary reactions might be expected to mimic some aspects of high-temperature chemistry. Aggregate grains bring together all components of the interstellar gas and dust, a unique situation outside planetary systems. As reaction products can remain trapped within aggregates and are shielded from radiation, conditions exist for the formation of larger molecules. The feedstock for this chemistry would be the ices accumulated during aggregation, together with light atoms and radicals from the ambient gas that diffuse into the interior of the aggregate. Sputtering of silicate and carbon solids and PAHs by cosmic rays will inject heavier atoms, ions and molecules into these ices. Examples of possible reaction products produced by sputtering within dust grain aggregates are shown in Table 12.4.

Primary results of this chemistry might be a variety of organometallic compounds, *e.g.* Si substitutions for C in hydrocarbons. Silane compounds are possible, including methyl, ethyl and phenyl derivatives, and in the presence

Table 12.4 Reaction products produced by sputtering within grain cavities. The reactant is liberated from the source and then reacts immediately with an adjacent condensate. M = Si, Mg, Fe. R = chemical group, R' = ring. Data from Duley (2000).[21]

Source	Reactant	Condensate	Products
Silicate	M, M$^+$	H$_2$O ice	M(OH)$_n$
		CO	H$_x$MOCH$_y$
		a-C:H	H$_x$MCR
		PAH	H$_x$MCR'
	O, O$^+$	H$_2$O ice	H$_2$O, H$_2$O$_2$
		CO	CO$_2$
		a-C:H	CO, H$_2$O, H$_2$CO
		PAH	HOR'
	MO, MO$^+$	H$_2$O ice	MO, H$_x$MO
		CO	MO
		a-C:H	MOCR
		PAH	MOR'
H$_2$O	OH, H$_2$O, H$_3$O$^+$	CO	COOH
	Clusters	a-C:H	Alcohols, aldehydes, complexes
		PAH	Clusters, phenols, aldehydes
a-C:H, PAH	C$_6$ ring, R'	H$_2$O ice	R'–(H$_2$O)$_n$
		CO	R, R'
		a-C:H	Hydrocarbons
		PAH	PAH
		Silicate	R'OR
Various	C, C$^+$	H$_2$O ice	H$_2$CO, HCOOH, H$_3$COH
		CO	C$_2$O, CH$_4$
		a-C:H	CH$_4$, hydrocarbons
		PAH	H$_x$CR', CH$_4$
		Silicate	H$_x$CSiR
	CH$_n$, CH$_n^+$	H$_2$O	CH$_3$OH, H$_2$CO, CH$_4$
		CO	H$_2$CO, CH$_4$
		a-C:H	CH$_4$, hydrocarbons
		PAH	CH$_4$, hydrocarbons, H$_x$CR'
		Silicate	CH4, H$_3$CSiR

of ammonia silanes with amine and amino groups. Sputtering of the hydro-carbon component in aggregate grains including PAH should lead to the cleaving of C–C single bonds with the liberation of volatile fragments. Injection of these into CO and H_2O ices may yield a variety of aldehydes, ketones, carboxylic acids and alcohols. Many chemical outcomes considered exotic for space chemistry are standard products in terrestrial laboratories (Duley 2000).[21]

Dust aggregates can be impulsively heated by collisions with other aggregates or grains and by cosmic-ray impacts. The heat released during a collision may lead to the vaporization of the ice content filling the cavities, in which radicals and molecules from the ice enter a transient, warm, high pressure gas phase. The triggered chemistry is then reminiscent of the one powered by chemically-induced explosions (see Section 9.5), with a few remarkable differences: (i) a solid-state chemistry involving elements such as Si, Mg and Fe sputtered from silicate particles; (ii) facilitated secondary reactions and rapid quenching of the reaction products; (iii) a hydrogen rich atmosphere inside the cavities. The resulting mixture is a reasonable analogue of the conditions that Stanley Miller[17] envisaged as plausible for the primitive Earth atmosphere in the famous experiment at the University of Chicago. Therefore, grain aggregates may represent in the interstellar medium the equivalent of terrestrial micro-laboratories containing raw materials of reducing chemical composition suitable for conversion into COMs. The final products are likely to be very similar to those obtained from laboratory chemistry under terrestrial conditions. Because of the reducing atmosphere in the cavities large organic molecules are allowed to form, with the inclusion of sputtered atoms from the grain substrate, such as Ca, Mg, Cu, Fe, P, *etc.* (*e.g.*, Solov'ev *et al.* 1999[22]). It is recognised that metals are necessary to many important organic constituents of living organisms, such as phosphorus, nucleic acids, copper and iron in blood pigments and magnesium in plant pigments. Recent numerical experiments[23] based on *ab initio* molecular dynamics simulations of aqueous systems subject to electric fields (*e.g.* describing lightning) and on metadynamics analysis of chemical reactions showed that glycine spontaneously forms from mixtures of simple molecules. Formic acid and formamide are key intermediate products of the early steps of the Miller reactions. Formamide represents the simplest molecule containing the peptide bond. It is therefore of great interest as an important precursor in the abiotic synthesis of amino acids, and of further prebiotic chemistry. When nitrogen compounds are present in the initial mixture, the processing of interstellar ice analogues produces formamide (see, *e.g.*, Jones *et al.* 2011[24]).

The chemical products contained in these aggregates would then be incorporated in planetesimals and comets, leading to the plausible connection between chemistry in cold protostellar clouds and that on the planets. The first solid particles were microscopic in size. They orbited the Sun in nearly circular orbits. Gentle collisions allowed small particles to stick together and make larger particles which, in turn, attracted more solid particles.

Planetesimals formed during this process acted as seeds for planet formation (Chapter 10). At first, planetesimals were closely packed. They coalesced into larger objects of up to a few kilometres across in a few million years. Eventually, a size distribution was established in which there were very many more small bodies than large. If life did originate on the surface of Earth, it could either have succeeded many times before the end of the major impact era or, if it started only after the conditions became less hostile, it happened rather quickly. Even if cometary and asteroidal impacts had destructive effects, refreshing the chemical conditions to raw materials both in their interiors and the primordial Earth, the gentle dusty rain of smaller dust particles provides a continuous injection of prebiotic materials.

During the construction of planetesimals, the chemical content of dust aggregate cavities was partially protected against the destructive effects of ultraviolet radiation. Such radiation has a significant degree of linear polarization. When linearly polarized waves hit the aggregate grains, the field within the cavities acquires the features of a circularly polarized wave. This fact may have connections to meteoritics and observed amino acid asymmetries, as we shall discuss in the next Section.

12.4   Chiral Selection in Space

Chiral molecules exist in two forms, with the same chemical make-up, each form being a mirror image of the other. Although laboratory experiments will tend to produce equal quantities of the left- and right-handed versions of a given chiral molecule (enantiomers), many of the chiral molecules found in living organisms come in only one variety. Biomolecules are in general chiral and almost exclusively homochiral. Of the 20 protein amino acids, 19 are chiral and of the L-conformation. Nucleic acid sugars are also chiral, but of D-conformation. The reasons for such asymmetry in living matter are the subject of a long standing debate. Some have argued that equal numbers of both versions of each chiral molecule were present at the onset of life and that it was only during biological evolution that the imbalance occurred. That view has become increasingly unpopular, however, with the realisation that the fundamentally important process of protein folding seems to require chiral imbalances (*e.g.*, Avetisov and Goldanskii, 1996[25]), while for nature to have selected the left- or right-handedness of each molecule during evolution would involve extraordinarily complex processes.

Since its discovery by Engel and Nagy in 1982[1] in the Murchison meteorite, enantiomeric enrichment in meteoritic amino acids has been widely studied, and there is now firm evidence of non-negligible excesses in the L-enantiomer for more than ten species in a dozen meteorites (Table 12.5).[26] Such findings demonstrate that it is possible to create asymmetrical molecules in space conditions from a mixture that does not initially contain any chiral substances.

A possible explanation for this symmetry breaking might be the production by weak interactions of a slight parity-violating energy difference among the enantiomers, *i.e.* a situation in which the true mirror image of an object or experiment is energetically not equivalent to the original. It has not been possible, to date, to determine experimentally the value of this energy difference. Such energy differences may be calculated by sophisticated *ab initio* quantum mechanical methods, and results generally give energy differences $\sim 10^{-14}$ J mol, corresponding to enantiomeric excesses $\leq 10^{-10}$. In almost all cases the calculations based on parity violation predicted the correct handedness for the biogenic amino acids. However, even if parity violation causes a left or right preference on the molecular level, this preference is not necessarily linked to the evolution of biomolecular homochirality. Other hypotheses include chirality transfer from spin-polarized electrons to chiral molecules,[27] chiral symmetry breaking upon spontaneous crystallisation,[28] and other rather *ad hoc* proposals such as chiral symmetry breaking during crystallization.[29]

For evident reasons, the highest astrophysical relevance has been awarded to processes involving radiation and magnetic fields. Among photochemical effects, only circularly polarized light and a static magnetic field collinear with a light beam are truly chiral systems and thus can potentially produce an enantiomeric enhancement within initially racemic mixtures. The interaction of circularly polarized light with an isotropic medium containing

Table 12.5 Detected enantiomer excess values of amino acids in carbonaceous meteorites. The negative sign tells indicates that the *R* enantiomer is the dominant one. Data from Myrgorodska *et al.* (2014).[26]

Amino acid	L-Enantiomer excess (%)	Meteorite[a]
Isovaline	−1 to +18.5	EET, LEW, LON, MN, MY, OR, QUE
Norvaline	−0.7 to +3.7	EET, LON, MN, OR, QUE
α-Methylnorvaline	1.4, 2.8	MN, MY
Valine	−0.4 to +43.6	EET, LEW, LON, MN, MY, OR, QUE
α-Methylvaline	1.0, 2.8	MN, MY
α-Methylnorleucine	1.8, 4.4	MN, MY
Isoleucine	3.6–50	EET, GRA LAP, ME, MIL, PCA, QUE
allo-Isoleucine	−60 to −2.2	EET, GRA LAP, ME, MIL, PCA, QUE
α-Amino-α-methylheptanoic acid	7	MN
α-Amino-α,β-dimethylpentanoic acid	1.4–5.2	MN, MY
allo-α-Amino-α,β-dimethylpentanoic acid	2.2–10.4	MN, MY
α-Methylglutamic acid	2–3	MN

[a]EET: EET 92042; GRA: GRA 95229; LAP: LAP 02342; LEW: LEW 90500; LON: LON 94102; MN: Murchison; MY: Murray; OR: Orgueil; ME: MET 00426; MIL: MIL 07525; PCA: PCA 91082; QUE: QUE 99177.

chiral molecules in the presence of a constant external magnetic field may be described by several phenomenological constants, relating to optical activity and dichroism, summarised by slightly different dielectric constants associated with right- and left-handed circularly polarized light. A difference in the refractive indices of a racemic mixture for (+) and (−) circularly polarized light the optical activity appears as a rotation of the plane of linearly polarized light traversing the medium. A difference in the absorption coefficients of a racemic mixture for left- and right-handed circularly polarized light the circular dichroism may cause an enantiomeric excess, if one of the enantiomers is more efficiently photosynthesised, photolysed or photocatalysed (see Figure 12.2). Among the terms arising in the interaction with the magnetic field, only magnetic circular dichroism and magnetochiral dichroism result from a breaking of the spatial mirror symmetry, either due to the chiral nature of the molecular constituents of the medium, or induced by the preferential direction introduced by the magnetic field (see Jorissen & Cerf, 2002,[30] for a detailed description of interaction of chiral molecules with light).

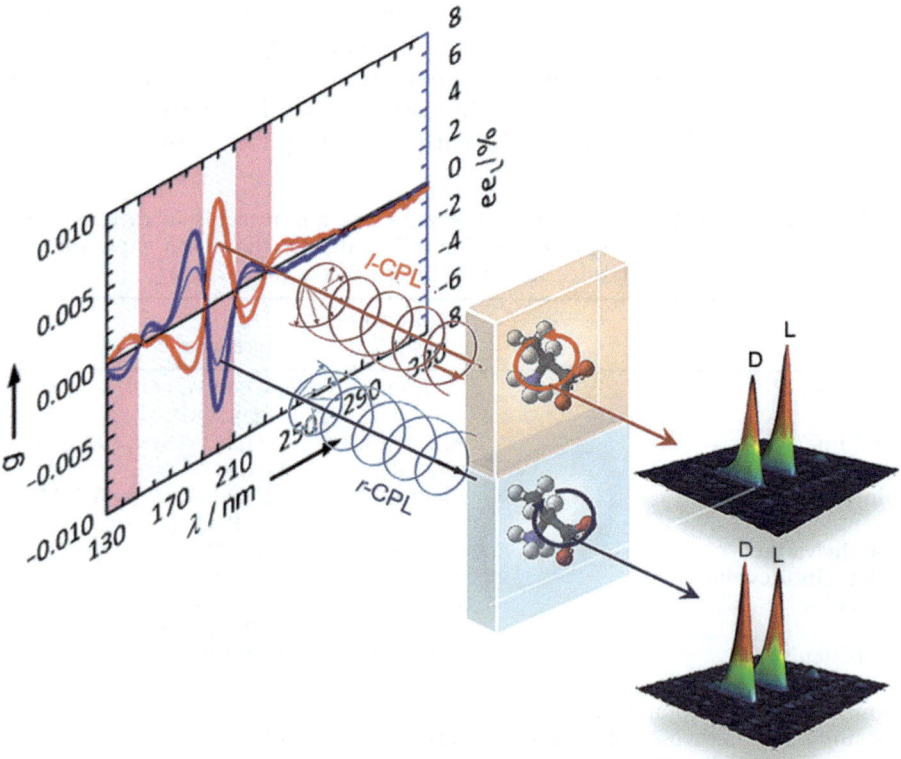

Figure 12.2   Asymmetric photolysis of a racemic chiral species. Circularly polarized light of a given sign is absorbed differentially by the molecule, giving rise to an enhanced photo-destruction of one of the enantiomers. Reproduced with permission from Myrgorodska *et al.* (2014).[26]

Most proposed extraterrestrial mechanisms postulated the enantiose-lective partial photolysis of racemic constituents in the organic mantles of interstellar grains in interstellar molecular clouds caused by the circularly polarized ultraviolet components of synchrotron radiation produced by relativistic electrons orbiting the rapidly rotating neutron star remnants of supernovae (Rubenstein *et al.* 1983[31]). In essence the suggestion is that ultra-violet circular polarized light from some astronomical source acted on chiral molecules in the molecular cloud from which the Solar System formed, to produce an excess of one enantiomer.

For the above scenario to work amino acids must form in space, and some source of ultraviolet circularly polarized light at a wavelength of about 200 to 230 nm needed to cause asymmetric photolysis in amino acids – must be pres-ent to irradiate the amino acids. As to the first condition, we have seen that a number of laboratory experiments imply that photochemical reactions take place on to and into icy dust grains. Concerning the second condition, ultra-violet circularly polarized light is a rare phenomenon in space. Observations of young stellar objects reveal the existence of circularly polarized light in the infrared, but a substantial degree of polarization (in the infrared) has been detected only in massive star formation regions. Synchrotron radiation from neutron stars is not predicted to be strongly circularly polarized. Very few such sources show optical synchrotron radiation and in the few that do circular polarization has not been observed. Radiation from white dwarfs and white dwarf binaries can be highly circularly polarized but any effect on molecular clouds and star formation regions must rely on rare chance encounters. Diffuse starlight in the ultraviolet region is either thermal or linearly polarized through dichroic extinction by interstellar grains.

Observations of young solar-type stars (having an age of less than 10 million years) in the Milky Way, considered to represent the early Sun, as well as meteoritic data, show that the early Sun was much more active than it is today, and it emitted copious amounts of energetic radiation, mainly cosmic ray particles and X-rays. In such conditions the abundance of amino acids, or any other COMs, is very limited both in the gas phase and on grain surfaces, unless the chemical species is well buried within the ice. On the other hand, chemicals stored inside grain aggregates would be partly shielded from the extremely unfavourable environmental condi-tions in the early stages of Sun life, and thus exposed to a lesser degree of photo-processing.

The effect of dust grain aggregates may not be, however, limited to the passive role of protection. The most general state of polarization of an elec-tromagnetic field is elliptic and the direction of propagation is given by the Poynting vector. The plane of the polarization ellipse is orthogonal to the Poynting vector, so that the state of polarization is conveniently described by the Stokes parameters constructed with the Poynting vector itself. In the interstitial material and in the cavities of the aggregate the direction of prop-agation of the field changes from point to point. The plane of the polariza-tion ellipse is no longer orthogonal to the Poynting vector. The net result is the creation of additional components of the fields that thus lose their original state of being linearly polarized, acquiring a degree of circularity. In technical terms, the fields depolarise (see Borghese *et al.* 2005[32] for the detailed theory). This effective ultraviolet circular polarization is generated *in situ*, and exposes to asymmetric photolysis the amino acids formed in the cavities of dust aggregates. However, the effect is purely geometric and in this sense is not chiral since it does not provide a symmetry breaking. In fact, the sign of the induced circular polarization changes with rotation of the cavities with respect to axes parallel to the propagation of the wave. The breaking of spatial symmetry is provided by dust grain alignment with respect to the stellar incident field. When embedded in a protoplanetary disk real aggregates can be efficiently aligned by interacting with the gaseous flow both in subsonic and supersonic regimes (see Chapter 2). The alignment arises from grains having irregularities that scatter atoms with different efficiency in the right and left directions. Although, the tendency for grains is to align with long axes perpendicular to the magnetic field, paramagnetic dissipation is not involved (Lazarian & Hoang 2007[33]), and the specific chemical composition of a dust aggregate is irrelevant.

Whatever is the mechanism producing enantiomer selection by asymmetric photolysis (if any) a second step is needed: in all the cases proposed, the enantiomeric excess generated among chiral molecules is very low, from 0.001% to about 10% in the best cases. Such tiny chiral prevalence can be increased by means of autocatalytic processes to homochirality. Autocatalysis may be driven by kinetics in processes far from equilibrium or be based on thermodynamics and equilibrium phase behaviour (for a review see Weiner *et al.* 2010[34]). At this stage, the interstellar dust legacy is nonetheless almost completely lost.

Further Reading

1. M. H. Engel and B. Nagy, *Nature*, 1982, **296**, 837.
2. J. R. Cronin and S. Pizzarello, *Science*, 1997, **275**, 951.
3. A. Belloche, R. T. Garrod, H. S. P. Müller and K. M. Menten, *Science*, 2014, **345**, 1584.
4. E. Herbst and E. F. van Dishoeck, *Annu. Rev. Astron. Astrophys.*, 2009, **47**, 427.
5. H. Campins, K. Hargrove, *et al.*, *Nature*, 2010, **464**, 1320.
6. S. Pizzarello and E. Shock, in *The Origins of Life*, ed. D. Deamer and J. Shostak, Long Island, NY, Cold Spring Harbor Laboratory Press, 2010, p. 89.
7. M. J. Mumma and S. B. Charnley, *Annu. Rev. Astron. Astrophys.*, 2011, **49**, 471.
8. J. E. Elsila, D. P. Glavin and J. P. Dworkin, *Meteorit. Planet. Sci.*, 2009, **44**, 1323.
9. H. Hidaka, M. Watanabe, A. Kouchi and N. Watanabe, *Astrophys. J.*, 2009, **702**, 291.

10. G. M. Muñoz Caro, U. J. Meierhenrich, *et al.*, *Nature*, 2002, **416**, 403.

11. M. P. Bernstein, J. P. Dworkin, S. A. Sandford, G. W. Cooper and L. J. Allamandola, *Nature*, 2002, **416**, 401.

12. M. Nuevo, U. J. Meierhenrich, *et al.*, *Astron. Astrophys.*, 2006, **457**, 741.

13. C. Meinert, J. J. Filippi, P. de Marcellus, L. d'Hendecourt and U. J. Meierhenrich, *ChemPlusChem*, 2012, **77**, 186.

14. A. S. Burton, J. C. Stern, J. E. Elsila, D. P. Glavin and J. P. Dworkin, *Chem. Soc. Rev.*, 2012, **41**, 5459.

15. A. Burton, J. Elsila, M. Callahan, M. Martin, D. Glavin, N. Johnson and J. Dworkin, *Meteorit. Planet. Sci.*, 2012, **47**, 374.

16. J. M. C. Rawlings, D. A. Williams, S. Viti, C. Cecchi-Pestellini and W. W. Duley, *Faraday Discuss.*, 2014, **168**, 369.

17. S. L. Miller, *Science*, 1953, **117**, 528.

18. S. L. Miller, *Biochim. Biophys. Acta*, 1957, **23**, 480.

19. J. Oró and A. Kinball, *Biochem. Biophys. Res. Commun.*, 1960, **2**, 407.

20. C. Kittel, *Introduction to Solid State Physics*, John Wiley & Sons, New York, 8th edn, 2005.

21. W. W. Duley, *Mon. Not. R. Astron. Soc.*, 2000, **319**, 791.

22. V. N. Solov'ev, E. V. Polikarpov, A. V. Nemukhin and G. B. Sergeev, *J. Phys. Chem. A*, 1999, **103**(34), 6721.

23. A. M. Saitta and F. Saija, *Proc. Natl. Acad. Sci. U. S. A.*, 2014, **111**, 13768.

24. B. M. Jones, C. J. Bennett and R. I. Kaiser, *Astrophys. J.*, 2011, **734**, 78.

25. V. Avetisov and V. Goldanskii, *Proc. Natl. Acad. Sci. U. S. A.*, 1996, **93**, 11435.

26. I. Myrgorodska, C. Meinert, Z. Martins, L. d'Hendecourt and U. J. Meierhenrich, *Angew. Chem., Int. Ed.*, 2015, **54**, 1402.

27. D. Campbell and P. Farago, *J. Phys. B*, 1987, **20**, 5133.

28. D. K. Kondepudi, R. J. Kaufman and N. Singh, *Science*, 1990, **250**, 975.

29. C. Viedma, *Phys. Rev. Lett.*, 2005, **94**, 065504.

30. A. Jorissen and C. Cerf, *Origins Life Evol. Biospheres*, 2002, **32**, 129.

31. E. Rubenstein, W. A. Bonner, H. P. Noyes and G. S. Brown, *Nature*, 1983, **306**, 118.

32. F. Borghese, P. Denti, R. Saija and C. Cecchi-Pestellini, *J. Phys. Conf. Ser.*, 2005, **6**, 59.

33. A. Lazarian and T. Hoang, *Astrophys. J.*, 2007, **669**, 77.

34. B. Weiner, W. Szymanski, D. B. Janssen, A. J. Minaard and B. L. Feringa, *Chem. Soc. Rev.*, 2010, **39**, 1656.

CHAPTER 13

Are We Nearly There Yet?

13.1 Introduction

"Are we nearly there yet?" is the oft-repeated cry of the young child in the back seat of the car, on a long journey. We can apply the same question to our long scientific journey: our destination is to obtain a broad understanding of the role of dust grains in astrochemistry. The cry from the back seat is—in effect—from astronomers who would like to use the results of the deliberations of astrochemists as an aid to understanding processes in the Universe. This book shows that the journey has certainly begun, is clearly heading in the right direction, and is still in progress. We are now on the high-speed motorway and are rapidly approaching—if not the end of our journey (*i.e.* the complete picture)—at least, a feeling that we are in the right locality so that we need to look carefully at the road signs for our destination. We can comfort that restless child in the back seat—the astronomer—with the good news that the journey's end is at least approaching.

Researchers tend to focus on the immediate problems they face and do not always remember to lift their heads from time to time to enjoy the view and appreciate how far they have already come. Yet progress in surface and solid-state astrochemistry, especially since the 1990s, has been very rapid indeed, and successful in obtaining data from laboratory experiments and theoretical investigations. The essential computational modelling that glues these data together with observational results draws useful insights and indicates future directions for study. Many of the ideas underpinning surface chemistry on interstellar dust grains had been discussed in the 1960s, especially by Salpeter and his collaborators (*e.g.* Gould & Salpeter 1963 [1]), but the next three decades justifiably resounded with the great successes of gas-phase astrochemistry. However, these successes were also

The Chemistry of Cosmic Dust
By David A. Williams and Cesare Cecchi-Pestellini
© David A. Williams and Cesare Cecchi-Pestellini 2016
Published by the Royal Society of Chemistry, www.rsc.org

accompanied by the recognition of significant failures that have definitively emphasized the necessity for surface processes to accompany processes in the gas phase. The discovery in the 1970s of interstellar ice mantles on dust grains in many regions of space[2] showed that the solid-state was also an essential subject for astrochemistry. Fortunately, the advances during the 1990s in laboratory techniques for the study of surfaces and ices and in fast computation of *ab initio* calculations—and the availability of people who had the imagination and skills to address these difficult problems—soon showed how fundamental progress could be made. The timing turned out to be ideal.

For those who flew the flag for surface and solid-state astrochemistry through these barren years, the long wait was over. Once humorously categorized as "the last refuge of a scoundrel" and topics only to be broached when all possible gas phase endeavours have failed (*cf.* Charnley *et al.* 1992[3]), grain surface chemistry and solid-state chemistry are now accepted as important components playing an essential role alongside gas-phase chemistries. Both are clearly necessary: one cannot choose one or the other: it is a total package. The acceptance of this position is welcome progress.

13.2   Stages on Our Journey

13.2.1   The Nature of Interstellar Dust: Where We are Now, and What We May Hope to See

Our journey began with *Section I: Defining the Chemical and Physical Nature of Interstellar Dust* and considered information from remote observations, from collected Solar System dust, from laboratory studies of dust grain analogues, and from the modelling that draws all this information together. Our motivation here is straightforward: we need to understand the physical and chemical nature of interstellar dust if we are to describe appropriately the chemistry that occurs on the surfaces of that dust.

Remote observations of dust (Chapter 2) through the extinction it causes in the visual regions of the spectrum have been made for about a century and the (by now) very detailed data have been probed exhaustively for signatures of the nature of interstellar dust. Remote observations are now much more extensive than extinction and polarization and range far beyond the visual. Depletion measurements provide important constraints on the elemental abundances locked in grains. Absorption and emission features and continuum emissions in various wavelength ranges contain important signatures and must be accounted for in any model of dust; the origins of some features, such as the ERE, remain to be reliably assigned. Another topic of major uncertainty that remains is the abundance and role of PAHs; the observational evidence is strongly suggestive that PAHs of some type are abundant in the interstellar medium, but precise identifications have yet to be made. The nature and abundance of PAHs is an area in which future progress may be expected to occur.

Solar System dust (Chapter 7) is the only material that can be directly examined in laboratories on Earth or on board space missions, and these data provide useful constraints on the precursors of Solar System dust, *i.e.* interstellar dust. Of course, Solar System dust has been through the star formation process involving physical conditions that are grossly different to those in interstellar clouds. Aggregation of interstellar dust grains certainly occurred in the denser regions of star formation, and dust processing by intense radiation from the proto-Sun certainly occurred. Therefore, the interpretation of data concerning Solar System dust to provide information about interstellar dust is dependent on our understanding of those key processes during the formation of stars; this understanding is currently rather rudimentary, but should improve. There has been a flood of data from NASA's STARDUST cometary coma sample return mission, launched in 1999; samples from that mission were returned in 2006. We shall see a flood of new data, especially from missions such as that of Rosetta and its lander, Philae, on comet 67P/Churyumov–Gerasimenko in 2014. Therefore, the information that we may infer from Solar System dust about interstellar dust is likely to grow rapidly in extent and reliability. It is important that we understand as thoroughly as possible the star- and planet-formation processes that we discussed in Chapter 10, so that we may obtain the most convincing picture of interstellar dust.

Models of interstellar dust (Chapter 3) attempt to develop a satisfactory description of dust from a chosen range of observational data. Such models must have predictive power to be convincing; simply fitting a selected set of data is not satisfactory. The optical and near-optical extinction and polarization measurements traditionally used as guides to the nature of interstellar dust cannot by themselves adequately constrain the models, as there are many free parameters available to the modeller. It is important that current models attempt to take into account all types of available observational data, and results from the work of Draine and his collaborators have been pre-eminent in that activity.[4] Recently, there has been considerable convergence among models of interstellar dust in diffuse clouds (discussion of the formation of ices on dust grains in dense clouds is deferred until Chapter 9). All the models assume a similar range of grain sizes of both carbons and silicates. A component of PAHs is included in most models. There are some differences in categorizing the structure of grains; some authors consider grains to be composed of one basic material such as silicate or amorphous carbon, while others allow grains to be composed of mixtures of these materials. Porosity is, at the moment, a free parameter. However, as well as these broad measures of agreement, there is also a fundamental disparity of view about grains: do they respond in some way to their environment, or are they essentially inert? In the former case, there is the possibility that one may understand why dust grains appear to be subtly different from one region to another (this time-dependent view is explored more fully in Chapter 6). In the latter case, the modelling is simply a rather complex curve-fitting problem in which model parameters are adjusted to make computations fit observational data, without much reference to the local physical conditions. We may hope to see this fundamental difference of view resolved in the near future.

13.2.2 Dust and Its Evolution in the Interstellar Medium

At this point in our journey towards a broad understanding of the chemical roles of interstellar dust we have deduced from observations and from collected interplanetary material some plausible models of interstellar dust. But where does the dust come from? Are the conclusions we have reached about interstellar dust in harmony with our views on the origin of dust in circumstellar regions and on the processing of circumstellar dust as it emerges into interstellar space? We discuss these issues in *Section II: The Formation of Dust and its Evolution in the Interstellar Medium.*

Stellar atmosphere gas that is initially warm and dense and is ejected by explosions or shocks from the central object into interstellar space, passes through a range of densities and temperatures to more tenuous and cooler states. Given our experience of soot-forming internal combustion engines, it is hardly a surprise that at some point during that expansion of stellar gas the physical conditions encourage the formation of solid nuclei on which other materials, perhaps less prone to nucleate, may be deposited, so that dust grains form (Chapter 5). Similarly, stars with atmospheres that go through this phase of nucleation are known to fall into two distinct classes, carbon-rich and oxygen-rich, so it is to be expected that dust grains formed in this way are of two main types: carbonaceous grains and metal silicates/ oxides. The scenario has been thoroughly investigated for many years, and the current picture for dust formation in stellar envelopes is presented in the very comprehensive monograph by Gail & Sedlmayr (2014).[5] Similar processes must form dust in the ejecta from novae and supernovae. In particular, supernovae may be very significant contributors to dust masses; recent observations show that dust masses around supernovae may grow significantly in the decades following the outburst.[6] However, the total dust mass that supernovae may contribute to a galaxy is currently unclear.[7]

In broad terms, it seems evident that dust from these near-stellar sources must be the precursor material to interstellar dust. However, significant processing occurs during the passage of this dust into the interstellar medium and during its residence there. We may expect to discover in the near future how the power-law size distribution—apparently near-universal in the interstellar media of many different galaxies—is achieved. More interesting, perhaps, is the response of these grains to various modifying influences in the interstellar medium (Chapter 6). In this book, we take the view—based on laboratory data—that this response may be rapid enough to compete with other relevant timescales in the interstellar medium. If so, then the optical behaviour and perhaps also the chemical properties of interstellar dust may be time-dependent. Regardless of whether this position turns out to be correct, this time-dependent approach allows a detailed determination of the chemical nature of the dust, especially of its outer surface layers. These are crucial for surface chemistry on dust. We may hope that developments in the near future will determine the various evolutionary rates of dust in the interstellar medium more precisely.

13.2.3   Chemically Active Interstellar Dust

Our journey has so far brought us to a satisfactory position. We have models of interstellar dust that are in harmony with the observational data and are consistent with available elemental abundances. The origin of interstellar dust grains appears to be happily accounted for as dust in near-stellar regions, assuming that the near-stellar dust evolves on its approach to and in the interstellar medium. The benefit of this approach is that it defines fairly precisely the nature of dust in different regions of interstellar space, and therefore we may reliably discuss chemistry operating on the surfaces of these grains (Chapter 8) and in and on the ices that may be deposited on these grains (Chapter 9). *Section III: Chemically Active Interstellar Dust* is the heart of this book on the astrochemistry of cosmic dust.

The pre-eminent problem of astrochemistry has been to understand the formation of molecular hydrogen in surface reactions on dust grains. Molecular hydrogen in the Milky Way galaxy (and others) cannot be made in the observed abundances by any known gas-phase reactions, and—by default—surface reactions were first invoked, for without molecular hydrogen the whole edifice of gas-phase astrochemistry would collapse. The introduction of surface reactions into astrochemistry has had a happy conclusion. Over the last two decades the problem of H_2 formation in surface reactions has attracted hundreds of publications based on fundamental *ab initio* calculations and on highly sophisticated experiments in several different laboratories. This enormous research effort has been expertly reviewed by Vidali (2013).[8]

The research programme has been outstandingly successful. We now have a good understanding how the H_2 formation reaction occurs on various types of surface at low temperatures, along the lines that have been discussed for over half a century. We know that the reaction can in some circumstances proceed with the high efficiency demanded by the observations. We understand in crude terms the energy budget in these processes; we know that the H_2 molecules are formed in internally excited states, leave the surface with some (non-thermal) kinetic energy and deposit some energy into the surface. We have some idea how these various processes differ on different materials.

There remain some concerns. The formation of H_2 at higher temperatures (~100 K) has been demonstrated for some materials, but not for all likely grain analogues. While it is clear that newly-formed H_2 should be internally excited, emission from these excited states is masked by other processes in the interstellar medium and has—so far—never been observed. There is no precise agreement on the nature of the excitation due to formation, and the energy deposited in the surface is still poorly constrained. The physical nature of the surfaces assumed in computations or adopted in experiments are clearly unlike the small-scale nature of interstellar dust, and it is unclear to what extent the morphology of dust grains, and indeed possible porosity, will have on the conclusion reached so far.

It has been speculated that chemisorption may lead to high H-coverage on some surfaces, and that this might be relieved by a catastrophic release of chemical energy. In other words, there might be a runaway explosion. If so, then H_2 formation by this means would occur in "puffs" rather than continuously.[9] Further study of this mechanism is required. Can it compete with the continuous formation mechanisms? What would be the consequences for internal excitation in the products H_2 molecules, and how much energy would be directed into the surface?

By contrast with studies of H_2 formation, rather little work has been carried out for formation of simple hydrides. Models suggest that these could be significant routes to interstellar chemistry, without adverse effects. Further work on the formation of OH and H_2O, NH, NH_2, and NH_3 in surface reactions would be desirable. Do carbon molecules form in such a way, or are they deposited as "soot" on the surfaces of dust grains? There are, of course, competing gas-phase routes for these hydrides.

As regards the deposition of ices on grain surfaces and the chemistry that may be promoted by UV, X-rays, or fast particles on their surfaces and in their bulk, there is a considerable amount of work in progress (see the excellent review by van Dishoeck *et al.* 2013 [10]) but the current situation as regards the chemistry occurring in these ices is not yet well-defined. The mechanisms observed to control the deposition of ices at a fairly well-defined depth into any particular cloud have yet to be definitively explored. However, laboratory work that shows convincingly that the formation of complex organics from simple ices can be promoted by ultraviolet or X-rays, or by fast particles, and released into the gas phase under TPD. Our understanding of the details of such conversions is, unfortunately, rather limited at present, but more quantitative measurements (for example, those of Öberg *et al.* 2009 [11]) have shown how to take us out of the realm of "cooking" (as Nigel Mason has termed the early work) to increase molecular complexity, into an era of precision measurements in solid-state astrochemistry. Extensive computer models that drive chemistry with radicals created by a weak cosmic-ray induced radiation in the interior of dark clouds have been explored in detail by Herbst and collaborators (*cf.* Vasyunin & Herbst 2013 [12]). Such models have no difficulty in predicting significant abundances of complex organics as the ices are warmed by a nearby star. While the predicted chemical richness is very pleasing, these models usually depend on adopted values for very large number of poorly-known physical data on mobilities, binding energies, *etc.* The highly speculative "explosions" model also seems capable to generating a wide range of complex organic molecules, with abundances that appear to be competitive with observations. Intensive study in the laboratory and through computational modelling will certainly need to continue before our understanding of the growth of molecular complexity is nearly complete.

13.2.4   Roles of Dust in the Universe

The purpose of the final section of this book: *Section IV: Roles of Dust in the Universe* is to display the important roles that dust grains play in the evolution of a galaxy, and to determine the extent to which those roles depend on the precise nature of the dust. We have learnt how the physical and chemical nature of dust may be affected by its environment, and so we can begin to consider whether processes that are familiar to us in our locality of the Milky Way galaxy operate in similar ways elsewhere in the Milky Way and in other galaxies.

In star- and planet-formation in the nearby Universe (Chapter 10) the essential role is to provide enough extinction in star-forming regions so that gas-phase chemistry is able to produce adequate numbers of simple molecular coolants. These coolants must not be frozen-out into ices too early, or—if they are—must be returned to the gas from the ices by effective desorption processes. These processes are not particularly dependent on the physical and chemical nature of the dust. All that is required is a dust mass relative the mass of gas that is not too dissimilar to that in the Milky Way, and also a range of grain sizes so that extinction is substantial in the UV as well as in the visual region of the spectrum; such a range of sizes ensures that the dust grains can also be effective radiators.

Dust grains are the raw material for the formation of terrestrial planets and the cores of massive gaseous planets, and they undergo significant processing during the star and planet forming process. None of these processes is significantly dependent on the nature of the dust grains.

The first star formation in the high red shift Universe proceeded without benefit of dust, and formed a short-lived population of massive stars (see Chapter 11). These Population III stars ended their lives as supernovae and seeded the early Universe with heavy elements. Dust became relatively abundant at a remarkably early epoch in the evolution of the Universe, and from that point on star formation occurred essentially as in the local Universe.

Section IV ends with a brief consideration of the relation of astrochemistry to astrobiology. It is clear that ice processing is capable of producing molecules of a higher complexity—COMs—than those produced purely by gas-phase chemistry. However, although these COMs are relevant to astrobiology, they must be considered essentially as feedstock for molecules of even greater complexity; *i.e.* biological molecules. The possible importance of cavity chemistry in generating this complexity is briefly considered, and the role of cavity chemistry in introducing chirality in product molecules is mentioned. Obviously, this is a field in which much work is required.

13.3   The Journey's Progress

Our journey towards an understanding of surface and solid-state astrochemistry is making rapid progress. The querulous child in the back seat—the astronomer who wishes to make use of the conclusions of the astrochemists—can be reassured. Section 13.2 lists some the astrochemical successes of the last two decades. It also makes clear some of the scientific areas that need to be visited or re-visited before our journey can reach its conclusion.

Further Reading

1. R. J. Gould and E. E. Salpeter, *Astrophys. J.*, 1963, **138**, 393.
2. F. C. Gillett, T. W. Jones, K. M. Merrill and W. A. Stein, *Astron. Astrophys.*, 1975, **45**, 77.
3. S. B. Charnley, A. Tielens and T. J. Millar, *Astrophys. J.*, 1992, **399**, L71.
4. B. T. Draine, *Annu. Rev. Astron. Astrophys.*, 2003, **41**, 241.
5. H.-P. Gail and E. Sedlmayr, *Physics and Chemistry of Circumstellar Dust Shells*, Cambridge University Press, 2014.
6. M. Matsuura, E. Dwek, *et al.*, *Science*, 2011, **333**, 1258.
7. T. Temim, E. Dwek, K. Tchernyshyov, M. L. Boyer, M. Meixner, C. Gall and J. Roman-Duval, *Astrophys. J.*, 2015, **799**, 158.
8. G. Vidali, *Chem. Rev.*, 2013, **113**, 8762.
9. C. Cecchi-Pestellini, W. W. Duley and D. A. Williams, *Astrophys. J.*, 2012, **755**, 119.
10. E. F. van Dishoeck, E. Herbst and D. A. Neufeld, *Chem. Rev.*, 2013, **113**, 9043.
11. K. I. Öberg, R. T. Garrod, E. F. van Dishoeck and H. Linnartz, *Astron. Astrophys.*, 2009, **504**, 891.
12. A. I. Vasyunin and E. Herbst, *Astrophys. J.*, 2013, **762**, 86.

Subject Index

Page numbers in *italics* refer to figures. Page numbers with T indicate a table.